Poppy Tushingham

Creating an Empire of Informers

Vigilanzkulturen / Cultures of Vigilance

Herausgegeben vom / Edited by
Sonderforschungsbereich 1369
Ludwig-Maximilians-Universität München

Band / Volume 10

Poppy Tushingham

Creating an Empire of Informers

Vigilance in the Assyrian Empire
and King Esarhaddon's *adê*-Covenant
of 672 BC

DE GRUYTER

Funded by the Deutsche Forschungsgemeinschaft (DFG, German Research Foundation) – Project-ID 394775490 – SFB 1369

This book is a revised version of a doctoral dissertation written at the University of Munich (Ludwig-Maximilians-Universität München) that was defended in April 2023.

ISBN 978-3-11-132341-1
e-ISBN (PDF) 978-3-11-132343-5
ISBN (EPUB) 978-3-11-132346-6
ISSN 2749-8913
DOI https://doi.org/10.1515/9783111323435

Library of Congress Control Number: 2023951180

Bibliographic information published by the Deutsche Nationalbibliothek
The Deutsche Nationalbibliothek lists this publication in the Deutsche Nationalbibliografie; detailed bibliographic data are available on the Internet at http://dnb.dnb.de.

Cover illustration: Copy of a treaty between Esarhaddon and Humbaresh, ruler of the city of Nashshimarta in Media, Tablet, seal-impression, clay, BM 132548, ND 4336, Nimrud.
© The Trustees of the British Museum
Printing and binding: CPI books GmbH, Leck

www.degruyter.com

Acknowledgements

This study is a lightly edited version of my dissertation, submitted to the History Department at the LMU Munich in January 2023 and defended in April of the same year.

There are many people to whom I owe my sincere thanks, both for making this dissertation possible and for shaping and improving its contents. In particular, I wish to thank my supervisor, Karen Radner, for her guidance during the writing of this dissertation and for her encouragement of my academic endeavours more broadly. So too, Jamie Novotny, my second supervisor, has been very generous with his time throughout this period, and I thank him for his invaluable feedback and corrections. For their participation in the examination process, I also thank Arndt Brendecke, who acted as third examiner, and Michael Jursa, who performed the role of external assessor.

I am grateful to all of the members of the DFG-funded collaborative research centre (CRC), 'Cultures of Vigilance (SFB 1369)', in which I undertook this project from 2019 – 2023. It has been a pleasure to work in such a productive and interdisciplinary environment, and I have benefitted greatly from my colleagues' expertise. I would especially like to thank the CRC speaker, Arndt Brendecke, for laying the foundations of the CRC's conceptual framework and for steering our joint endeavours. I am also indebted to everyone who participated in the discussions when I presented my work, as well as those who read and provided comments on my drafts in our writing workshops. The two CRC working groups in which I took part, on 'responsibilization' and 'semantics', were wonderful environments for the exchange of ideas. I also thank my fellow members of the CRC graduate school, as well as our coordinators: Alina Enzensberger and Benjamin Steiner. Martina Heger of the CRC publication office also guided me through the publication process, and Thomas Seidler kindly prepared the index.

My LMU Munich colleagues in Ancient History and in Assyriology have provided me with a great deal of help and support, both academic and otherwise, during this period. Andrea Squitieri in particular helped me to get to grips with QGIS and shared his materials with me, which allowed me to create the maps found in this study. Tonio Mitto, meanwhile, kindly corrected the German summary that I submitted with my thesis. I have also very much enjoyed teaching undergraduate students over these past years, who frequently provided me with food for thought, especially during the course on the Assyrian Empire that I taught together with Mary Frazer in WS 20/21. In addition to this, it was a pleasure to participate in the various activities on digital cuneiform studies organized within the LMU-

 https://doi.org/10.1515/9783111323435-001

UCB Research in the Humanities programme, led by Karen Radner and Jamie Novotny in Munich and Niek Veldhuis and Laurie Pearce at Berkeley from 2017–2022.

Beyond LMU Munich, I owe my thanks to Shuichi Hasegawa for reading the sections of this dissertation that concern the Hebrew Bible and the *Story of Ahiqar* and providing me with insightful comments and useful corrections. Similarly, I would like to thank Eckart Frahm for discussing my dissertation with me, and for providing me with interesting ideas and valuable materials, particularly concerning *The Underworld Vision of an Assyrian Prince.* In the summer of 2019, I had the good fortune to take part in the *Heidelberger Forschungslabor Alter Orient*, led by Stefan Maul and Betina Faist, whom I would like to thank as well as thanking my fellow participants. The experience was a valuable learning opportunity in itself, and the sections of my thesis on Ashur in particular greatly benefitted from our joint work.

Despite the constraints of the COVID-19 pandemic, I was able to make use of the student rooms in the British Museum and the Vorderasiatisches Museum, for which I thank the staff of both. I have also been able to make ample use of many digital resources, in particular the various text corpus projects hosted by the Open Richly-Annotated Cuneiform Corpus (Oracc) and the photographic materials of the Cuneiform Digital Library Initiative (CDLI), and extend my thanks to all who have given their time to contribute to these efforts.

Finally, I thank my family – Lorna, David, Ralph, Ollie, Alice and Eddie – for their support and love throughout, as well as my husband, Pablo.

Contents

List of tables —— XI

List of figures —— XIII

Chapter 1: Introduction —— 1
1.1 Historical overview —— **3**
1.2 Terminology used in this study —— **8**
1.3 Source material —— **16**
1.4 Structure and contents of this study —— **27**

Part 1: **The royal call to vigilance in 672 BC**

Chapter 2: King Esarhaddon and his empire in the covenant composition —— 33
2.1 The king and his heir as the linchpin of empire in the covenant —— **34**
2.2 The legal parties and the composition of empire in the covenant —— **38**
2.3 Using the gods to access place in the covenant —— **43**
2.4 Conclusions —— **53**

Chapter 3: Directing vigilance in the covenant composition —— 55
3.1 Scenarios of potential danger in the covenant stipulations —— **63**
3.2 Mandated reactions in the covenant stipulations —— **72**
3.3 Forbidden reactions in the covenant stipulations —— **82**
3.4 Directing vigilance in the covenant oath —— **89**
3.5 Conclusions —— **93**

Chapter 4: Laying the ideological groundwork for enacting the covenant —— 95
4.1 Promoting covenant and delegitimizing usurpation in *Esarhaddon's Apology* —— **97**
4.2 Reframing the role of provincial subjects in royal narratives —— **106**
4.3 Conclusions —— **113**

Chapter 5: Putting the covenant into practice —— 116
5.1 Scaling vigilance at the covenant ceremony —— **117**
5.2 The covenant ceremony as a staggered event —— **120**

5.3 The covenant ceremony as an iterative event — **132**
5.4 Conclusions — **137**

Part 2: **Responses to the call to vigilance**

Chapter 6: Responses to covenant at Esarhaddon's court — 141
6.1 The succession covenant in letters from royal advisors — **142**
6.2 Scholarly musings on covenant and vigilance — **152**
6.3 Conclusions — **164**

Chapter 7: Responses to Esarhaddon's covenant in the provinces — 166
7.1 Responses from Ashur — **168**
7.2 Responses concerning the western provinces — **183**
7.3 The covenant of the king as the punisher of private legal parties — **194**
7.4 Conclusions — **200**

Chapter 8: Responses to Esarhaddon's covenant across the client states — 203
8.1 Responses from Babylonia — **204**
8.2 Letters from Itti-Šamaš-balaṭu, royal delegate in the Phoenician client state of Arwad — **216**
8.3 Reframing the duty of vigilance in Judah: Deuteronomy 13 — **223**
8.4 A literary response: wrongful accusation in the *Story of Ahiqar* — **233**
8.5 Conclusions — **239**

Chapter 9: Conclusions. Creating an empire of informers — 241
9.1 Space: geographical, administrative, social — **244**
9.2 Responsibilization: successes — **245**
9.3 Responsibilization: limitations and misunderstandings — **248**
9.4 Responsibilization: failures — **251**
9.5 Spatial dynamics of responsibilization — **253**

List of tables

Table 1: Structure of the covenant stipulations
Table 2: References to the succession covenant in Esarhaddon's court correspondence
Table 3: References to the succession covenant in the royal correspondence from Ashur
Table 4: References to the succession covenant concerning the western provinces in the royal correspondence
Table 5: References to the succession covenant concerning Babylonia in the royal correspondence
Table 6: References to the succession covenant from western client states in the royal correspondence

 https://doi.org/10.1515/9783111323435-002

List of figures

Figure 1: Map of find locations of manuscripts of Esarhaddon's succession covenant
Figure 2: Manuscript of Esarhaddon's succession covenant, Kalhu version. ND 4327; Iraq Museum, Baghdad, reproduced from Wiseman 1958, pl. 1
Figure 3: Schema of the vigilant diamond
Figure 4: Map of locations of responses to Esarhaddon's succession covenant from the provincial extent of the empire
Figure 5: Map of locations of responses to Esarhaddon's succession covenant from client states of the empire
Figure 6: Map of locations of responses to Esarhaddon's succession covenant from the court, provinces and client states

 https://doi.org/10.1515/9783111323435-003

Chapter 1: Introduction

By the early seventh century BC, the Kingdom of Assyria was the dominant power in the Middle East. It was a mature state: its basic structure had been in place for some seven centuries, and it had been an empire, in modern terms, for approximately two hundred years.[1] Its dominion spread far beyond the capital, Nineveh, located in what is now northern Iraq, and encompassed most of the habitable Middle East. Administratively, the empire employed a two-tier system: direct control over Assyria's approximately seventy provinces,[2] and indirect control over the client states.[3] Nonetheless, the clear distinction between these two administrative categories belies the diversity of the empire. Recent studies, particularly in the field of archaeology, have demonstrated that Assyrian imperialism was enacted in a deeply heterogenous fashion across its holdings.[4] As such, the nature and degree of a subject's interaction with the state would have varied greatly depending on their physical and, of course, social, position within the empire. The state, for instance, would have had more access to those living in lowland urban settlements, or farming in their surroundings, than it would have had to those living beyond such environments. This notwithstanding, the Assyrian state can be broadly characterized as making relatively uniform basic demands of its subjects across the ex-

1 On the status of Assyria as an empire, see for instance Radner 2014b.

2 On the two-tier system, see Postgate 1992. On Assyrian imperial structure, see also Liverani 2017. The number of Neo-Assyrian provinces was established by Karen Radner in RlA 11, 42–68 s.v. Provinz. C. Assyrien. Each Assyrian province was headed by a provincial governor appointed directly by the king. In the Assyrian language, the governors were called *pāhutu* or *bēl pāhete*, literally meaning 'proxy' (CAD P, 367–369 s.v. *pīḫatu* in *bēl pīḫati*; AHw, 862 s.v. *pīḫātu(m)*). On this terminology, see also Postgate 1995, 2 and RlA 11, 42f. s.v. Provinz. C. Assyrien.

3 As noted by J.N. Postgate (1992, 252), the term 'client' state is preferable to the frequently used 'vassal' state, as the latter has feudal connotations. One might object to the term 'client' as having implications of reciprocity and voluntary participation. In some respects, this is accurate, as client rulers could expect benefits from their relationship to Assyria, such as military support in case of invasion or revolt. However, they were often forced to become or remain so on pain of their own death and the annexation of their country. While client kings were theoretically independent rulers, the Assyrian monarch would sometimes install a puppet ruler. King Sennacherib (r. 704–681 BC), for example, boasts in his royal inscriptions of installing a local ruler in Babylonia (modern southern Iraq) named Bel-ibni, "a scion of Šuanna (Babylon) who had grown up like a young puppy in my palace, as king of the land of Sumer and Akkad (i.e. Babylonia)" (RINAP 3/1 nos. 1: 54, 2: 13 and 3: 13 and RINAP 3/2, no. 213: 53).

4 For a recent argument that the Assyrian Empire should be characterized as 'dynamic and variegated', rather than 'institutionalist and schematic', see Düring 2021. See also the important studies Parker 2001 and Bagg 2011.

 https://doi.org/10.1515/9783111323435-004

tent of the empire. On the one hand, subjects were expected to contribute material goods, in the form of tax from the provinces and tribute from the client states. On the other, they were required to supply manpower, in the form of conscripted labour and military support.[5]

It is the central thesis of this study, however, that under Esarhaddon, king of Assyria (r. 680 – 669 BC), the Assyrian state made a further demand of its subjects: it required them to contribute their attention on behalf of the crown. In the year 672 BC, Esarhaddon imposed a succession covenant, a binding agreement witnessed by the gods, on his subjects across the entire extent of his empire. This covenant required the empire's population to monitor one another for signs of disloyalty to the monarch and especially to his newly appointed crown prince, Ashurbanipal. They had to swear to report suspicious activity to the crown, and, in certain cases, even to intervene physically on the crown's behalf, taking action against dissidents. In addition to scrutinizing one another, they were required to monitor and regulate their own attitudes towards the crown. They had, for example, to love the new crown prince like they loved their own lives, and to act on his behalf with their whole hearts. In this way, Esarhaddon attempted to harness the attention of Assyria's population for the state's own ends, ensuring his own safety and – in particular – that of his chosen successor to the Assyrian throne. While the covenant was imposed on subjects by the state in a vertical fashion, it was not designed to create a system of top-down surveillance. Instead, it sought to provoke a bottom-up dynamic of mutual monitoring that this study terms 'vigilance'.

Esarhaddon's succession covenant was not mere empty rhetoric. Rather, the monarch and his advisors took steps to implement the covenant throughout the Assyrian Empire, from the Zagros to the Levant, and across social groups. The demand to report plots against a treaty partner had been a feature of Ancient Near Eastern treaties since the third millennium BC, and as such was not new.[6] Indeed, Esarhaddon's covenant utilized existing legal tools and expressed its demands in ways that that would have been familiar to many of Assyria's subjects to some degree. This notwithstanding, Esarhaddon's succession covenant constitutes the first well-documented attempt by an imperial power to compel its population to monitor one another on its behalf across its holdings. The available evidence allows the modern scholar to examine the phenomenon from the royal point of view, both by looking at the covenant text itself and also at sources that provide insights into its

5 J.N. Postgate's monograph (1974b) remains the most thorough study of taxation and conscription for the Neo-Assyrian period. More recently, see also Radner 2007. On recruitment and logistics in the Assyrian army, see also Dezső 2016.

6 See discussion in Koch 2008, 24 f. So too, of course, the act of relaying information about peers is well attested in earlier source material.

practical implementation across the empire. So too, written sources of various types shed light on the possible ways in which Assyria's subjects interpreted, acted on, or even possibly rejected, these royal demands. In this way, the source material from the final years of King Esarhaddon's reign provides a unique opportunity to examine an early empire's attempt to establish a duty of vigilance among its subjects, and to measure the extent of its success in doing so.

1.1 Historical overview

By the year 672 BC, the Kingdom of Assyria was at the zenith of its power. During the period of Esarhaddon's rule and the early reign of his son and successor, Ashurbanipal, the empire reached its maximum geographical extent. In contrast to his predecessors, Esarhaddon was largely free from the threat of attack to his holdings from external forces. He was even able, in 674 BC, to campaign as far afield as Egypt, the first such invasion by an Assyrian monarch. While he was not initially successful on his campaign, he succeeded in returning to Egypt and taking Memphis, the capital, in 671 BC.[7] It is in part for these reasons that the Assyrian specialist Eckart Frahm rightly terms this period Assyria's 'imperial heydays'.[8]

Nonetheless, ruling a large and heterogenous state such as Assyria in the early seventh century BC was not without considerable challenges. From a dynastic standpoint, the family to which Esarhaddon belonged had maintained its hold on the Assyrian throne for an impressive duration.[9] The Assyrian throne was, however, no stranger to bloodshed and usurpation. Indeed, both Esarhaddon's

7 The Assyrians went to significant lengths to ensure the success of their 671 BC campaign (Radner 2008a).

8 For a survey of the history of the Neo-Assyrian period (ca. 1000–609 BC) and a characterization of the period of the reigns of Sennacherib, Esarhaddon and Ashurbanipal (r. 704–ca. 631 BC) as Assyria's 'imperial heydays', see Frahm 2017b, esp. 183–190. Other surveys include Radner 2015a and 2017c, 93–108, as well as Cancik-Kirschbaum 2015, 75–95. This period can also be considered a high point in the history of Assyria and Babylonia from a cultural standpoint. Indeed, many of the cuneiform scholarly and literary works known to us today were found as a result of the excavation of Assyrian libraries, particularly the royal collections at Nineveh, which were created and expanded during first half of the seventh century BC. Much of the evidence comes from what is frequently termed 'Ashurbanipal's Library', although this is a misnomer for the Assyrian royal library. For discussion, see Robson 2019, esp. 10–48, as well as Reade 1986, Pedersén 1998, 158–165 and Robson 2013.

9 On the special status of the Assyrian royal family and its ramifications for succession, see Radner 2010a, esp. 26 f.

grandfather, Sargon II (r. 721–705 BC) and his great-grandfather, Tiglath-pileser III (r. 744–727 BC), had usurped the throne, deposing their brother and father respectively.[10]

Esarhaddon's own accession had been difficult. His father, Sennacherib (704–681 BC),[11] appears to have nominated him to the position of crown prince in around 683 BC.[12] He likely swore his subjects to a succession covenant to this effect, although the only surviving fragments of this document do not preserve the name of the son being elevated to the position of crown prince, instead leaving a blank space where this information should be.[13] While this succession covenant bears some points of comparison with that of Esarhaddon, there is as of yet no contemporary evidence that it was disseminated in anything like the same way. If this lack of evidence does indeed reflect the reality and the covenant was imposed (if at all) only on a select group of subjects, likely in Ashur, then there are various possible explanations for this. Firstly, it seems entirely possible that Sennacherib did not consider it necessary, a sentiment that would apparently have been consistent with those of his forebears. Alternatively, Sennacherib's final years were very likely characterized by political turmoil, at least among the ruling family, and it may not have been feasible to implement the covenant more widely, whether or not this was what the monarch desired.

Esarhaddon was probably only able to enjoy his new status as crown prince for a brief period. The precise details of events are impossible to reconstruct with certainty, in particular because the most detailed source on Sennacherib's death and Esarhaddon's succession is a royal inscription commissioned by Esarhaddon himself, first disseminated in 673 BC, several years after the events took

---

10 For a survey of Assyrian usurpations, see Mayer 1998, as well as Radner 2016. On Tiglath-pileser III see PNA 3/2, 1329–1331 s.v. Tukultī-apil-Ešarra no. 3, as well as Garelli 1991 and Zawadzki 1994, on Sargon II see PNA 3/2, 1239–1247 s.v. Šarru-kēnu, Šarru-kīn, Šarru-ukīn no. 2, as well as Thomas 1993 and Vera Chamaza 1992.

11 For a brief overview of Sennacherib's reign, see Frahm 2017b, 183–186, as well as RINAP 3/1, 1 and RINAP 3/2, 1–30.

12 Contemporary sources confirm that Esarhaddon was indeed Sennacherib's official crown prince, in particular a royal inscription in which Esarhaddon is mentioned by his other name, Aššur-etel-ilani-mukin-apli, and is described as the 'the senior son of the king, who (resides in) the House of Succession' (RINAP 4, no. 13: 1). Many have argued that the Esarhaddon was appointed as crown prince at some point around 683 BC (see for instance SAA 6, xxix–xxxiv, Frahm 1997, 18, Melville 1999, 20, PNA 3/1, 1115 s.v. Sīn-aḫḫē-erība and Šašková 2010b, 152). Nonetheless, as both Andrew Knapp and Auday Hussein have pointed out recently, this dating is not entirely certain (Knapp 2015, 303–304, fn. 7; Hussein 2020, 82). For a broader discussion of Sennacherib's family, see also RINAP 4, 26–30.

13 SAA 2, no. 3 and Frahm 2009a, nos. 67–69.

place. By his own admission, however, Esarhaddon lost his father's favour towards the end of the latter's reign and was forced into exile, something that he blames on his older brothers, whom he accuses of plotting against him.[14] If Esarhaddon and the short account in 2 Kings and Isaiah are to be believed, it was these brothers who murdered Sennacherib in 681 BC.[15]

Esarhaddon was able to overcome this rival faction and ascended to the Assyrian throne some two months after the death of his father.[16] Once he took the throne, he imposed a loyalty oath on his subjects and put many of his brothers' co-conspirators to death. Meanwhile, Esarhaddon's brothers themselves, whom the biblical version of events identifies as Urdu-Mullissu and a brother with a name ending in 'šarru-uṣur', managed to escape beyond Esarhaddon's reach.[17] Esarhaddon's quick accession to the throne, and the apparent lack of major resistance to his rule in the early years of his reign, suggest that he was able to assert his claim to the throne fairly decisively. Nonetheless, it seems reasonable to assume that this episode would have made him acutely aware of the fragile position of the monarch and his crown prince.

Despite this, it is not clear that Esarhaddon immediately chose a successor on coming to the throne.[18] Whatever his earlier arrangements, he officially nominat-

---

14 RINAP 4, no. 1: i 26–27; RINAP 4, no. 1: i 29–31 and i 39. Note that it has been suggested that Esarhaddon may have taken refuge in the city of Harran, classical Carrhae, near modern Şanlıurfa (Leichty 2007; RINAP 4, 2).

It is unclear whether Esarhaddon was still crown prince at the end of Sennacherib's reign, seeing as he fell out of favour with his father before the latter's death. When Esarhaddon was elevated to the status of crown prince, Sennacherib gave him gifts and a new name, as well as imposing a succession covenant (Hussein 2020, 81). There is no evidence to suggest he did this for a new successor, but it is possible that Esarhaddon was no longer his preferred successor.

15 2 Kgs. 19:37; Isa. 37:38. Note also that, while some have chosen to see the biblical account as external confirmation of Esarhaddon's account of his succession (a theory put forward by Simo Parpola (1980) which has found wide acceptance), it has been argued that the biblical account may have been based on Esarhaddon's own version of events (most recently, see Knapp 2020, 116, fn. 3, 168 f., fn. 16 and 178).

Most scholars accept the theory that Sennacherib was murdered by Esarhaddon's brothers (as in Parpola 1980, Frahm 2009b, 2014, 219 and 2017, 186, Karen Radner 2003c, 116–167 and 2016, 53, and Šašková 2010b. Nonetheless, some others have questioned this version of events, instead suggesting that Esarhaddon may have been involved in the assassination. Recently on his subject, see Knapp 2015, 301–307 and 2020, as well as Dalley and Siddall 2021.

16 As recorded not only in Esarhaddon's royal inscription describing the event (RINAP 4, no. 1: i 87), but also in a Babylonian chronicle (RINAP 4, 6–7; see also Grayson 1975, no. 1: iii 34–38).

17 2 Kgs. 19:37; Isa. 37:38. Recently on Urartu see Tsetskhladze (ed.) 2021; RINAP 4, no. 1: i 84.

18 There are several references to a *mār šarri* (literally 'son of the king') in archival documents dated to the early years of Esarhaddon's reign. Theodore Kwasman and Simo Parpola have argued that the term always refers to the crown prince (SAA 6, xxvii–xxix), which would mean that Esar-

ed Ashurbanipal, one of his younger sons, as his successor to the Assyrian throne in the year 672 BC. At the same time, he took the unprecedented step of naming his eldest living son,[19] Šamaš-šumu-ukin, as crown prince of the region of Babylonia, modern southern Iraq. Esarhaddon was the direct ruler of Babylonia, and thus by selecting two different sons as his successors to the thrones of Assyria and Babylonia, Esarhaddon split his holdings. In effect, however, Esarhaddon was choosing Ashurbanipal as the heir to Assyria and the empire, and placing Šamaš-šumu-ukin in the future role of a client ruler to Assyria, albeit one of high status due to Babylonia's cultural and religious significance in Assyrian eyes.

In order to ratify his new succession plan, Esarhaddon imposed his succession covenant, demanding the vigilance of all of Assyria's subjects in the service of the crown. Esarhaddon's public selection of a successor at this time may have been prompted by the sense that his position as monarch had become less secure, possibly due to a combination of difficult events that occurred in the lead up to 672 BC. These include his unsuccessful campaign against Egypt in 674 BC,[20] the death of his wife, Ešarra-ḫammat, in the year 673 BC,[21] and his struggle with his chronic illness.[22]

Whether or not Esarhaddon's health was a factor in his decision to name his successor, he died only slightly over three years after making these arrangements. Esarhaddon died of natural causes in the eighth month (Arahsamna, i.e. October–November) of 669 BC, while on his way to campaign in Egypt for a third time. Thus, during his own reign at least, his covenant was in force for some three and a half

---

haddon had appointed a crown prince – whose identity is unknown – almost immediately after coming to the throne. Despite this, various scholars have argued against this interpretation of the term *mār šarri*, suggesting that it means 'prince' more generally, and therefore cannot be taken as evidence of the existence of a crown prince at this time (Hussein 2020, 69–76, see also Knapp 2015, 305, fn. 12). Furthermore, it is clear that Esarhaddon at the very least considered appointing a crown prince in 677 BC. One extispicy query (a question to the sun god, Šamaš, answered by a diviner specialized in inspecting the organs of sheep and goats for divine messages) shows that he attempted to promote his eldest son, Sîn-nadin-apli, to the position in late 677 BC. The fate of Sîn-nadin-apli is unknown, presumably he had died or fell out of favour with his father by 672 BC, or both. On the topic, see PNA 3/1, 1138 s.v. Sīn-nādin-apli, Novotny and Singletary 2009, 169, as well as Hussein 2020, 82 f.

19 On Esarhaddon's children and their relative ages, see Novotny and Singletary 2009.

20 As Andrew Knapp has recently argued (2016).

21 Grayson 1975, 60–87 and 125–128, nos. 1 and 14. The two chronicles differ slightly on the date of the death of the king's wife: the first states that she died on the 5th day of month Addaru (XII), while the second states that she died on the sixth. However, as these two chronicles frequently differ on the dates of particular events by several months, the relative closeness of these dates can perhaps be taken as an indication of its wider importance.

22 As suggested in Radner 2003c, for instance.

years in total. It is worth stressing that these years were largely characterized by military and diplomatic successes. In 674 BC, Esarhaddon concluded a treaty with Elam in modern Iran, a longstanding antagonist of the Assyrian Empire. He succeeded in maintaining this during the final years of his reign.[23] As already mentioned, in the year 671 BC Esarhaddon succeeded in invading Egypt and taking Memphis, the capital. This was a significant victory and brought both people and booty to Assyria.[24]

Despite these triumphs, Esarhaddon faced the most significant challenge to his internal control of the empire since his accession in these final years of his reign: plotters located in the core region of the empire, in Harran to the west, and with ties in Babylonia schemed to overthrow the monarch at this time, and intended to replace him with an insurgent named Sasî.[25] Esarhaddon discovered this conspiracy and crushed it in 670 BC. The plot was apparently sufficiently widespread, and its suppression sufficiently brutal, that the latter was recorded in two Babylonian chronicles covering this period.[26] In an attempt to avoid such internal threats, Esarhaddon made regular oracular queries concerning the trustworthiness of particular officials.[27] He also undertook the 'substitute king ritual' multiple times throughout his reign, a procedure whereby the monarch sought to avoid harm by changing place with a substitute for up to 100 days.[28] He also had monuments erected in the provinces showing him with his two new crown princes,[29] and may have imposed additional loyalty oaths on certain groups.[30] Perhaps related to all of this, the king's private correspondence reveals that Esarhaddon continued to struggle with his health during this period.

After Esarhaddon's death, his two crown princes took their places on the thrones of Assyria and Babylonia respectively, apparently without opposition. Esarhaddon's mother, Naqi'a, outlived her son: a covenant she imposed in the wake of her son's death on various members of the royal court, including Esarhaddon's older sons, swearing them to loyalty to Ashurbanipal, their new monarch, has survived until the present day. Like most known treaty documents, it is likely either a

---

23 Radner 2019, 319.

24 On Egyptian scholars at the Assyrian court in the wake of the 671 BC invasion, see Radner 2009.

25 See discussion in Nissinen 1998, 135–150, Radner 2003c, Frahm 2010, 120–126, Radner 2016, 52f.

26 RINAP 4, 6–8; Grayson 1975, no. 1: iv 29 and no. 14: 27'. See also discussion in Radner 2003c.

27 SAA 4, lxiii.

28 Radner 2003c, 171–176; Parpola 1983, xxii–xxxii.

29 Such monuments survive from the provincial capitals Til Barsip (modern Tell Aḥmar) and Sam'al (modern Zinçirli), published as RINAP 4, nos. 97 and 98 respectively.

30 See SAA 2, nos. 7 and 14. The latter is not a treaty, but seeks to reinforce the message that the imposition of Esarhaddon's succession covenant was important.

draft or a chancellery copy, and thus it is not possible to be completely sure if it was enacted, although most scholars presume that it was.[31] In the short term, then, Esarhaddon's succession arrangements can be considered to have been successful. It is worth noting, however, that Šamaš-šumu-ukin revolted against his brother's rule in 652 BC, some twenty years after his father imposed his succession covenant. This prompted a war between Assyria and Babylonia that lasted for four years and likely accelerated the Assyrian Empire's decline: by the end of the seventh century, the Kingdom of Assyria was no more.

1.2 Terminology used in this study

1.2.1 Vigilance

It is the aim of this study to examine the imposition of and responses to Esarhaddon's succession covenant through the lens of 'vigilance', applying this concept to the source material as an ideal type. The term 'vigilance' as employed in this study differs from colloquial English use, and refers to 'a linking of individual attentiveness to goals set by others'.[32] In this way, the concept stresses the potential role of the human both as a subject and an object of attention. This study thus conceives of attention as a resource that can be directed by others. Across time and space, groups and organizations have attempted to harness the resource of attention and use it to accomplish objectives that they have set.

The term 'vigilance' used in this sense was first coined by the historian Arndt Brendecke, in a context that is particularly relevant and instructive for this study. Brendecke utilized the concept of vigilance in his 2009 study of the Spanish Empire, a revised version of which appeared in English translation in the year 2016.[33] The English title of this work is *The Empirical Empire*, and in it Brendecke traces the relationship between 'information and the exercise of sovereignty' in the Spanish Empire.[34] As Brendecke argues, the task of the Spanish monarch – overseeing a vast territory from which he was separated by the Atlantic Ocean – constituted a new kind of challenge of state governance, one that centred on issues of knowledge. The sovereign was limited both in terms of his knowledge about

31 SAA 2, no. 8.
32 As defined in the website 'SFB 1369 "Vigilanzkulturen"': https://www.sfb1369.uni-muenchen.de/forschung/index.html.
33 Brendecke 2009 and 2016.
34 Brendecke 2016, 1.

these 'very distant and frequently unfamiliar' places *per se*, as well as about the actions of those located there.[35]

Vigilance, according to Brendecke, was one of the key tools used to overcome this problem. Subjects of the Spanish crown were encouraged to report to the king, thus exercising the *libertad de escribir* 'freedom to write', granted to all subjects of the Spanish crown.[36] This rhetoric, coupled with a postal system that made the claim credible, meant that any subject could conceivably report to their far-away monarch. The result of this was that, at least in theory, any two subjects who came into contact with one another could potentially relay to the crown any disloyalty that they observed on the part of the other, resulting in negative consequences for the latter party. Brendecke suggests that the knowledge of this led subjects to anticipate external monitoring, prompting them pre-emptively to adjust their behaviour to ensure that others would have nothing negative to report. As such, when effective, the *libertad de escribir* was an important instrument by which the Spanish crown was able to maintain control over the colonies both actively and passively.

While the Assyrian monarch was not separated from the lands of his dominion by an ocean, there are clear parallels between the case of the Assyrian Empire and that of early-modern Spain. The empire during the reign of Esarhaddon was at its greatest extent so far, and in particular its provincial holdings had ballooned in size over the preceding seventy years. The territories of the empire were spread across a wide variety of terrains, many of which were difficult to access or indeed inaccessible at particular times of the year. While the Assyrian king was very mobile, going with his army on campaign, for instance, he would certainly not have been intimately familiar with all of the regions under his direct or indirect rule, and relied on written correspondence to keep abreast of what was going on in his empire. Furthermore, in the case of Esarhaddon specifically, some letters sent to him by his physicians seem to indicate that he was prone to bouts of illness which left him either unable or indisposed to leave his chambers.[37] Meanwhile, as in the Spanish instance, the empire was controlled on the ground by a relatively small number of people. These individuals had access to substantial wealth and manpower, and were frequently positioned at a substantial geographical remove from the monarch, potentially allowing them to increase their own power or even launch a coup attempt without Esarhaddon's immediate knowledge.

35 Brendecke 2016, 1.

36 Brendecke 2016, 117.

37 Radner 2003c, 169, with additional bibliography.

While this study does not take a comparative approach, by exploring the Assyrian Empire under Esarhaddon through the lens of vigilance, I hope to add to the understanding of pre-modern imperial attempts to harness attention as a means of 'remote control'.[38] Overall, the importance of lateral and bottom-up watchfulness as a central tenet of control in pre-modern empires has been too little considered in scholarly discourse. In the field of Assyriology, the fact of the frequent reports to Esarhaddon and evident watchfulness of subjects towards one another has been commented upon many times,[39] but remains in need of systematic examination and more detailed theoretical exploration. This study seeks to further examine the attempt in Assyria to guarantee the stability of the empire by bridging the knowledge gap between the monarch's awareness and the locations that he was not able to observe directly. It also seeks to highlight a previously underappreciated aspect of the manner in which the crown attempted to achieve this: harnessing the attention of Assyria's subjects for its own purposes.

Brendecke rejects the use of the term 'surveillance' to describe what he considers to be 'vigilance'. He states that the former describes a generalized '*sur*-veiling (watching *over*) that never attains the intensity and sharpness of detail characteristic of local watchfulness'.[40] Surveillance is something that can be carried out remotely over large areas and, as such, at least in a pre-modern setting, can only ever be a diffuse watchfulness disseminating from the top downwards. Vigilance, meanwhile, is intimately linked with the human senses and the attention of individuals. It is immediate, direct, and exists as a series of peaks and troughs, dictated in large part by the possibilities and limitations of an individual's own attention span. As a basic concept, it also contains the possibility for greater diversity in terms of its directionality than does surveillance. Beyond this, the term 'surveillance' is so closely linked with the technologies of the contemporary or twentieth-century 'surveillance state' that it risks anachronism when applied to the distant past. The term 'vigilance', in contrast, has no such associations.

Various successful studies have examined topics that, for the purposes of this investigation, are very similar to vigilance, albeit through the lens of surveillance. As early as 1927, the historian Ernst Kantorowicz, for instance, explored the topic of 'mutual surveillance' in his study of the reign of the Holy Roman Emperor Frederick II (1194–1250).[41] More recently, in his examination of a similar topic in the context of the United States, *Citizen Spies*, Joshua Reeves used the term 'lateral surveil-

38 As Brendecke describes it (2016, 113).

39 See the discussion in 1.3.1 below.

40 Brendecke 2009, 114.

41 As discussed by Brendecke (2016, 113; Kantorowicz 1985, 251–255). The German term used by Kantorowicz is 'gegenseitige Überwachung'.

lance' to describe the monitoring of private US citizens by one another from the eighteenth century until the present.[42] While both of these studies are highly successful in many respects, the terms themselves leave something to be desired. Reeves and Kantorowicz, despite having written these monographs almost a century apart, and in different languages, are faced with a similar challenge, namely that of circumventing the strong top-down associations of words that, in both languages, refer to watching over someone. Their respective solutions lead to somewhat oxymoronic phrases that, in part, highlight these associations: the term 'surveillance' with no qualifications means top-down surveillance, and anything else requires an adjective disclaiming that fact. In contrast, 'vigilance' does not presuppose an institutional or hierarchical direction, and can thus be 'top-down' in the sense of a superior monitoring a subordinate, but it can just as easily be 'bottom-up' and 'lateral'. It can also be 'mutual', and – crucially – it can be directed both towards others and towards the self. This latter means that watchfulness of the self, along with the adaptation and self-censorship that come with it, do not have to be divided into their own separate conceptual category, but can be analyzed as part of the broader phenomenon of vigilance. In a field such as the study of the Assyrian Empire, where many of the historical investigations are – primarily due to the nature of the available evidence – very strongly focused around top-down state interventions and the person of the king, it is of particular value to employ concepts that allow for a reading of the sources in ways that challenge this model of the workings of empire.

This study is written within the context of the DFG-funded Collaborative Research Centre, 'Cultures of Vigilance. Transformations – Spaces – Practices', the core objective of which is to 'research the historical and cultural foundations of vigilance'.[43]

1.2.2 Space

It is important to note that, while there are various similarities between the Spanish and Assyrian cases as described thus far, there are some key differences between the territorial realities of these empires. So too, the focus and framing of this study diverge in various ways from those of Brendecke's *The Empirical Empire.* The most obvious difference between the spatial dynamics of the two cases, as al-

42 Reeves 2017, 14–15.

43 A description of the Collaborative Research Centre and its aims can be found on the website: 'SFB 1369 "Vigilanzkulturen"': https://www.sfb1369.uni-muenchen.de/index.html.

ready indicated, is that the early modern Spanish court was separated from most of its holdings by a vast ocean. Thus, the monarch, on one side of the divide, was required to collect information on what was happening on the other side, a place in which he was 'blind'.[44] Naturally, Brendecke frames his analysis around this dichotomy: the court in Madrid, a familiar location for the monarch, and the colonial territories, an unfamiliar zone about which knowledge was collected. As such, Brendecke's focus is on vigilance as a method of long-distance control of the colonies.

While some zones of the Assyrian Empire would certainly have been unfamiliar to the Assyrian monarch, this study seeks to move away from the centre/periphery model. Geographical distance from the crown is an important metric when thinking about the implementation of and reaction to a duty of vigilance in the form of the succession covenant. Geographical features, such as waterways and mountains, are also important to understanding the spatial dimensions of vigilance within the Assyrian Empire.[45] Nevertheless, this is certainly not the only category of space that is relevant when considering the distribution of vigilance across Assyria's holdings.

The rulers of the Assyrian Empire themselves, for instance, seem to have conceived of the state primarily as being divided into a series of administrative units. One of the primary distinctions between these units was their status as either Assyrian province or client state. There is frequently correlation between geographical and administrative space in the Assyrian Empire: provinces are often either geographically closer or more easily accessible from the capital than client states, which are located further away. Nonetheless, this is by no means always the case and the two concepts, geographical distance from the crown and administrative category, are to be clearly delineated from one another. In addition to this, it is also important to differentiate between individual units, even within the same category. Some provinces, for instance, had belonged to Assyria since before the founding of the empire, while others had been annexed only a few years prior to the imposition of Esarhaddon's succession covenant. Even within individual provinces or client states, one finds other administrative categories, such as the town and the countryside. It is one of the tasks of this study to explore how geographical and administrative differences interacted and affected the imposition and enactment of vigilance across the empire.

44 Brendecke 2016, 42.

45 On the geography of the Ancient Near East, see Liverani 2014a, 17–33. On the Assyrian heartland more specifically, see Ur 2017.

Finally, this work also focuses on social distance from the crown. Whereas in the Spanish case, the desire to collect reports was focused on the colonies, in the Assyrian case, any interaction at which the king or his crown prince were not personally present (and indeed some where they were), was considered potentially relevant to the task of protecting the dynasty. This study seeks to draw conclusions about the degree of interest that the crown had in imposing the duty of vigilance on those at different degrees of social distance from the crown itself. This includes members of the king's 'inner circle', who had direct access to the monarch,[46] high-ranking officials in the provinces or client states, far away but able easily to write to the king, and the ordinary inhabitants in various urban and non-urban locations throughout the empire.

In thinking about social distance, this study draws in part on Eleanor Robson's recent monograph, which she terms a 'social geography of cuneiform scholarship'.[47] In this work, Robson establishes the value of thinking about ancient Assyrian and Babylonian 'knowledge networks' in spatial terms. While the dynamics at play in this study are significantly different to those of scholars exchanging knowledge, some of the concepts that Robson employs are useful for my purposes. Firstly, her discussion of the relevance of Latourian Actor-Network Theory (ANT) to first millennium cuneiform culture can be applied here. In contrast to other network models, Bruno Latour argued for the inclusion of the non-human in the analysis, allowing – as Robson notes – for the incorporation of divinities and clay tablets in the analyses of Assyriologists.[48] Such notions are useful when attempting to reconstruct the social distance between the king and his subjects. While the king himself may have been at a substantial remove from a subject when conceptualizing them as nodes made up of a string of human contacts, it is perhaps correct to view these distances as effectively being collapsed by a belief in a mutual relationship with a particular deity, or by a cuneiform tablet written on behalf of the king or in his name and presented to or read aloud to a subject. The reverse can also be true, and to this effect Robson quotes the geographer David Livingstone, who comments that: 'People close together physically may be 'miles apart' in terms of social distance or cultural space, living, as it were, in totally different worlds.'[49] Thus, a monarch can be socially fairly close to a provincial governor living several

46 The phrase 'inner circle' is used by Simo Parpola (SAA 10, xxv–xxvii). See also his analysis of the scholarly milieu in Parpola 1983, xv–xxi.

47 Robson 2019, 28–42.

48 Robson 2019, 40; Latour 1987.

49 Robson 2019, 40; D. N. Livingstone 2003, 6.

days' journey away, exchanging frequent correspondence, while living in at a vast social remove from some of his subjects living in the same city as him.[50]

1.2.3 Responsibilization

In conceptualizing vigilance as defined above, it is necessary also to think about responsibility and its transmission, which is here termed 'responsibilization'. In order for vigilance to work, the subjects of the Assyrian Empire had to be made responsible for directing their watchfulness towards targets and topics dictated by the crown. In addition to this, it was important that they be responsible for taking the correct course of action in response to observing something of relevance. In the US context, Joshua Reeves refers to this as 'seeing/saying responsibility'.[51] Thus, using the senses to gather information is one side, and action – in this case 'communicative action' – is the other side of the civic responsibility of the American citizen in Reeves's analysis.[52] Reeves correctly identifies a failure on the part of those interested in surveillance over recent years to adequately stress the latter element in this dyad: in his analysis this is limited to communication, and other more direct forms of action as a consequence of what is seen or heard are not considered. This is a key difference between the model of vigilance described by both Brendecke and Reeves in their respective analyses and the focus of this study. While communicative action was a key component of the responsibility of Assyrian subjects to the crown – 'microphysical acts of communication', as Reeves describes it in the US context[53] – other types of action were also required of the Assyrian subject. Thus, a kind of state-sponsored vigilantism can be considered as part of the responsibility of Assyria's subjects. This fits well with the conceptual framework introduced by Reeves in his analysis of 'the body as surveillance technology'.[54] Reeves highlights the human capacity to use the senses (particularly sight and hearing), as well as the communicative capacity of the mouth. In addition to these, though, other bodily abilities must be considered in the Assyrian context, such as the ability of the body to attack or physically hinder others from acting against the crown.

The dual responsibilities inherent in vigilance, attention and reaction, are united both by their connection to the human body, but also by the challenge of

50 On the wide range of people who sent letters to the Assyrian monarch, see Radner 2015b.
51 Reeves 2017, 3.
52 Reeves 2017, 5–8.
53 Reeves 2017, 8.
54 Reeves 2017, 8–11.

directing them in a manner consistent with the goals that they are intended to fulfil. It is necessary for the would-be vigilant person to know where they are to direct their attention and when, thus scaling their attention as appropriate. It is also vital for them to know how to react to the things that they perceive. It is not, however, sufficient for the potentially vigilant person to know what they are required to look or listen out for and how they are supposed to react when they notice it. Rather, they must have sufficiently accepted and internalized these things in order to participate in vigilance. In this way, these people are required to conceive of themselves as responsible for directing their attention towards the relevant object and for acting based on their observations. The transfer of responsibility in this context is here described as 'responsibilization'.

The term 'responsibilization' has its roots in criminology and sociology.[55] At its most basic, responsibilization is a social transfer by which a societal expectation is conveyed from a primarily responsible party to a secondarily responsible party. In the context of vigilance, the responsibility that is being transferred is the 'enhanced performance of alertness'.[56] Relevant factors when analyzing responsibilization include the function and source of the primary responsibility, the transmitter, the reason for the transfer of responsibility, the mode of the transfer, as well as the actor with secondary responsibility, the degree of compulsion that they are under, the scope of the internal effects of transmission, and its external effects.[57]

In the Assyrian Empire, the responsibility for monitoring the self and others on behalf of the Assyrian crown was to be transferred by Esarhaddon to all subjects of the empire. The chosen mode of transfer was the succession covenant on behalf of Ashurbanipal, which was designed both to give subjects the sense that they were irreversibly bound to their agreement to do so, on pain of divine punishment, and also to describe and explain what they were responsible for, namely what exactly they were required to watch and listen out for and how they were required to react based on their perceptions.

55 For a discussion of the use of the term and bibliography, see Kölbel et al. 2021, 4f. See also Gadebusch Bondio et al. 2023. Other examples of the use of the concept include Hinds and Grabosky 2010 and Böschen 2018.

56 These definitions follow those given in Kölbel et al. 2021, 5: 'Responsibilisierungen im Kontext von Vigilanz beziehen sich auf eine erweiterte Wachsamkeitsleistung, die an sich einem primärzuständigen Akteur in einem übergeordneten Interesse obliegt.'

57 These concepts are discussed in detail in Kölbel et al. 2021.

1.3 Source material

1.3.1 Esarhaddon's succession covenant

Esarhaddon's succession covenant is the central source for this study. The Assyrian term for covenants was *adê*, which is also translated by some modern scholars as 'treaty' or 'loyalty oath'.[58] The covenant was written in a mixture of the Assyrian and Babylonian languages using the Assyrian variant of the cuneiform script on large clay tablets, measuring approximately 30 x 45 cm.[59] While it is by no means impossible, or even unlikely, that the covenant was translated into other languages, there is no direct evidence for this.[60] The decision to compose the covenant mostly in the Assyrian language, although some portions are in Babylonian,[61] portrays the linguistically-diverse Assyrian Empire as a place in which Assyrian is the central language.

Some eleven manuscripts of the text are preserved: this number is not entirely certain because each of the exemplars is smashed into fragments, and, as some

58 As I have discussed the topic of the definition and translation of the term *adê* elsewhere (Tushingham 2023), I will not rehash my arguments here. While the phrase 'Esarhaddon's Succession Treaty' (EST) is commonly used in the secondary literature, I consider 'covenant' a more accurate translation than 'treaty', particularly as applied to Esarhaddon's *adê* of 672 BC. As such, I generally use the term 'covenant', unless the context demands otherwise.

59 Different manuscripts have slightly different measurements (see SAA 2, xlviii and Lauinger 2012, 90).

60 A good example of a case in which this certainly did happen is the Bisitun inscription of the Achaemenid king Darius I (r. 522–486 BC). The inscription is engraved on Mount Bisitun in the Old Persian, Elamite and Babylonian languages and their respective forms of the cuneiform script (for recent discussion see Brosius 2021, 48 f.). An Aramaic version of the inscription, in alphabetic script, was found at Elephantine in Egypt (Kratz 2022; the papyri of Elephantine are further discussed in Chapter 8.4).

61 Watanabe 1987, 43 f.; Watanabe 2017, 473 f. Kazuko Watanabe calculates the relative percentages of Assyrian and Babylonian lines of the covenant composition as 81 % and 19 %, respectively. She notes that the vast majority of the Babylonian lines are found in the traditional curse portions. While Watanabe (2017, 487) suggests based on this observation that the covenant was 'addressed to Babylonians in the first place… And the main purpose of the ESOD was evidently to place Babylonia in the position of puppet-state under Assyria (see Watanabe 2014, 165)'. The fact that the overwhelming majority of the covenant's lines were composed in Assyrian, however, undercuts this assertion. In particular, the oath itself seems to suppose that those repeating it will be speaking Assyrian, as it is written in that language. As such, it seems clear that the use for the traditional curses of Babylonian, which was frequently used in literary texts composed in Assyria, was intended more to convey a sense of their connection to the history of the Mesopotamian curse tradition than it was to appeal to a specifically Babylonian audience. In this way it worked similarly to the use of the ancient cylinder seals, to convey a sense of permanence.

of the manuscripts were discovered as a group and in other cases the precise find location is lost, it is not always clear to which manuscript a fragment belongs. That the tablets were designed to be disseminated across the empire and displayed is evident from both their content and design: they had a horizontal piercing intended to allow them to be hung up on a string or chain.[62] Nine of the known manuscripts were found in 1955 in the Nabû temple, Ezida,[63] at the central Assyrian city of Kalhu, modern Nimrud.[64] Three fragments of the same composition were also found at Ashur, one of which was published in 1939/40, while the second and third were published only recently, in 2009.[65] While they can securely be stated to come from Ashur, their precise findspot is unknown.[66] Also in 2009, a new manuscript was discovered at the Assyrian provincial capital of Kullania, modern Tell Tayinat in the Turkish province of Hatay (see Figure 1).[67]

Until this most recent discovery, the status of Esarhaddon's succession covenant was the subject of much debate. As the treaty partners of all of the manuscripts found at Kalhu were eastern city-lords, clients of Assyria, it was hypothesized that the covenant had been composed specifically for them, and that this was an indication that they had special status at the Assyrian court.[68] In contrast, others argued that the identity of the legal parties in these manuscripts was more likely an accident of preservation, with the covenant having been imposed on all subjects of the Assyrian king.[69] The discovery of the Kullania manuscript confirmed the latter viewpoint, as it was imposed on the provincial governor of Kullania, his administration and the people of that province.[70]

The thrilling discovery of the new manuscript from Kullania has led to a flurry of new publications on Esarhaddon's succession covenant. The bulk of modern

62 As noted in Lauinger 2011, 11 and discussed further in Harrison and Osborne 2012, 137 and Lauinger 2012, 90.

63 This is the Sumerian name of the temple, meaning 'true house' (A. George 1993, 160, no. 1239). Sumerian was a *lingua sacra* in Assyria and Babylonia in this period.

64 Kazuko Watanabe is the leading authority on the Kalhu fragments of Esarhaddon's covenant, and reconstructs 'at least nine copies' (Watanabe 2019, 238). The text editions of the Kalhu version of the covenant are: Wiseman 1958, Watanabe 1987 and SAA 2, no. 6.

65 For a brief overview, see Watanabe 2019, 237 f. The text editions are Weidner 1939/40 and Frahm 2009a, nos. 70 f.

66 Frahm 2009a, 135.

67 Lauinger 2012.

68 Liverani 1995.

69 Thus Radner 2006. Writing in the same edited volume, see Steymans 2006, 349 for the argument that it is 'unthinkable' that King Manasseh of Judah was not required to enter into an *adê* with the Assyrians concerning Ashurbanipal's succession. Note too that Hans Ulrich Steymans had argued elsewhere that the covenant tablets were to be displayed and worshipped as 'icons' (2003).

70 For the text edition, see Lauinger 2012.

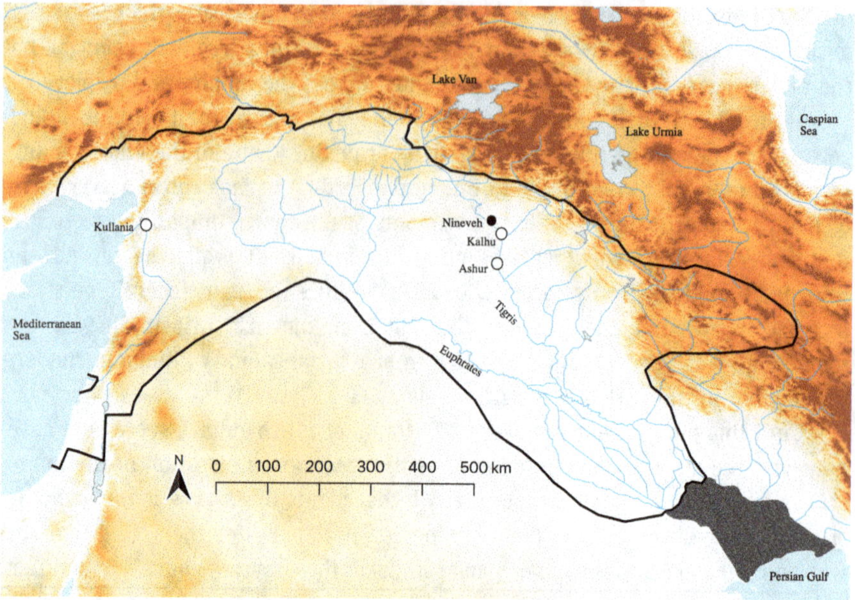

Figure 1: Map of find locations of manuscripts of Esarhaddon's succession covenant: Ashur, Kalhu and Kullania. The capital, Nineveh, is marked in black for reference. The shaded area indicates the ancient coastline.

scholarship on the subject has addressed the evident link between the covenant composition and the biblical Book of Deuteronomy. As the kingdom of Judah was a client state of Assyria at the time of the covenant's imposition, the Tell Tayinat manuscript has reignited the debate on the subject of whether those who composed Deuteronomy would have had access to the covenant text, in what form, and whether and in what way the book was influenced by it.[71] These debates, of course, tie into larger questions in the field of biblical studies concerning Deuteronomy's ideology, chronology and tradition history. Assyriologists, meanwhile, have addressed a variety of topics such as the religious and ideological implications of the *adê*-documents, the insights that the text provides into the structure of the provincial administration, and Iron Age textual mass-production, as at least 110 and possibly more like 200 individual manuscripts of the *adê* were drawn up within

71 As in, for instance, Levinson 2010, Steymans 2013 and Crouch 2014. The debate on this topic has continued apace, as illustrated by the recent publication of an issue of the journal *HeBAI* dedicated to the topic of 'the treaty framework of Deuteronomy', with contributions referencing Esarhaddon's succession covenant including Edenburg and Müller 2019, Lauinger 2019, Morrow 2019, Pakkala 2019 and Steymans 2019.

a short timeframe.[72] Further studies concerned the logistics of the covenant's imposition and its use in ceremonial contexts.[73]

The crown's very clear demands that Assyria's subjects monitor themselves and others, reporting on those around them, have certainly not been lost on modern scholars.[74] Nonetheless, no sustained study of the communication, enactment, and outcome of this has yet been made. It is the aim of the present study to offer a thorough examination of Esarhaddon's succession covenant through this lens, using the concept of vigilance outlined above.

The contents of Esarhaddon's succession covenant can be broken down into sections or categories in various ways, but the basic structure is fairly simple:[75]

1. Seal identification (i–iv);
2. Introduction of legal parties (§ 1: 1–12);[76]
3. Divine witnesses and adjuration (§§ 2–3: 13–40);
4. Stipulations (§§ 4–36: 41–413);
5. Curse section I (§§ 37–56: 414–493);
6. Oath (§ 57: 494–512);
7. Curse section II (§§ 58–106: 513–663);
8. Date and colophon (§ 107: 664–670).

These eight sections each fulfil a distinct purpose, and it is clear that the text of the covenant tablets was carefully composed, with particular focus on its intended audience and its objectives. While the composition can be viewed as a fairly het-

72 On the religious and ideological implications, see Fales 2012, Lauinger 2013, Watanabe 2014, 2017, 2020 and 2021. On mass production, see Watanabe 2015, Lauinger 2015 and Lauinger 2021. On the provincial administration, see Ponchia 2014.

73 This work has been carried out by Jacob Lauinger (most recently in Lauinger 2019), as well as Cristina Barcina (2016; see also Barcina 2017).

74 Various studies have highlighted the importance of related topics, such as that of A. Leo Oppenheim in his 1968 study on 'The Eyes of the Lord', who states that 'a sort of secret information service' may have been one of the novel innovations of the Assyrian Empire compared with prior Mesopotamian states (Oppenheim 1968, 174). Other scholars have written, for instance, on watchfulness as a key component good service to the Assyrian crown (Baker and Gross 2015; Fales 2011) and on the importance of informants under Esarhaddon and the fact that this was connected to some degree to the imposition of the succession covenant (SAA 16; Frahm 2010; Fales 2017a; Radner 2019).

75 These designations seek in particular to allow for the analysis of the covenant composition from the perspective of its intended practical use. For alternative typologies, see recently Piccin 2020, whose identification of the structural elements of the *adê* draws on SAA 2, xxxv. Kazuko Watanabe also provides a slightly different interpretation of the structure in Watanabe 2017, 473.

76 Note that two different versions of the introduction of legal parties are known to modern scholars, one for the client states (found in the Kalhu versions) and one for the provinces (found in the Tell Tayinat version).

erogenous 'patchwork',[77] the segments combine to produce a complex but cohesive impression of the duty of vigilance that Esarhaddon wished to convey to his subjects, as well as its projected impact and the consequences of failure to adhere to it.

Figure 2: Manuscript of Esarhaddon's succession covenant, Kalhu version. ND 4327; Iraq Museum, Baghdad.[78]

77 As the Neo-Assyrian *adê* is termed in Piccin 2020.

78 Reproduced from Wiseman 1958, pl. 1.

1.3.2 Other royally-commissioned sources

The succession covenant is by no means the only source that sheds light on the call to vigilance made by Esarhaddon in 672 BC. In particular, various texts written in the cuneiform script and Assyrian language that can be described as 'royally-commissioned' sources are useful in this regard. This category encompasses all texts that were composed on behalf of the Assyrian crown by royal advisors working at least nominally under the supervision of the king. Beyond the *adê*-covenants themselves, the two main royally-commissioned texts discussed in this study are *Esarhaddon's Apology* and *Esarhaddon's Letter to the God Aššur.*[79]

Esarhaddon's Apology is not an independent text, but rather the prologue to Esarhaddon's longest known royal inscription, known today as Nineveh A.[80] Royal inscriptions were typically written in the first person singular, from the perspective of the Assyrian king. Stylistically, the Assyrian royal inscriptions vary substantially from monarch to monarch, in a way and to a degree that cannot be wholly explained away by factors such as changes in the staff responsible for drafting the inscriptions. As such, it is likely that individual kings had significant influence on the content and style of their royal inscriptions, even though they certainly did not compose the texts themselves.[81]

Besides the gods, who judged the king's actions, the 'target audience' of a royal inscription was posterity, specifically future kings of Assyria. It is for this reason that royal inscriptions were frequently built into the fabric of palaces and temples as foundation deposits. The latter is the case for the Nineveh A inscription, a building narrative concerning the *ēkal–māšarti*, 'Review Palace', in Nineveh. The prisms on which it was inscribed were intended for use as foundation deposits there.[82] Subsequent monarchs were expected to renovate such buildings and thus find the royal inscriptions of their forebears.[83] Nevertheless, these are by no means the only people who would have been familiar with compositions of this genre. Royal inscriptions were displayed in a wide variety of locations, including temples

79 Published as RINAP 4, nos. 1 and 33. On the royal inscriptions generally, see recently Frahm 2019, as well as RlA 6, 65–77 s.v. Königsinschriften. B. Akkadisch, Grayson 1980, esp. 149–171 and Fales 1999.

80 RINAP 4, no. 1: i 1 – ii 11. Note that the Apology was also used as a prologue to two later prisms bearing a very similar inscription to Nineveh A, known as Nineveh F/S and Nineveh D/S (RINAP 4, nos. 5 and 6).

81 This characterization of authorship in the royal inscriptions follows Hayim Tadmor's argument (1997, 328 f.).

82 On this palace, see most recently Maul and Miglus 2020. See also Kertai 2015, 150–153.

83 Radner 2005, esp. 203.

and palaces, but also in public areas in central and provincial cities, as well as on rock surfaces in non-urban environments, and therefore some of them at least would have been known to contemporary audiences.[84]

The process of drafting, re-drafting and editing the royal inscriptions – and indeed all royally-commissioned texts – would have served as an opportunity for the king and his advisors to reflect on the events and decisions of the monarch's reign thus far, and the manner in which they should be framed. Such an exercise would surely have also informed their plans for the future and served to create in real time a shared narrative about the world in which they were operating.[85] As such, the royal inscriptions can be seen as instrumental in both recording and determining the course of a monarch's reign. Seen in this way, the king himself, along with his inner circle, perhaps constituted the most significant audience for this genre of texts.[86]

Esarhaddon's Apology is termed thus by modern scholars because it chronicles in considerable detail the difficult circumstances of Esarhaddon's succession to the Assyrian throne already briefly discussed above, presenting him as the rightful successor to his father, Sennacherib. Esarhaddon presents the support of the gods and the people of Assyria as keystones of his legitimacy. He also stresses the importance of the imposition by his father of a succession covenant on his behalf, thus choosing to portray the use of succession covenants as a vital component of Assyrian legitimacy in a composition disseminated mere months before he enacted his own succession covenant.

Esarhaddon's Letter to the God Aššur, meanwhile, is often described as a royal inscription, but the text is framed as a letter from the monarch to his divine master, the god Aššur. Aššur was the head of the Assyrian pantheon and his cult centre

---

84 Frahm 2019, 141, with relevant literature.

85 Indeed, the royal inscriptions did not always stick strictly to narrating what had already happened during a monarch's reign, they also sometimes described planned events as though they had already occurred. As discussed in Novotny 2014, Esarhaddon himself did this when describing his renovations at the Aššur temple: 'In true Assyrian style, the work is described as having already been completed, although little, if any, actual work apart from the demolishing of the old temple and the fashioning of bricks had likely taken place' (Novotny 2014, 95).

86 Eckart Frahm (2019, 142) states the following on the subject: 'It seems, then, that Neo-Assyrian royal inscriptions were primarily addressed to the king himself, later Assyrian rulers (who would find the inscriptions of their predecessors in foundation deposits), and the leading circles of Assyria's political and religious establishment. Exposing these "inner" audiences to the royal *res gestae* was a way of reinforcing among them the aggressive ideology on which the Assyrian state was founded. A secondary, and clearly less important purpose of the texts may have been to instill fear in foreigners by showing them the terrible things Assyrian kings would do to those who did not submit to them.' See also Liverani 2014b.

was located in the city of Ashur, modern Qal'at Sherqat, located in the Assyrian heartland.[87] A particularly important element of the distinction between this text and a royal inscription is the fact that such 'letters to gods' were not used as foundation deposits, but rather were written on tablets, and likely read aloud in public places.[88] *Esarhaddon's Letter to the God Aššur* specifically may well have been read out in Ashur, Assyria's religious capital.[89] *Esarhaddon's Letter to the God Aššur* recalls his campaign of 673 BC to the small client kingdom of Šubria, located in the Upper Tigris region of modern Turkey, and was likely composed in that year. The justification for this campaign is the Šubrian king's failure to uphold his *adê*-treaty with Esarhaddon. In particular, Esarhaddon accuses his client of harbouring Assyrian fugitives, and it has been suggested by some modern scholars that these runaways included Esarhaddon's own treacherous brothers and their supporters.[90] In either case, the *Letter* links the concept of *adê* to the claim that not only client kings, but also the inhabitants of the provinces, will inevitably be punished for any disloyalty to the Assyrian crown.

1.3.3 Who composed royally-commissioned sources?

I have referred to 'royally-commissioned' texts and to the monarch's 'advisors', or 'inner circle', but it is worth dwelling at greater length on the identity of the people who wrote these sources, including the succession covenant. The composer, or composers, of the covenant text are not mentioned explicitly in any of the available sources. Despite this, I consider it most probable that the *rab–ṭupšarri* 'chief scribe' would have been responsible for the text's composition. The chief scribe was a powerful court official and prominent scholar, with many responsibilities including that of overseeing the composition of royal inscriptions, as well as managing the royal libraries.[91]

At the time of the covenant's imposition in the month of Ayyaru (II, i.e. April/May) of 672 BC, the chief scribe was a man named Issar-šumu-ereš. He came from an old and influential family of royal scholars, descended from one Gabbu-ilani-ereš, who had been chief scribe under the kings Tukulti-Ninurta II (r. 890–884 BC) and Ashurnasirpal II (r. 883–859 BC). Issar-šumu-ereš's father, Nabû-

87 The source is published as RINAP 4, no. 33.
88 On this genre of text, see Pongratz-Leisten 1999, 210–265.
89 On the subject of such texts being read aloud, see A. Leo Oppenheim in 1960, as well as Frahm 2019, 141 and Pongratz-Leisten 1999, 273f. and 2013.
90 For instance Leichty 1991 and Fuchs 2012.
91 On the role of the chief scribe, see Luukko 2007. More recently see also Gross 2020, esp. 55–56.

zeru-lešir, had himself been chief scribe under Esarhaddon, while his paternal uncle, Adad-šumu-uṣur, was that king's personal *āšipu* 'healer'.[92] Prior to his promotion to the position of chief scribe, which likely occurred upon his father's death in 673 BC, Issar-šumu-ereš had been engaged at court and worked in tandem with other members of the king's inner circle – including his father and uncle – to advise the king.[93] Before his promotion, it has been suggested that he was the *ṭupšar–ēkalli* 'palace scribe', the official in charge of the chancery.[94] While this is conjectural, it is certain that he was an important man even before he was elevated to the position of Esarhaddon's chief scribe and *ummānu* 'chief scholar', a station for which he had doubtless been prepared by his close work with his father.

One must therefore picture Issar-šumu-ereš as someone who belonged to a rarified group of highly-educated scholars, a man whose family had a long and proud history of proximity to the crown and who had been prepared for a career in scholarship and royal service from an early age. Despite this, the model of royal patronage of scholars and advisors in place at the Assyrian court was a precarious one, with the loss of royal favour a constant and real threat. Issar-šumu-ereš would have known this well from his own family history. His grandfather, Nabû-zuqup-kenu, had been an influential scholar and royal advisor under King Sargon II (r. 721–705 BC), but at some point he lost the favour of Sargon's successor, Sennacherib; as a result, Nabû-zuqup-kenu left the royal court at the capital, Nineveh, and decamped to the city of Kalhu. When Esarhaddon acceded to the throne the family's fortunes changed for the better, but this was not to be permanent, as both Issar-šumu-ereš's brother and his cousin later fell out of favour with Esarhaddon's successor, Ashurbanipal.[95]

It is not entirely clear when Issar-šumu-ereš's father and predecessor as Esarhaddon's chief scribe, Nabû-zeru-lešir, died, nor is it certain how long it would have taken to compose the text of the covenant inscription. As such, it seems pos-

92 The Akkadian term *āšipu* has been translated as 'exorcist', as well as 'incantation priest' or '*āšipu*-healer'. For further discussion see Chapter 6.1.2.

93 Luukko 2007; see also Robson 2019 on the Gabbu-ilani-ereš family, esp. 113.

94 Luukko 2007, 253.

95 On the topic of the royal patronage of scholars in this period, see Westbrook 2005, Radner 2011a, Radner 2017b, 221–223 and Robson 2019, esp. 49–97. On Nabû-zuqup-kenu, see May 2018 and Frahm 1999. Issar-sumu-ereš's brother, Šumaya, fell out of favour with Ashurbanipal while the latter was still crown prince, as is clear from two extant letters to Ashurbanipal complaining of his financial situation and seeking to be reinstated (SAA 16, nos. 34–35). At some point in the first years of Ashurbanipal's reign, meanwhile, one of Issar-šumu-ereš's cousins, Urdu-Gula, son of Adad-šumu-uṣur, fell out of favour with the new king. Both Urdu-Gula and his influential father, Adad-šumu-uṣur, wrote to the king attempting to rectify the situation, apparently without success (SAA 10, nos. 224 and 294).

sible that Nabû-zeru-lešir was involved in its composition, at least at the early stages. As both men were in close contact with Esarhaddon, advising him on a range of topics, as the royal correspondence shows, it seems likely that Issar-šumu-ereš, and possibly his father before him, sought Esarhaddon's opinions and approval regarding the covenant's composition as it was in progress. Esarhaddon often solicited opinions from several of his advisors regarding one topic. As such, one may conjecture that other members of the king's 'inner circle' of advisors would have been invited to give their opinions on the composition and perhaps to contribute their ideas.[96] Issar-šumu-ereš himself may also have sought aid from his fellow scholars in Nineveh, and perhaps those in Ashur whom he would have encountered during his frequent trips to the city to oversee ritual activity. He may also have made use of his various contacts in Babylonia for such purposes.[97] This last possibility is particularly interesting due to the inclusion of material in the Babylonian language in the text.

1.3.4 Non-royally commissioned sources

In order to examine the effects of the imposition of Esarhaddon's succession covenant, and, relatedly, the duty of vigilance throughout the Assyrian Empire, it is necessary to identify the various responses to the covenant and its duty of vigilance. The majority of the sources included in this study as potential responses to the covenant and its call to vigilance are cuneiform texts inscribed on clay tablets. Significant textual genres within this category include letters, legal documents, and literary works.

The letters relevant to this topic come from the royal correspondence of the Assyrian crown, and thus are all missives sent to the king or crown prince. Nonetheless, the identities of the people sending these letters vary significantly, as do their locations and the contents of their letters. Assyrian letters were typically written on clay tablets using the cuneiform script and either the Assyrian or Babylonian language, although one letter survives that demonstrates the use of different media along with other languages and scripts, in this case Aramaic written in

96 SAA 10, xxv–xxvii.

97 It is possible to reconstruct such activities and ties from Issar-šumu-ereš's dossier of letters to the king. For an overview, see PNA 2/1, 577–579, s.v. Issar-šumu-ērēš no. 3. That Babylonians could have been involved in the composition of the covenant may be an explanation for the mixing of the Assyrian and Babylonian dialects within the covenant composition (Radner 2021, 180).

alphabetic script.[98] Letters written on clay tablets were encased in clay envelopes bearing seal impressions and the initial lines of the letter, which included the identity of the sender and intended recipient.

Assyrian royal correspondence was transported using a postal system that was likely put in place under King Shalmaneser III (r. 858–824 BC), who undertook various administrative reforms.[99] The long-distance communication network was known as the *hūl šarri* 'king's road', and was maintained across the provincial extent of the empire.[100] Many letters dating to the reign of Esarhaddon have survived until the present day. In contrast to the correspondence that survives from the reign of his grandfather, Sargon II, much of which was authored by high-ranking state officials in the provinces, a high proportion of the letters dating to Esarhaddon's reign come from advisors, scholars and priests, many of whom were members of the court. After the imposition of Esarhaddon's succession covenant, various members of the monarch's entourage, as well as individuals living in the provinces and some client states, made allusions to it in their letters to the king or his crown princes. These documents, then, provide compelling insights into the ways in which letter-writers chose to respond to the covenant, and how they decided to present these responses to the Assyrian crown.

The legal documents discussed in this study, meanwhile, generally recorded private transactions between individuals, although it is possible that some of these texts were of relevance to local officials for tax-collection purposes. Various Neo-Assyrian legal documents from the reign of Esarhaddon onwards reference the *adê ša šarri* 'covenant of the king' in penalty clauses designed to ensure that no one contravenes the legal agreement.[101] These documents are found at various locations across the provincial extent of the empire, and frequently date several decades after the imposition of Esarhaddon's succession covenant, up until the late seventh century.[102]

The literary material incorporated in this study includes two Assyrian literary compositions, *The Sin of Sargon* and *The Underworld Vision of an Assyrian Prince*.[103] Both of these narratives explicitly reference covenants, and appear to reflect in different ways on the role of covenant throughout the empire and its mean-

98 Recently on this issue, see Radner 2021, 173–176, discussing the evidence of the so-called Ashur Ostracon.

99 Radner 2014a, 71; Radner 2008b.

100 Radner 2014a, 71–77.

101 As discussed recently in Barcina 2016, 39–41 and Radner 2019, 322–325.

102 See Chapter 7.3.

103 They are published as SAA 3, nos. 33 and 32 and discussed in Chapters 6.2.1 and 7.1.4 respectively.

ing for particular groups of Assyria's subjects. I argue that they were both in all likelihood composed in the years after the imposition of Esarhaddon's succession covenant and can therefore be regarded as evidence that their authors were reflecting on the implications of such covenants more broadly. While one or both of these compositions may have been written at court, this is not certain. Likewise, in contrast to the royally-commissioned source material discussed above, and regardless of where they were first produced, they were not necessarily written with the oversight or even the knowledge of the monarch.

In addition to cuneiform sources, I discuss the possible influence of Esarhaddon's succession covenant on the Book of Deuteronomy, and in particular on the duty of vigilance evident in Deuteronomy 13. Despite the striking parallels between portions of Esarhaddon's succession covenant and Deuteronomy, the exact dynamics that led to these similarities are the subject of much debate. As such, it is necessary to consider a range of possible reconstructions of the relationship between these two texts and the people behind them.

Finally, I consider a famous Aramaic literary text, *The Story of Ahiqar.* Found on a papyrus in Elephantine that dates to the fifth century BC, this manuscript dates to more than a century after the fall of the Assyrian Empire. Nonetheless, the narrative of the wise sage Ahiqar is explicitly set during the reign of King Esarhaddon. In contrast to the two Assyrian literary narratives discussed above, the *Story of Ahiqar* does not explicitly mention a covenant, but it seems to reflect enduring memories of the climate of vigilance that existed at Esarhaddon's court.

1.4 Structure and contents of this study

This study is divided into two halves. Part 1, 'The royal call to vigilance in 672 BC', analyzes the text of Esarhaddon's succession covenant of 672 BC, as well as other royally-commissioned documents written around that time, and the practical enactment of the covenant viewed through the lens of vigilance. It also examines the spatial dynamics of the duty of vigilance as expressed in the written sources and as may have existed when the covenant was put into practice.

Chapter 2, 'King Esarhaddon and his empire in the covenant composition', sets out the main actors presented in the covenant, as well as examining its depiction of the empire as a spatial entity. The analysis in this chapter argues that, while the covenant composition is to some degree a reflection of the empire and the people who inhabited it, it also seeks to create a stylized and idealized version of these things, that is designed to promote and promulgate a duty of vigilance according to the crown's priorities.

Chapter 3, 'Directing vigilance in the covenant composition', focuses on the sections of the covenant composition that most explicitly seek to set out a programme for vigilance to be followed across the empire. These sections are the stipulations and the oath section of the text. These portions of the composition highlight both what and who are to be considered as potentially dangerous to the crown and how subjects ought to react in a variety of different scenarios. I argue that these scenarios are structured in such a way as to impress upon the subject the impossibility of the failure of the covenant's aims. I also conclude that the various interesting disparities between the framing of the stipulation and oath sections hint that they were intended to be used differently.

Chapter 4, 'Laying the ideological groundwork for enacting the covenant', shifts away from the covenant composition itself, focusing on two compositions commissioned shortly before the covenant's imposition. This chapter argues that these royal narratives present covenants in the Assyrian Empire as functioning as Esarhaddon wished his own succession covenant to function. I also attempt to draw conclusions about the likely target audience of these inscriptions, and suggest that these were groups of particular interest to Esarhaddon when it came to ensuring vigilance.

Chapter 5, 'Putting the covenant into practice', meanwhile, focuses on the practicalities of the covenant's imposition, arguing that the ceremonies in which it was enacted would themselves have been instances of heightened vigilance. I present the covenant's imposition as having taken place in the form of a staggered event and, following Jacob Lauinger's characterization, argue that it may also have been an iterative process, intended to be repeated regularly across the empire.[104]

Part 2 of the study turns its attention to the responses to Esarhaddon's call to vigilance across three spatial categories: the court, the provinces, and the client states. Primarily, I seek to compare what can be discerned about the reality of the aftermath of the covenant's composition with the royal rhetoric as laid out in Part 1.

Chapter 6, 'Responses to Esarhaddon's covenant at the Assyrian court', focuses on the responses to the covenant among Esarhaddon's entourage, specifically Esarhaddon's 'inner circle' of advisors. Members of this group appear to have reflected on the significance of Esarhaddon's succession covenant, as well as the significance of covenants more broadly, in various ways. Two of Esarhaddon's personal healers, Urdu-Nanaya and Adad-šumu-uṣur, wrote letters to the monarch mentioning his succession covenant. So too, members of the monarch's entourage at this time may have been responsible for the literary composition, *The Sin of Sargon*,

104 As Lauinger argues in his 2013 and 2019 articles.

which explores the relationship between the Assyrian monarch and the gods, and mentions a covenant in relation to the death of King Sargon II. Finally, a prophecy compilation that may have been drawn up in the years following 672 BC includes two prophecies that reference a covenant.

Chapter 7, 'Responses to Esarhaddon's covenant in the provinces', expands the geographic scope of my analysis to the Assyrian provincial system. While this chapter seeks to examine the evidence from the provincial system as a whole, I note that the available evidence was found predominantly in the city of Ashur. Other locations that are particularly strongly represented include the capitals of provinces to the west of the Assyrian heartland, such as Harran (near modern Şanlıurfa) and Guzana (modern Tell Halaf). Several letters pertain to the provincial extent of the empire, while private archives from the provinces have also yielded documents that mention covenant, including literary documents and legal texts.

In Chapter 8, 'Responses to Esarhaddon's covenant across the client states', I examine immediate responses to the covenant's imposition. Several letters mentioning covenant were sent from Babylonia in the years following the imposition of the succession covenant. Babylonia was an administrative special case among Assyria's client states, as it was directly ruled by King Esarhaddon. Just one letter comes from a different client state, Arwad. Beyond these direct responses, I also discuss the possible evidence of a response to the covenant in one client state, Judah, in the form of the parallels to Esarhaddon's covenant found in Deuteronomy, as well as the implications of the references to false accusation in the *Story of Ahiqar*.

Finally, the concluding chapter of this study is dedicated to determining the extent to which responsibilization can be concluded to have been successfully carried out in the case of Esarhaddon's succession covenant. In seeking to answer this, it is also possible to draw broader conclusions about continuity and disjuncture between this early empire's pretensions to power over its subjects and what it is possible to determine about the actions and opinions of Assyria's subjects themselves.

Part 1: **The royal call to vigilance in 672 BC**

Chapter 2: King Esarhaddon and his empire in the covenant composition

This chapter examines the portrayal of the king, Esarhaddon, and his empire in the covenant composition. The covenant reflects the manner in which its composers sought to depict geographical, administrative and social space within the empire. The composers of the covenant, building on and adapting the *adê* genre, linked the composition to pre-existing administrative and geographical structures, perhaps in part to ensure the successful penetration of the covenant's message into the areas of the empire that they considered most relevant to its demands. In this sense, the covenant composition can be taken to some degree to reflect the structure of the empire as it existed in the conception of those writing it. Equally, however, the depiction of the empire in the text also ties in with the covenant's aims. In this way, the text portrays an idealized version of the empire, which was to be made into a reality by the enactment of the covenant and its duty of vigilance. The description of the empire in these documents thus ties in with the dynamics of the Assyrian Empire and its power structures as the composers saw them, but also constitutes an act of creation, in which the composers attempt to shift those very dynamics.

In order to trace these overlapping and, at times, opposing threads within the covenant's text, I have divided the present chapter into three main sections based on the empire's construction within the composition. I argue that, according to the covenant, space is frequently defined by social distance from the king, as well as by administrative category, something that is also conceived of in terms of the relationship to the crown. As such, the chapter's first section deals with the Assyrian crown, which is presented as the linchpin of the empire in the covenant text. While the monarch is viewed in this fashion in many surviving Assyrian texts,[1] the covenant stresses that the Assyrian crown is composed not only of the king himself, but also of his chosen successor, Ashurbanipal. The second section deals with the description of the legal parties of the covenant, which are separated primarily by administrative zone, but also to an extent by social distance from the monarch. Despite this, an effort is made in the composition to universalize the status of these subjects within the empire, detaching them from their geographical context. The final section is dedicated to an examination of the use of deities in the composition, positing that they were utilized in part as a means to access

1 On the topic, see, for instance, the recent edited volume *The King as a Nodal Point of Neo-Assyrian Identity* (Bach and Fink 2022).

 https://doi.org/10.1515/9783111323435-005

place. The deities listed in the covenant show a clear focus on the empire's urban populations, particularly those living in Assyria's historic and economic heartland provinces, as well as in the western provinces and possibly some of the western client states. In this way, the covenant composition appears to attempt to tread a fine line between presenting the empire as a complete entity, unified by its relationship to the crown, and also stressing the importance of its duty of vigilance to particular target groups.

2.1 The king and his heir as the linchpin of empire in the covenant

King Esarhaddon is the first person mentioned in the text of the covenant, only the deity Aššur is cited before him. The covenant tablets are all presented in the manner traditional for the genre, as an agreement between two main parties.[2] King Esarhaddon is framed as the covenant's initiator and proprietor, imposing it on the other party. The covenant is described in this initial section as the *adê ša Aššur-ahu-iddina* 'covenant of Esarhaddon', and this statement is followed by a brief royal titulary. In two of the extant manuscripts, the section that introduces the legal parties concludes with the statement *adê issikunu iškunūni* 'he (Esarhaddon) established the covenant with you (pl.)'.[3] Esarhaddon is thus the subject of this section, and all other agents are mentioned in relation to him. Indeed, the subjects with whom Esarhaddon concludes the covenant are described as *ammar Aššur-ahu-iddina šar Māt–Aššur šarrūtu bēlūtu ina muhhišunu uppašūni* 'all those over whom Esarhaddon, king of Assyria, exercises kingship and lordship'.[4] So too, in the third section of the covenant, Esarhaddon is described as having *udanninūni iṣbatu iškunūni* 'confirmed, made and concluded' the *adê*-covenant in the presence of the gods.[5] Here, then, Esarhaddon is presented as the generating force behind the covenant.

2 Unfortunately, the oldest extant Assyrian covenant compositions are broken and do not preserve the initial lines. Compare, however, Esarhaddon's covenant with Ba'al, the king of Tyre (SAA 2, no. 5), imposed in ca. 677 BC. On the bilateral nature of Assyrian *adê* texts, see Radner 2019.

3 The manuscripts are the Kullania version (T 1801, § 1: i19) and one of the Kalhu tablets (ND 4327, § 1: 12).

4 SAA 2, no. 6, § 1: 7–8.

5 The Kullania manuscript confirms that the text reads *udanninūni iṣbatu iškunūni*, which some had previously attempted to restore as *udanninūni issikunu iškunūni* (Lauinger 2012, 114, § 2: i 28). The subjunctive predicates indicate that these verbs refer to an omitted *adê ša*, here to be translated as 'covenant that' (SAA 2, xxxvi).

In terms of the conceptualization of Esarhaddon in the text, the monarch's name is always accompanied by the title *šar Māt–Aššur* 'king of Assyria' and,[6] while this is not unusual for Assyrian written documents concerning the king, the repetition of Esarhaddon's name with this title emphasizes his power and status. It also serves to link him inexorably with one place, Assyria (literally 'the land of Aššur'). The composers do not include the more elaborate titulary used for Esarhaddon in, for instance, his royal inscriptions. These titles often emphasize the king's relationship with the gods, as well as his heroism and lineage.[7] In Esarhaddon's case, he also took on the titles of *šakkanak Bābili šar māt Šumeri u Akkadi* 'governor of Babylon and king of Sumer and Akkad (i. e. Babylonia)' in various documents.[8] In the covenant, however, the titles used to introduce Esarhaddon are the following: 'king of the world, king of Assyria, son of Sennacherib, (likewise king of the world), king of Assyria'.[9] This is the longest section of titulary relating to Esarhaddon in the text. Once more, the titles emphasize a particular place, Assyria, above all else. At the time, this would probably have been understood to refer to the provincial extent of the empire.[10] The statement that Esarhaddon is *šar kiššati* 'king of the world (lit. totality)', meanwhile, highlights the sheer all-encompassing scale of his dominion.[11] One could argue that this choice of titulary therefore reinforces the notion that the empire is structured using a two-tiered

6 Barbara Cifola refers to this epithet as the '"national" title' (Cifola 1995, 76 and passim).

7 For a brief recent discussion on the topic of Neo-Assyrian royal titulary, see Liverani 2017, 107–115. See also Ellie Bennett's recent study of 'masculinities' in Neo-Assyrian royal inscriptions (2019).

8 As in RINAP 4, passim. The 'Sumer' of the title refers specifically to southern Babylonia, while 'Akkad' refers to northern Babylonia (Bagg 2020, 404–408, s.v. Māt-Šumeri u Akkadî). By referring to Babylonia in this way, Esarhaddon uses the traditional terminology and presents himself as a true Babylonian king, rather than a foreign overlord.

9 SAA 2, no. 6, § 1: 1–2.

10 As established in J. N. Postgate's discussion of the provincial and client state systems (Postgate 1992).

11 This introduction is consistent with that of Esarhaddon and Ba'al, king of Tyre (SAA 2, no. 5). The statement that Esarhaddon is 'king of the world' is present in five of the eight exemplars that preserve this section and that Sennacherib was 'likewise king of the world' is present in four of the eight. This statement is not present in Esarhaddon's treaty with Ba'al, and one may perhaps wonder whether the composers added this clause in order to stress that this, in contrast to that covenant, was not to be seen merely as a bilateral treaty between the kings of two distinct regions. Instead, it stresses Esarhaddon's hegemony over the entire region, including the client states. For this reason, it is perhaps interesting that one of the exemplars that does not include this phrase is the only extant copy of the provincial version of the covenant (i. e. the Kullania version), as the provinces were considered to be part of 'Assyria'. However, this may simply be a coincidence.

system: the land of Aššur, on one hand, and the rest of the world, 'totality', on the other. The chosen titles do not emphasize any other portion of the empire in particular, as, for instance, the inclusion of Esarhaddon's Babylonian titles would have done. These two broad spatial categories are linked in the world as depicted in the covenant only by their relationship to Esarhaddon himself. This notion of a direct tie between the king and his subjects is reinforced by the use throughout the composition of the statement *bēlkunu* 'your (pl.) lord'. Thus Esarhaddon is regularly described in terms of his relationship to and authority over his subjects, who are addressed directly. Esarhaddon's titulary in the covenant composition is not innovative, but rather relies on long-established royal epithets. As with the reference to Esarhaddon's father, these epithets highlight Esarhaddon's claims to legitimacy on traditional grounds, and in this way can be seen as positioning him in time as well as space.

Esarhaddon's chosen crown prince of Assyria, Ashurbanipal, is first mentioned several lines after his father, in either line eleven or line forty-three of the text, depending on the manuscript.[12] In both lines, it is stated that Esarhaddon has concluded the covenant *ina muhhi Aššur-bāni-apli mār šarri rabiu ša bēt rēdûti mār Aššur-ahu-iddina šar Māt–Aššur* 'concerning Ashurbanipal, the great crown prince of the House of Succession, son of Esarhaddon, king of Assyria'. Thus, while the covenant is stated to be about Ashurbanipal, he is not included in its instigation, nor is he one of the parties bound by it. Instead, Ashurbanipal is to be understood as the object of the covenant, he is the matter about which the legal agreement is being concluded by Esarhaddon, his subjects, and the gods.

The name used most frequently in the covenant text is that of *Aššur-bāni-apli* 'Ashurbanipal'. He is mentioned by name some sixty-two times. Esarhaddon, meanwhile, is evoked by name on forty-five occasions, and in nineteen of these cases his name is given as an aspect of Ashurbanipal's titulary, as in the above citation. Ashurbanipal also has the longest titulary in the covenant composition. It always includes the statement that he is the *mār šarri rabiu ša bēt rēdûti* 'great crown prince of the House of Succession'. The identity of his father is mentioned in addition to this in some 33% of cases.[13] The attention of a reader or listener to the covenant is thus continually redirected to the person of Ashurbanipal. Esarhaddon's status is frequently used to promote that of his son, linking their posi-

12 The two manuscripts that include lines § 1: 11–12 (i.e. the Kullania version and Kalhu manuscript ND 4327) first mention Ashurbanipal in line 11. All the other exemplars that preserve the section first mention Ashurbanipal in § 4: 43.

13 Note that in SAA 2, no. 6, this formulation is translated as 'great crown prince designate'. This is certainly a more elegant English phrase than the literal translation, but it negates the importance of a location, the Succession Palace, in the definition of the crown prince.

tions and underlining their relationship to the crown and to Assyria. This highlights the fact that Ashurbanipal's position is conferred on him by his father, the king of Assyria. That Ashurbanipal will soon be the king of Assyria is suggested by the mention of his paternity, just as Esarhaddon's legitimate kingship is underlined by the mention of his father Sennacherib. The repeated statement in this context that Esarhaddon is *bēlkunu* 'your (pl.) lord', hints at the fact that Ashurbanipal himself is soon to assume this role and will have a similar relationship to the subjects of the Assyrian crown. This is even stated explicitly in § 17, with the stipulation:

> On the day that Esarhaddon, king of Assyria, your lord, passes away, (on that day) Ashurbanipal, the great crown prince of the succession [palace], son of Esarhaddon, king of Assyria, your lord, shall be your king and your lord.[14]

Here Esarhaddon is described as *bēlkunu* 'your lord' twice, while the statement is also made that Ashurbanipal will be the king and lord of the subjects of Assyria. Here, then, Esarhaddon's projected transference to Ashurbanipal of his own kingship and lordship over Assyria, on the one hand, and over 'you (pl.)' on the other, is made explicit. Thus, Ashurbanipal is shown as holding in the future the position currently occupied by his father. That Ashurbanipal is termed the *mār šarri rabiu*, literally the 'great son of the king',[15] serves both to highlight once more his status as the son of Esarhaddon, while also stressing that he is a special son of the king, differentiating him from and elevating him above his brothers.

The rest of Ashurbanipal's titulary associates him clearly with a place, the *bēt rēdûti*, literally the 'House of Succession', located in Nineveh.[16] In spatial terms, therefore, Ashurbanipal is positioned in the heartland of Assyria. His identification with a *bētu* 'house' or, in this case 'palace', is perhaps also significant.[17] The palace was the central node of Assyrian imperial rule: subjects and foreigners alike were

---

14 SAA 2, no. 6, § 17: 188–191.

15 The precise meaning of the phrase *mār šarri* (literally 'son of the king') is contested. Theodore Kwasman and Simo Parpola have argued that the term always refers to the crown prince (SAA 6, xxvii–xxix). While some have accepted this interpretation, other scholars have argued that there is insufficient evidence to assert that this is universally the case (Hussein 2020, 69–76, see also Knapp 2015, 305, fn. 12).

16 As is clear from Ashurbanipal's inscriptions: see, for instance, RINAP 5/1, no. 11: x 51.

17 In this context, the term *bētu* can certainly be taken to refer to a 'palace' by modern standards, although it is worth noting that the Assyrian title of *ēkallu* 'palace' was, at least in royally-commissioned texts, reserved for the residence of the king (RlA 10, 218 s.v. Palast. A. V. Mittel- und Neuassyrisch).

received there, as was correspondence. In addition to this, administrative matters were overseen there, and tribute and booty could be housed in them, with some palaces also serving as military centres. Governors, the *bēl pāhiti* or simply *pāhutu* 'proxy', ruled from palaces located in their provincial capitals. The association of a crown prince with a palace, therefore, may have served to signal that Ashurbanipal was going to participate in the running of the empire.

It is also worth addressing the role of Šamaš-šumu-ukin, the crown prince of Babylon, in the covenant composition, as well as the role of Esarhaddon's other sons. The covenant's short subscript (colophon) explicitly states that Šamaš-šumu-ukin's own position as crown prince of Babylon was established at the same time as that of Ashurbanipal as crown prince of Assyria.[18] It is therefore striking that Šamaš-šumu-ukin is largely conspicuous by his absence from the covenant's text. Aside from the colophon, this second crown prince is mentioned just once, in a stipulation which limits his future power as much as it seems to promote it. The main body of the covenant does not state that it has been imposed on his behalf. In the composition, Šamaš-šumu-ukin is associated with Babylonia alone, which perhaps serves to stress his lack of authority over all other portions of the empire, which were the dominion of Esarhaddon and Ashurbanipal.[19] Other sons of Esarhaddon are also mentioned in the composition, but never by name, and their authority is clearly limited. As such, the Assyrian crown is portrayed as being comprised of the king and his chosen heir, Ashurbanipal, who alone constitute the administrative and social centre of the empire.

2.2 The legal parties and the composition of empire in the covenant

The initial lines of the covenant (§ 1: 1–12) are concerned with establishing its legal parties. In addition to introducing Esarhaddon, the imposer of the covenant, and Ashurbanipal, concerning whom the covenant is composed, this initial section also introduces those who are required to enter into the covenant. This is the only part of the covenant composition that had multiple known versions.[20] Thus

18 SAA 2, no. 6, § 107: 666–670: 'The treaty of Esarhaddon, king of Assyria, conclu[ded] on behalf of Ashurbanipal, the great crown prince designate of Assyria, and Šamaš-šumu-ukin, the crown prince designate of Babylon.' Compare Lauinger 2012, viii 63–71.

19 See further discussion in Chapter 3.

20 While there are slight differences between the curse sections of the Kalhu and Kullania versions, these can be disregarded as small errors on the part of the scribes, hardly surprising

far, two distinct variations of the section are known: one for the client kingdoms (the manuscripts of which were found at Kalhu) and one for the provinces (the manuscript of which was found at Kullania). One cannot rule out the possibility that other versions of this section of the composition were also used, for the subjects of Babylonia, for instance. Nevertheless, it is worth noting the two known versions of the covenant have the potential to encompass all subjects of the Assyrian crown, considering that from 672 BC, Babylonia can be viewed as a client state, albeit an anomalous one.[21] The central dichotomy of the empire as a spatial entity in the covenant is therefore that of the administrative distinction between provinces and client states.

The version of the introductory section used for the client kingdoms is framed as a covenant between Esarhaddon and the named ruler of the region, termed a *bēl āli* 'city lord':

> Covenant of Esarhaddon king of the world, king of Assyria, son of Sennacherib, (likewise king of the world), king of Assyria, with Humbareš (etc.), city-ruler of Nahšimarti (etc.), his sons, his grandsons, with all the Nahšimarteans, the men in his hands great and small,[22] as many as there are from sunrise to sunset, all those over whom Esarhaddon, king of Assyria, exercises kingship and lordship, (with) you, your sons and your grandsons who will be born in days to come after this covenant.[23]

The formulation of this section of the covenant in this version is fairly similar to earlier treaties between the Assyrian king and the ruler of a less influential polity.[24] The treaty between Esarhaddon and Ba'al, king of Tyre, which was probably concluded in 676 BC, begins in a similar fashion: '[with Ba'a]l, king of Tyre, with [..., his son, and his other sons and grandsons, with a]ll [Tyrians], great and small'.[25] The tablet is only fragmentarily preserved, but it is clear that the treaty begins, as in the case of this version of Esarhaddon's succession covenant, with the personal name of the current ruler of the polity. The list of covenant partners then continues down through the generations of the male family line, with the scope then widening. Thus, the initial portion of Esarhaddon's succession covenant is framed like a typical bilateral covenant between two states, something that is

given the logistical challenge of drawing up so many identical tablets over a short period of time (Tushingham 2023).

21 See discussion in Chapter 8.1.

22 Note that this formulation *ṣehru* <u> *rabiu* could also be translated literally as 'young and old' (as it is, for instance, in RINAP 4).

23 SAA 2, no. 6, § 1: 1–10.

24 SAA 2, no. 5. This is discussed in, for instance, SAA 2, xxx.

25 SAA 2, no. 5: 1–3.

belied by the fact that so many identical ones were concluded simultaneously. In this way, the covenant composition confirms the status of these places as client states, reinforcing this conception of the structure of empire. The covenant composition also aligns itself with what has gone before, namely prior covenant agreements, which were commonly conducted between the Assyrian king and a new client ruler.[26]

In the Kalhu versions of this section, the first legal partner to be named is the city lord, for instance Humbareš, followed by his descendants and presumed successors, his sons and grandsons. These parties are not named, nor is it stated whether they have yet been born. The composition then widens its scope to include all Nahšimarteans, a sentiment that is then expanded upon in the following lines, in which it is emphasized that the covenant applies to both 'great and small' and indeed with every individual over whom Esarhaddon exercises 'kingship and lordship'. It is important to stress that the agreement is framed as existing directly between King Esarhaddon and each of the people mentioned in the section – from the city lord to anyone under Esarhaddon's rule. The repetition of the word *issi* 'with' highlights the direct nature of the connection between the king and each treaty subject.

Despite this direct connection with Esarhaddon, however, the covenant partners listed in this version of the text are, at least initially, defined through their relationship with the local city ruler and not the Assyrian king. As such, the covenant composition uses a single named person, linked to a specified place, as the point of entry to state that the covenant is universally binding. By following the male line of the ruling family of these client state rulers, the composition is applied to the future rulers of these locations. This initial section can perhaps be taken to reflect the nature of Assyria's authority over its the client states, which relies on the Assyrian king's influence over the local ruler. It is through him that the Assyrian king can claim influence over the client state and its inhabitants as an administrative unit of the empire.

The composition then switches to the second person, stating that it is *issikunu <issi> mar'ēkunu mār mar'ēkunu ša urki adê ana ūmē ṣâti ibbaššûni* 'with you (pl.), with your sons and your grandsons (lit. the sons of your sons) who will be born in the days after this covenant'.[27] Here, the implied reference to the future at the beginning of the clause is made explicit with the statement that the covenant also applies to those who have not yet been born. So too, it appears that, despite the apparent traditional and practical necessity of accessing the covenant's intended

26 As illustrated recently in Radner 2019, 309–312.
27 SAA 2, no. 6, § 1: 9–10.

audience through their location and their local ruler, the composition also attempts to negate this fact by addressing these people directly and stating that Esarhaddon has imposed the covenant on them.

The second extant version of the covenant composition is known from one manuscript, found in a Neo-Assyrian shrine in modern Tell Tayinat, ancient Kullania (also Kunalia and Kunulua), located in the northern stretches of the Orontes Valley in southeast Turkey.[28] Kullania, the capital of the eponymous province, was annexed by the great conquering king of Assyria, Tiglath-pileser III (r. 744–727 BC), in the year 738 BC.[29] This manuscript's introductory section differs significantly from that of the client states in a manner that clearly reflects the contrast in the Assyrian crown's governance of these two types of administrative unit:

> The covenant of Esarhaddon, king of Assyria, son of Sennacherib, king of Assyria, with the governor of Kullania, with the deputy, the major-domo, the scribes, the chariot drivers, the third men (on the chariot), the village managers, the information officers, the prefects, the cohort commanders, the charioteers, the cavalrymen, the exempt, the outriders, the specialists, the shi[eld bearers (?)], the craftsmen, with [all] the men [in his hands], great and small, as many as there a[re] – [wi]th them and with the men who are born after the treaty in the [fu]ture, from sunrise [...] to sunset, all those over whom Esarhaddon, king of Assyria, exercises kingship and lordship.[30]

Unlike the version of the covenant composition used for the client states, the introduction of the provincial governor, administration and subjects is not found in any other known Assyrian covenant. Despite this, the formulations used here may well have drawn on former compositions, such as the succession covenant of Sennacherib.[31] One key difference between the client version and the provincial recension is the omission of the governor's name. He is referred to solely by his title: *pāhat Kunulia* 'governor of Kullania'. This wording seems designed to reflect the nature of Esarhaddon's control over the provinces. On the one hand, the client states were overseen by a local ruler, thus rendering it necessary to reinstate and possibly renegotiate the arrangement between that polity and Assyria whenever power changed hands. As such, for Esarhaddon the point of entry in terms of claiming authority over the inhabitants of that area was the individual currently reigning there. On the other hand, provincial governors were directly chosen by

---

28 Bagg 2007, 141 f.

29 RlA 11, 61 s.v. Provinz. C. Assyrien.

30 Lauinger 2012, § 1: i 1–17a.

31 As the fragments of Sennacherib's succession covenant (Frahm 2009a, nos. 67–69) do not preserve this portion of the tablet, it seems perfectly possible that the 'Langfassung', as Frahm terms it (i. e. nos. 67 and 68), has a similar initial section.

Esarhaddon,[32] thus rendering it unnecessary to link the covenant, which was imposed upon all royal subjects, to a particular individual as opposed to a title. Moreover, the post of provincial governor was not life-long, in contrast to that of client ruler, and the framing thus allows the covenant to withstand changes in personnel.[33]

Following the provincial governor come sixteen professional titles, linked with the civil and military provincial administration. The title of *šaniu* 'deputy governor' and *rab bēti* 'majordomo' (a high military official) are both found in the singular.[34] Here, yet again, the covenant references specific people but does not do so by name, rendering the covenant binding even after new individuals have taken up these roles. The remainder of the professional titles follow in the plural, illustrating, as in the case of the Kalhu version, the gradual widening in the scope of the bond in this section of the covenant. The list shifts from those at the very top of the provincial administration, ergo those responsible for taxation and conscription, to those who were subject to it, and in doing so it appears to move down the provincial hierarchy.[35] That the *ṭupšarrē* 'scribes' are mentioned directly after the three individual heads of the provincial administration is interesting, given that they were not necessarily the highest-ranking members of the provincial administration. It seems probable that these individuals were considered particularly relevant to the aims of the covenant and as such they were featured prominently. The logic behind this decision is easy to follow: written information was of particular significance both in ensuring the implementation of the covenant, and in guaranteeing the ability of the empire's subjects to report seditious actions and speech, as this required letter-writing. In general, the more detailed list of the provincial version also implies that the intended changes to be wrought by the covenant were rather more concretely planned out for the provinces than they were in the client states.

Once more, then, the formulation of this introductory section prioritizes depth of penetration through the social layers of the administrative zone to which it pertains, as well as the longevity of the bond produced. The manner in which this is done differs for the client state and provincial versions, but in both cases it follows the local hierarchies through which Esarhaddon could claim his authority over the people in that region. At the same time, these passages serve to highlight the obligation of these people to Esarhaddon and Ashurbanipal, and thus establish a di-

32 Radner 2003a, 888.

33 As pointed out in Lauinger 2012, 113 f.

34 For a detailed discussion of the provincial administrative structure according to the Kullania copy of the *adê*, see Ponchia 2014, 513–516.

35 Ponchia 2014.

rect, personal relationship between subjects and the crown, circumventing the local authority figures through whom the connection was initially made.

Beyond this introductory section of the text, the composition addresses the people entering into the covenant directly, by means either of the second-person plural, 'you', or – in the oath section – in the first-person plural, 'we'. The oath section constitutes some 2.5 % of the whole covenant composition,[36] and was intended to be repeated verbatim by those entering into the covenant. The subjects would presumably have been expected to listen to the remainder of the covenant read aloud. Thus, after the introductory section, all subjects entering into the covenant are collapsed into a single group. The defining factor of this group, which encompasses all subjects of the Assyrian Empire, is their bond of loyalty to Ashurbanipal imposed by means of Esarhaddon's covenant. The hierarchical relationship between them is not stressed here; rather, all subjects are required to follow the same stipulations and can expect the same terrible consequences should they, in the words of the composition, *haṭû* 'sin against' the covenant.[37] In this way, while the text begins with the strict distinction between the inhabitants of the empire according to the client state or province in which they live, the framing of the remainder of the covenant presents the empire as one in which social space is flattened and geographical distance is, to a certain degree, disregarded. The king and his successor are located at the centre of the empire, with all subjects of the Assyrian crown existing at one degree of removal from him. The crown (consisting of the king and his successor, and excluding his other offspring) is therefore cast as the central node, with every other subject across the empire directly attached to him.[38] This contrasts, of course, with the reality of social networks in the Assyrian Empire, which was much more convoluted, with many more degrees of separation between the crown and most of the people over whom it exercised power.

2.3 Using the gods to access place in the covenant

2.3.1 The universality of Aššur, king of the gods

The first line of the text, before the king and his subjects are introduced in the main body of the composition, is a seal description: 'seal of the god Aššur, king

36 Lauinger 2019, 92.

37 CAD H, 156–158 s.v. *haṭû*.

38 This ties into the concept of direct rule, as well as the accessibility of the Assyrian king, linked to the concept of the 'king's word' *abāt šarri* (Radner 2003a, 887; more recently Frahm 2010, 106, with bibliography).

of the gods, lord of the lands – not to be altered; seal of the great ruler, father of the gods – not to be disputed.'[39] Each of the manuscripts of the covenant was impressed with three cylinder seals: the earliest of these seals was over one thousand years old by 672 BC, another was likely some four hundred years old and the one made most recently had been cut in the time of Sennacherib. These were divine seals belonging to the god Aššur, and signify his presence at the time that the covenant was drawn up.[40] On the mythological level, the sealing of a tablet by a god elevated it to the status of a 'Tablet of Destinies', a document whose contents were bound to occur.[41] That this was understood to be the meaning of Aššur's seal is made clear by the statement in the seal description that it is *lā šunnê* 'not to be altered' and *lā paqāri* 'not to be disputed'.

From a legal point of view, the use of a god's seal in this way can be interpreted as showing that Aššur was a witness to the tablet being draw up, and perhaps even that he had a degree of secondary responsibility for the successful outcome of the legal agreement.[42] The sealing and dating of cuneiform legal documents in order to render them binding was practiced across the provincial extent of the empire. As such, the legal significance of these seal impressions would have been clear to those who encountered the tablet throughout this administrative zone.[43] Whether the status of the tablet as a 'Tablet of Destinies' would have been equally clear to the subjects of this heterogenous, multicultural and multilingual empire is rather

39 SAA 2, no. 6: i–iv; Lauinger 2012, i–iv.
40 See the recent discussion of the seals and their significance in Watanabe 2021, as well as Watanabe 1985. Sennacherib's seal is published as RINAP 3/2, no. 212; on the Middle Assyrian seal, see recently Wallenfels 2022. That seal impressions are to be taken as proof of the presence of a party is clear from comparison with Neo-Assyrian legal practices (Radner 1997, 33). Note that, as Radner argues, the comparison between sealing a tablet and the modern practice of signing a document only partially applies, as a tablet was sealed before it was inscribed. As such, the sealing indicates the authenticity of the document by proving the presence of a party, as opposed to proving that they have read and agreed to everything written in the document.
41 The argument that the seal of Aššur renders the covenant a 'Tablet of Destinies' was first put forward by Andrew George (1986). His analysis has been widely taken up by Assyriologists working on the covenant. Indeed, Jacob Lauinger (2013; 2019) has argued more recently that the term *adê* is best understood as meaning 'destiny', so central does he consider the mythological status of the *adê* tablets. For a recent overview of the Old Assyrian period, see Veenhof 2017, esp. 72–74.
42 The sealing practice found in the *adê* bears several interesting points of comparison to Neo-Assyrian land sales composed in Ashur, which were often sealed by Assyrian officials, seemingly to take on some degree of secondary responsibility for the transaction (Tushingham 2019, 34 f.; see also Faist and Klengel-Brandt 2010). The Ashur link is particularly interesting here, as the practice of using the seal of the god Aššur also originates there.
43 On the structure of Neo-Assyrian conveyances, see Tushingham 2019.

more doubtful.[44] Rather, the mythological import of the seal impressions may only have been clear to a select subset of those who came into contact with the covenant. The fact that it was possible to interpret the seal impressions on various levels, however, would probably have rendered them meaningful to Assyria's subjects across a variety of locations and social positions.

The god Aššur, in contrast to deities originating in Babylonia, was equated with a geographical feature, the rocky outcrop upon which the citadel of the city Ashur was built.[45] The deity was so closely aligned with the city of Ashur, in fact, that it was not theologically feasible to build temples dedicated to him in other locations. Thus, in contrast to most important Assyrian deities, Aššur had only one temple.[46] The ancient seal of that god highlights the long history of the city of Ashur, led by its eponymous deity, and the god's involvement in Assyrian statecraft. The oldest seal used on the covenant tablet had been used since the Old Assyrian period (ca. 1900–1700 BC), when it was employed by the main governmental body of the small city-state of Ashur, the city *puhrum* 'assembly' in the *bēt ālim* 'city hall'.[47] All three seals of Ashur would certainly have been kept in that city.[48] More recently, Esarhaddon's father, Sennacherib, had undertaken various religious reforms at the temple. These changes were calculated to raise Aššur's status, establishing him as the head of the pantheon, instead of Marduk, the city god of Babylon and erstwhile head of the Babylonian pantheon.[49] Esarhaddon's use of a seal created under his father, in conjunction with the two older seals, stresses both the ancient and contemporary importance of Aššur, something that Sennacherib had been at pains to establish even at the expense of other deities. As these seals were also inexorably linked to the city of Ashur, they can also be interpreted as doing something similar for that city, stressing its ideological and religious centrality from the crown's perspective. Hence, each manuscript of the covenant not only bore the mark of the empire's central deity, but also evidence of its origins at what by its own telling was the empire's religious and historical core.

---

44 On the linguistic profile of the Assyrian Empire, see recently Radner 2021. See also Fales 2023b on the cultural heterogeny of the empire.

45 On the god Aššur and his status both as a geographical feature and a *deus persona*, see Lambert 1983.

46 Radner 2015a, 8–13.

47 Veenhof 2017, 72. For an extensive study of the city hall, see Dercksen 2004.

48 A. George 1986.

49 On Sennacherib's religious reforms, see Machinist 1984, Frahm 1997, 20 and 282–288, Vera Chamaza 2002, 111–167 and Pongratz-Leisten 2015, 416–426. On Esarhaddon's reaction to these reforms and their legacy, see *inter alia* Novotny 2014.

Nevertheless, while the seal description draws attention to Aššur and his supremacy, its statement that Aššur appears here in his capacity as *šar ilāni* 'king of the gods' also indicates that his actions here are being taken on behalf of all deities, while his title *bēl mātāti* 'lord of the lands', stresses their universal applicability across multiple realms. Thus, this initial line demonstrates the all-encompassing nature of the divinely-bestowed power of the covenant, while also physically connecting the tablets to the empire's heart. Indeed, by sealing each of the covenant tablets with the seals of the god Aššur and disseminating them throughout the empire, the covenant can be seen as bringing the empire's central deity to the provinces and client states. The stipulations of the composition also contain the demand that those who are bound to the covenant *naṣāru* 'guard' the tablet *kî ilikunu* 'like your (pl.) god'. This shows that the covenant tablet, imprinted with the divine seal impressions of Aššur, was intended for worship. In this way, the dissemination of the covenant was seen as connected to the worship of an object associated with the god Aššur. Particularly relevant in the present context is the use of the term 'your (pl.) god', which acknowledges that Aššur is not necessarily the local deity of the addressees of the covenant. Despite this, it demands that the tablet be treated as holy universally, throughout the entire empire.[50]

2.3.2 The specificity of local gods

In addition to the god Aššur, the covenant includes many other deities, several of whom are associated with particular locations. The main sections of the text that mention these gods are the divine witness section (§ 2: 13–40), as well as the two lists of divine curses (§§ 37–56: 414–493 and §§ 58–106: 513–663). The curses alone constitute some 34 % of the composition, showing just how important this element of the covenant was considered to be.

The divine witnesses feature significantly earlier in the composition than the curses. As such, this is probably the list of deities that those swearing to the covenant would have encountered first. This section states that the covenant is concluded *ina pān* 'in front of' the heavenly bodies Jupiter, Venus, Saturn, Mercury, Mars and Sirius,[51] as well as a list of seventeen named deities worshipped in Assyria: 'the gods dwelling in heaven and earth, the gods of Assyria, the gods of Sumer and Akkad, all the gods of the lands.'[52] The adjuration, meanwhile, demands

50 See also discussion in Radner 2017a, 81.
51 On the planets, see Hermann Hunger's RlA article (RlA 10, 589–591).
52 SAA 2, no. 6, § 2: 21–24; Lauinger 2012, i 24–28.

that those entering the covenant 'sw[ear contin]ually by Aššur, father of the gods, lord of the lands!'[53] They are also instructed to swear by each of the named deities mentioned in the list of divine witnesses. Furthermore, it is also mandated that they swear by various unnamed gods, whom the covenant lists by place:

> Ditto (i. e. swear continually) by the gods of the Inner City! Ditto by the gods of the Nineveh! Ditto by the gods of Kalhu! Ditto by the gods of Arbela! Ditto by the gods of Kilizu! Ditto by the gods of Harran! Ditto by the gods of Babylon, Borsippa and Nippur! Ditto by the gods of Assyria! Ditto by the gods of Sumer and Akkad! Ditto by all the gods of the lands! Ditto by all the gods of heaven and earth! Ditto by all the gods of one's land and one's district![54]

In the adjuration, therefore, after listing the highest deities worshipped in the Assyrian Empire by name, deities are explicitly associated with places, specifically with settlements. Each of the first five cities mentioned were important settlements located in the heartland of Assyria and were associated with significant cultic centres. Ashur was the symbolic centre of Assyrian power, while Nineveh was the royal capital.[55] Kalhu was the former capital of the empire, as well as the location of various important temples, including the primary temple of Nabû in Assyria, in which the nine manuscripts of the client ruler version of the covenant were found.[56] Arbela, meanwhile, had long been an important religious centre, particularly due to the presence there of the temple of Ištar of Arbela.[57] Located on the route between Kalhu and Arbela, the city of Kilizu was home to a temple of Adad.[58] In contrast, the sixth city mentioned in the list, Harran, was located substantially to the west of the Assyrian heartland. Harran was the site of an important and ancient religious centre, a temple to the moon god Sîn.[59] This temple was, however, by no means the only local cult centre of consequence in the region. The city of Harran, as well as its temple, had been promoted under Esarhaddon's grandfather, Sargon II, and its inclusion in this section of the composition can be taken as evidence of its continued and growing importance under Esarhaddon.[60]

---

53 My interpretation of the Gtn verbal form of *tamû* 'to swear' differs here from that of previous editors of the covenant: see discussion in Chapter 5.3.

54 SAA 2, no. 6, § 3: 31–40; Lauinger 2012, i 29–45'.

55 Bagg 2017, 77–80 s.v. Aššur (Stadt) and 456–466 s.v. Ninua (Ninive).

56 Oates and Oates 2001, 111–119. Bagg 2017, 277–283 s.v. Kalhu.

57 On Ištar of Arbela, see Porter 2004. On Arbela's position in the empire's core, see Radner 2011c and Bagg 2017, 53–57 s.v. Arbail.

58 RlA 11, 46–47 s.v. Provinz. C. Assyrien; Bagg 2017, 284–285 s.v. Kalzu.

59 Novotny 2020, 73–76. See also Gross 2014 and Hätinen 2021, 384–415.

60 On the promotion of Harran from the reign of Sargon II (r. 721–705 BC) until the end of the Assyrian Empire in 609 BC, see Novotny 2020. See also Radner 2003c, 173 and Leichty 2007, which present potential reasons for Harran's significance under Esarhaddon. Karen Radner sug-

The only other locations not in the Assyrian heartland mentioned by name are all listed together in one line and were all located in Babylonia. These cities were Babylon, the cult centre of the god Marduk and the royal seat of the Babylonian king; the neighbouring city of Borsippa, the Babylonian cult centre of the god Nabû; and Nippur to the south, the cult centre of the god Enlil, one of the most important deities in the Babylonian pantheon, decider of fates and appointer of kings.[61] As these locations are the last to be mentioned by name in the list and are also condensed into one line, the implication seems to be that these locations are subordinate in importance to the previous ones. The decision to include these places in such a way is perhaps indicative of a desire on the part of the covenant's composers to include the deities of these religiously highly significant places in the adjuration, while at the same time reflecting the status of these locations as lesser than the cities preceding them in the list.

The locations mentioned in the adjuration section of the covenant composition, then, are all urban sites, perhaps revealing a particular focus on large settlements as opposed to rural areas in the portrayal of empire in the text. The majority of the listed settlements were located in the Assyrian heartland. The locations not in the Assyrian heartland are all prominent religious centres, one of which was located at a strategically important point in a northwestern but well-integrated province some 300 km from the Assyrian capital. The other three were important Babylonian religious centres, whose temples had been subject to Assyrian building projects and whose gods were worshipped as some of the uppermost deities of the Assyrian pantheon. At the time that the covenant was imposed, these cities had been under Assyrian rule once more for some twenty-seven years, and were theoretically part of the Assyrian provincial system. In reality, however, Babylonia was never fully integrated into the provincial system: key markers of Assyrian provincial rule, such as the Assyrian legal system, were not implemented in Babylonia and it held a special status in the empire.[62] While these locations were not all in the Assyrian heartland, none of their deities can be considered foreign from an Assyrian point of view, as they had long been worshipped in Assyria by 672 BC.

The decision to frame the deities of these locations in this way, not naming them but rather naming their city, stands in sharp contrast to the only other As-

---

gests that Esarhaddon relied on Sîn to cure his chronic illness, while Erle Leichty posits that Esarhaddon may have found refuge in Harran during his exile prior to Sennacherib's murder.

61 Black and Green 2004, 76 s.v. Enlil, 128 s.v. Marduk and 133–134 s.v. Nabû, as well as the website: 'ORACC: Ancient Mesopotamian Gods and Goddesses'. On these locations, see Bagg 2020, 80–89 s.v. Bābili, 98–100 s.v. Barsip, 436–437 s.v. Nippur.

62 For a recent overview of Babylonia under Assyrian control, see Frahm 2017a. The most comprehensive study on the subject is Frame 1992.

syrian adjuration list of an earlier Assyrian covenant that has been preserved until the present day: that between Aššur-nerari V and Mati'-ilu, king of Arpad.[63] In that composition, each deity is named individually. While it cannot be known whether adjuration by deity location is an innovation of this particular covenant composition, those working on the text would likely have considered this wording carefully and chosen it because it was useful for their purposes. One may wonder whether their objective was to highlight not only the locations of significant deities, but also important populations: the followers of these particular deities or the residents of these particular locations. This would imply that the covenant was particularly intended to effect change in the cities of the Assyrian heartland and in Harran (or perhaps the wider zone in which Sîn of Harran was worshipped), as well as in the major cities of northern Babylonia, a difficult but high-status neighbour of the Assyrian heartland.

Although the populations of these locations are highlighted in the adjuration list, the final clause of the composition shifts its focus: 'Ditto (i.e. swear continually) by all the gods of one's land and one's district!' (literally *mātīšu nagīšu* 'his land and his district'; possibly a reference to the first covenant partner cited in each manuscript). This clause is also not known from older covenant compositions, although this may be due to the lack of extant sources. The decision not to reference specific locations by name here, but rather to use *-šu* 'his', serves to include all locations from which the legal parties came in the composition of the covenant. In this way, while the named locations chosen highlight not only the gods resident there but also their local populations as particularly relevant to the covenant, the final clause emphasizes the universal applicability of the covenant composition across all areas of the empire.

The use of deities in the curse sections, the first of which follows the stipulations and the second of which follows the oath section of the covenant, differs from the divine witness section, although some of the effects are similar. Across both sections, the curses can be divided into two basic types, described by Simo Parpola and Kazuko Watanabe as 'traditional' and 'ceremonial' curses.[64] The former category is made up of curses in which it is stated that a named deity will wreak destruction and dispense punishment, while the latter refers to curses inflicted collectively by all deities *ašibūtu kibrāti mala ina ṭuppi annīe šumšunu zakrū* 'who inhabit the (four) quarters (of the world), as many as are mentioned by name in this tablet'.[65]

63 SAA 2, no. 2.
64 SAA 2, xlii.
65 SAA 2, no. 6, § 56: 472–473; Lauinger 2012, vi 52–54.

The first category of curses includes individual deities from various locations. Hans Ulrich Steymans, who in 1995 published a monograph on the curses of Esarhaddon's succession covenant and their relationship to Deuteronomy, and who has continued to publish on the subject,[66] refers to the deities listed at the beginning of the first curse list (§§ 37–53) as 'Mesopotamian deities'.[67] As he also notes, the order in which they are mentioned appears to follow to a certain degree the hierarchy of the Assyrian pantheon.[68] This is significant, because it implies that they should be viewed in this context primarily as Assyrian (i. e. Northern Mesopotamian) deities, referencing the involvement of the Assyrian king and Assyrian people, as opposed to seeking primarily to refer to Babylonia (i. e. Southern Mesopotamia).

Nonetheless, while the order of the 'Mesopotamian' deities listed in the curses does indeed follow a hierarchical order to a degree, this cannot fully explain the decision to place the gods in the sequence in which they are found in the covenant. As Spencer Allen has noted, the fact that these deities are not found in the same order as they are listed in other compositions demands explanation. His study on the subject concludes that the deviations from tradition found in the composition, as modern scholars have seen them, sometimes serve thematic purposes.[69] Thus, the scribes swapped the order of the curses of Šamaš, the sun god, and Sîn, the moon god. This way, the leprosy-related Sîn curse followed the curse of Anu, the king of the gods, which also concerned disease, rather than the less clearly related Šamaš curse about blindness.[70] While the scribes evidently sought to maintain a sense of the Assyrian divine hierarchy, then, they apparently chose in particular to prioritize logical coherence within the text. Thus, rather than simply seeking to appeal to local literary or religious traditions, the scribes attempted to create an affecting, intuitive sequence of terrible, divinely-wrought consequences that would befall anyone who dared contravene the covenant. Such a message would presumably have been broadly understandable.

Following the curses based on members of the Assyrian pantheon, the geographical focus of the curse list moves further west.[71] The first curse list includes five curses inflicted by specific western deities, either individually or in pairs.

66 Steymans 1995.
67 As in Steymans 2013, 4.
68 Steymans 1995, 29.
69 Allen 2013, 21.
70 Allen 2013, 11. Note that Steymans sees this swap as a sign of the importance of the western settlement of Harran, rendering the moon god Sîn higher up in the hierarchy (Steymans 1995, 176).
71 Steymans 2013, 4.

These are the final curses wrought by named deities in the list. The gods cited are Aramiš (or Aramis) of Qarnina and Aza'i, Šarrat-Ekron, Bethel and Anath-Bethel, Adad and Šala of Kurba'il and Kubaba and Karhuha of Carchemish.[72] As Aramiš was likely the head of the western pantheon, it is probable that these deities are also ordered more or less hierarchically. Interestingly, while the settlement of Kurba'il has not been located, it seems likely that it was located within the Assyrian heartland, and it was certainly the primary cult centre of Adad in Assyria.[73] Despite this, Hans Ulrich Steymans argues convincingly that Adad of Kurba'il functions in this position in the list as a deity who connects the Assyrian pantheon with the many storm gods of the Levant and Armenia, through his equation with deities such as Ba'al, Hadad and Teššub.[74] In this way, a deity that would have been recognized and venerated by subjects in the Assyrian heartland is utilized to appeal to subjects in or from the western and northern zones of the empire as well. While these deities would not only have been worshipped in the provincial extent of the empire, the curses that mention deities associated with locations outside Assyria or Babylonia refer predominantly to deities worshipped in Levantine provinces that were integrated into the Assyrian Empire in the mid-eighth century, under King Tiglath-pileser III (r. 744–727 BC) and King Sargon II (r. 721–705 BC).[75] The deities of other areas, such as the eastern client states named in the Kalhu manuscripts, are not included. This could be taken as evidence that the curses were drawn up with more focus on the provinces than on the client states, as well as that, even within these two administrative categories, some specific locations were considered more relevant than others.

In contrast to the 'traditional' curses, 'ceremonial' curses generally rely on similes and are not attested in any other Neo-Assyrian covenant composition in the manner used in Esarhaddon's succession covenant. As the precursors of such curses are western rather than Assyrian or Babylonian,[76] their inclusion can be interpreted once again as evidence that the composers of the covenant sought to include in the text the traditions of the western portion of the empire.

---

72 Lauinger 2012, 119. Note that, in contrast to Jacob Lauinger's interpretation, Ariel M. Bagg considers Qarne and Qarnina to be distinct settlements Bagg 2007, 193 f.

73 Schwemer 2001, 595–600.

74 Steymans 2013, 4.

75 On the integration of the provinces, see RlA 11, 42–68 s.v. Provinz. C. Assyrien. On this section of the covenant and the locations with which it is associated, see Steymans 2013.

76 As Kazuko Watanabe notes, however, while there are parallels between these curses and Aramaic, Hebrew and Hittite traditions, it sometimes seems possible that similar similes have occurred to the composers of such curses independently, due to common experience: 'Ähnliche Vergleiche können aus gleichen Lebenserfahrungen stammen.' (1987, 33).

The only other Neo-Assyrian covenant composition that preserves such curses is the one between Aššur-nerari V and Mati'-ilu, king of Arpad. It describes one ritual associated with the conclusion of the agreement: the sacrifice of *hurāpu anniu* 'this spring lamb'.[77] As the use of the demonstrative pronoun suggests, the composition seems to be designed to accompany the actual sacrifice of the animal at the covenant ceremony, with the various dire consequences for King Mati'-ilu of Arpad if he breaks the covenant linked to the various stages of the lamb's dissection. In the first instance, for example, the lamb's death and in particular its separation from its fold are set out as akin to Mati'-ilu's potential punishment: if he breaks the covenant, he, his sons, his magnates and the people of his land will be ousted from his country, never to return or see it again.

In stark contrast to this, the ceremonial curses of Esarhaddon's succession covenant do not only contain one ritual image. The 'great gods of heaven and earth' in the first curse list are described as enacting hideous punishments that do not always contain obvious references to ceremonial actions.[78] In the second list, meanwhile, the formulation *kî ša* 'just as' is employed in several curses apparently also to be enacted by 'all the gods mentioned by name in this treaty tablet'. In the case of the Mati'-ilu covenant, therefore, the composition seems to imply that the ceremony was to include the ritual killing of an animal, with the text of the tablet accompanying and reinforcing the significance of the sacrifice. The great variety of imagery used in Esarhaddon's succession covenant composition, meanwhile, as well as the difficulty of imagining that some of the imagery used could be recreated for a ritual in a literal way (the *šamê ša siparri* 'brazen heaven' or literally 'sky of bronze' from which rain does not fall, for instance) resist an identical interpretation here. Several of the ceremonial curses do not contain demonstrative pronouns, and thus the composition itself does not portray the objects mentioned in those curses as physically present in the same location at the time of reading the composition. These curses are perhaps most accurately described as simile curses,[79] as they are not literal descriptions of the ceremonial context in which the covenant was concluded. While they may well have been accompanied by ritual actions, of course, they contrast clearly with references like those made to the lamb in the Mati'-ilu treaty. Nonetheless, these curses succeed in directing the attention of the hearer towards themselves, conveying the message that the curses apply universally to all subjects bound by the covenant. Indeed, the direct nature of these similes and the fact that they do not mention deities by name would have

77 SAA 2, no. 2. CAD H, 245 s.v. *hurāpu*.
78 SAA 2, no. 6, § 56: 472–493; Lauinger 2012, vi 52–76.
79 On simile curses, see Kitz 2007, 616 and 624 f.

meant that they were likely viscerally evocative to those forced to swear to them, regardless of location.

2.4 Conclusions

The first section of this chapter serves to illustrate that the covenant composition portrays the king and his crown prince as the linchpins of the Assyrian Empire. In this sense, it is the very person of the king and his successor, not a location, that constitute the centre of the empire in this composition. The text deliberately stresses the authority of these two figures as universal, contrasting this with the limited and locally-specific power wielded by others actors, such as the crown prince of Babylon, Šamaš-šumu-ukin.

The second and third sections of this chapter argue that the text of Esarhaddon's succession covenant presents the Assyrian Empire as a complex web of different administrative zones, social hierarchies, settlement structures and cultural regions. The composition highlights these elements of the empire by dividing its territories into provinces and client states, mentioning particular settlements by name and drawing on deities and curse traditions from various regions at the expense of others. These elements of the composition illustrate both the manner in which the covenant's composers sought to access particular groups through their specific location within this complex administrative, social, cultural and geographical network, as well as their priorities in addressing some groups more frequently and explicitly than others. Specifically, the composition seems most clearly to address city-dwellers of high status or belonging to the administration, living in the provincial extent of the empire. It focuses particularly on those in the Assyrian heartland and also, perhaps to a lesser degree, on those in the western extent of its dominion. The mention of Babylonian cities in the composition is significant, although the inclusion of deities worshipped in Babylonia need not exclusively be interpreted as pertaining only to Babylonia, as they were frequently also worshipped in Assyria. Assyrian deities take precedence in the composition, but the incorporation of Levantine deities and curse traditions seems to suggest a desire to address groups to whom these would have been familiar. These people may have lived in western provinces, in client states or indeed in the core region.

Despite this, while the composition uses specific allusions to place in order to assert its relevance across various diverse regions of the empire, it simultaneously seeks to flatten the distinctions between these places. Regardless of their position in geographical, administrative or social space, those listening to the covenant composition are addressed as 'you (pl.)' a single group with a direct relationship to the empire's living centre, the king. The list of divine witnesses to which one must swear ends with the exhortation that those entering the covenant swear by their local deity. Thus, all the people of the empire are addressed through a single demand. Even the curses appear to allow for ways to approach them that do not rely on a single cultural understanding of their contents: they are arranged in a thematically intuitive manner, and many of them rely on similes that would have been familiar from the hearer's own experience. It seems, then, that the scribes who composed Esarhaddon's succession covenant used the reality of the administrative and cultural divisions of the empire to access its subjects, but then sought to create another, much simpler conception of the empire: a group of subjects united and, regardless of physical distance, positioned in close social proximity to the Assyrian crown.

Chapter 3: Directing vigilance in the covenant composition

This chapter is dedicated to the analysis of the stipulations and oath section of Esarhaddon's covenant through the lens of the practical effects that they were intended to achieve. It argues that these sections sought primarily to create a duty of vigilance. At three hundred and seventy-two lines, the stipulations of the covenant make up some 56 % of the total composition. They are far more exhaustive than any stipulations found in earlier covenant compositions,[80] and can likely be considered one of the major innovations made by the covenant's authors. The stipulations constitute the portion of the composition that most explicitly lays out a vision of the duty of vigilance it sought to impose. Many of the stipulations attempt to steer the attention of the covenant parties in the service of the Assyrian crown, focusing in particular on the smooth succession of the crown prince of Assyria, Ashurbanipal. In contrast to the long stipulation section, the oath portion of the text is a fairly short segment at 2.5 % of the overall composition.[81] Despite this, it was a very important element in the communication of the covenant's intended consequences to those entering into its bond. The oath, too, demands vigilance on behalf of the crown, but does so in a somewhat different way.

For present purposes, the composition stipulations and, to a lesser degree, the oath, can be divided into three key elements: scenarios (the description of circumstances that would require a reaction on the part of those bound by the covenant), mandated reactions to the scenarios, and forbidden reactions to the scenarios. An overview of the scenarios and reactions in the stipulations is set out in Table 1 (see below). The summary I offer here is by no means the first, and is informed in particular by those of Donald Wiseman and, more recently, Frederick Mario Fales.[82] My summary differs from these, however, in that it divides the stipulations into the tripartite structure that I have outlined above. This system is designed to facilitate the in-depth analysis of the practical implications of the stipulations that this portion of the study aims to provide, something that was previously largely lacking.

---

80 The most interesting of these in this context is probably Sennacherib's succession covenant, as it can be most easily compared to that of Esarhaddon (SAA 2, no. 3; Frahm 2009a, nos. 67–69). In particular, the fragments of the long version of the covenant seem to indicate interesting parallels to Esarhaddon's succession covenant, as well as several points of departure (Frahm 2009a, nos. 67–68).

81 As pointed out by Jacob Lauinger (Lauinger 2019, 92).

82 Wiseman 1958, 23–24 and Fales 2012, 139–142.

 https://doi.org/10.1515/9783111323435-006

It is necessary briefly to comment on the language of the stipulation and oath sections of the covenant. The mandated and forbidden reactions to the covenant's stipulation scenarios are stated in the second person plural, while the oath uses first person plural. The sentences frequently begin with the particle *šumma* 'if' and use the subjunctive mood.[83] When phrased like this, the mandated reactions are expressed negatively, for example *šumma... lā tanaṣṣarāni,* literally 'if... you (pl.) should not guard', while the forbidden reactions are articulated positively, *šumma... tanaṣṣarāni,* literally 'if... you (pl.) should guard'. Parpola and Watanabe have argued that these formulations are to be translated 'you shall guard' and 'you shall not guard', respectively, with the former being a 'more solemn and binding' equivalent for the positive indicative and the latter for the negative indicative phrase.[84] The argument that these constructions are not to be considered simply as conditional clauses is convincing, and is strengthened in particular by the occasional interchangeable use of the indicative and subjunctive formulations in different manuscripts.[85] Nevertheless, the fact that this grammatical form was an established way of introducing an oath does not mean that the literal meaning, which surely originates in oath-bound conditional clauses,[86] would have been entirely obscure to those using them. As such, it seems reasonable to assert that, while these phrases were indeed a way to convey gravity, the implicit threat of the phrasing of *šumma... lā tanaṣṣarāni* 'should you not guard...', would have been apparent to those entering the covenant.[87]

83 CAD Š/3, 275–278 s.v. *šumma.*

84 SAA 2, xxxviii–xli. On this form, see also von Soden 1995, 293. See also Faist 2015 on oaths in Neo-Assyrian legal practice. In contrast to her argument in SAA 2, Kazuko Watanabe's translations in subsequent publications acknowledge this, as in, for instance, Watanabe 2019, passim: 'If, unlikely though it may be, you should not...'.

85 SAA 2, xl.

86 As is acknowledged by Parpola and Watanabe (SAA 2, xl).

87 See also Hans Ulrich Steymans's discussion on the topic (1995, 34–37), in which he concludes that the debate around the interpretation of the *šumma*-clauses pertains more to the purpose and role of the translation than it does about the meaning of the Akkadian, a conclusion with which I agree.

Table 1: Structure of the covenant stipulations.

	Scenario (When/If…)	Mandated action	Forbidden action
1. §§ 4–5: 46–72	**Event:** Death of Esarhaddon	(1) **Action:** Seat Ashurbanipal on throne, protect him in country and town, die for him. Speak the truth with him and advise loyally (**Attitude**), smooth his way.	
			(2) **Action:** Depose him or seat one of his brothers on the throne. Change or alter the word of Esarhaddon.
		(3) **Action:** Serve him.	
		(4) **Action:** Protect him.	
			(5) **Action:** Sin against him, bring a hand against him, revolt, anything not good and proper, oust him by helping the one of his brothers seize the throne. Set another king or lord over you, swear an oath to another king or lord.
2. § 6: 73–82	**Hearing** of opposition to succession	(1) **Report** it to Ashurbanipal.	
			(2) **Conceal** it.
3. §§ 7–9: 83–107	**Event:** Death of Esarhaddon in minority of his sons	(1) **Action:** Help Ashurbanipal to take the throne of Assyria and Šamaš-šumu-ukin to the throne of Babylon.	
			(2) **Action:** Hold back even one of the gifts given to Šamaš-šumu-ukin by Esarhaddon.
		(3) **Action:** Keep absolute honesty with Ashurbanipal and his brothers by the same mother. Serve them in a true manner (**Attitude**), speak the truth to them, protect them in country and town.	
			(4) **Action:** Sin against Ashurbanipal or his brothers by the same mother. Do evil against them, make an insurrection or do anything not good against them.

Table 1: Structure of the covenant stipulations. *(Continued)*

	Scenario (When/If…)	Mandated action	Forbidden action
4. §§ 10 – 11: 108 – 29	**Hearing** of treason against Ashurbanipal	(1) **Report** it to Ashurbanipal.	(2) **Conceal** it. (3) **Action:** Do anything evil or improper to Ashurbanipal. Seize him and put him to death, hand him to his enemy, swear an oath to another king or lord.
5. § 12: 130 – 46	**Spoken to** about treason against Ashurbanipal	1st preference: (1) **Action:** Seize the perpetrators and bring them to Ashurbanipal. If possible, seize and put to death, destroying their name and seed from the land. 2nd preference: (2) **Report** them to Ashurbanipal, then (3) **Action:** help him to seize them and put them to death.	
6. § 13: 147 – 61	**Hearing** (of anything) by coming into contact with insurrectionists	(1) **Report** it to Ashurbanipal, being loyal to him (**Attitude**). (3) **Report** it to Ashurbanipal and (4) **Action:** Seize and put to death the perpetrators.	(2) **Action:** Participate in oath.
7. §§ 14 – 16: 162 – 87	**Event:** Open rebellion **Event:** Taken hostage by rebels	(1) **Action:** Take a stand and (**Attitude**) wholeheartedly protect Ashurbanipal, defeat those who have revolted and rescue Ashurbanipal and his brothers by the same mother. (3) **Action:** Flee and come to Ashurbanipal.	(2) **Action:** Make common cause with one who may revolt against Ashurbanipal. (4) Set in your mind an unfavourable thought (**Attitude**). **Action:** Revolt against him, make rebellion, or do anything to Ashurbanipal which is not good.

Table 1: Structure of the covenant stipulations. *(Continued)*

	Scenario (When/If...)	Mandated action	Forbidden action
8. § 17: 188 – 97	**Event:** Esarhaddon dies, Ashurbanipal becomes king	(1) **Action:** Listen to whatever he says and do whatever he commands.	(2) **Action:** Seek any other king or any other lord against him.
9. §§ 18 – 21: 198 – 236	**Event:** Palace revolt against Esarhaddon **Event:** Suspicious messenger comes to the prince with a message from the king		(1) **Action:** Obey him. (2) **Action:** Listen to him or let him go away.
		(3) **Action:** Guard the prince strongly until one of you, who loves his lord and feels concern over the house of his lords (**Attitude**), goes to the palace and ascertains well-being of the king. Only afterwards go to the palace with the prince.	
			(4) **Action:** Hold an assembly to adjure one another and give kingship to one of you. (5) **Action:** Help any of Ashurbanipal's male relatives, in Assyria or those who have fled to another country, or anyone, to seize the throne of Assyria, nor hand over the kingship and lordship of Assyria.
		(6) **Action:** Help Ashurbanipal to seize the throne of Assyria and he will exercise kingship and lordship over you. (7) **Action:** Fall and die for Ashurbanipal and seek to do for him what is good.	
			(8) **Action:** Do for him what is not good, give him improper counsel or direct him in an unwholesome course.
		(9) **Action:** Continually serve him in a true and fitting manner.	
10. §§ 22 – 25: 237 – 301	**Event:** Death of Esarhaddon and murder of Ashurbanipal		(1) **Action:** Make common cause with the murderer and become his servant.

Table 1: Structure of the covenant stipulations. *(Continued)*

Scenario (When/If…)	Mandated action	Forbidden action
	(2) **Action:** Break away and be hostile (to the murderer), alienate all lands from him, instigate a rebellion against him, seize him and put him to death and help a son of Ashurbanipal to take the throne of Assyria. (3) **Action:** Wait for a woman pregnant by Esarhaddon (or) for the wife of Ashurbanipal (to give birth). After a son is born, bring him up and set him on the Assyrian throne. Seize and slay the perpetrators of rebellion, destroy their name and seed from the land, shedding blood for blood, avenging Ashurbanipal.	
		(4) **Action:** Give Ashurbanipal a deadly drug to poison him, practice witchcraft against him, make gods and goddesses angry with him.
	(5) **Attitude:** Love Ashurbanipal like yourselves	
		(6) **Action:** Slander Ashurbanipal's brothers, his mother's sons, before him, speak anything evil about them, lift your hands against their houses, or commit a crime against them, or take anything away from the gift which their father has given them, or the acquisition that they have made.
	(7) **Action:** Speak well of them before Ashurbanipal. (8) **Action:** Speak to sons and grandsons to be born in the future, give them orders to guard the covenant and not sin against it (verbatim).	

Table 1: Structure of the covenant stipulations. *(Continued)*

	Scenario (When/If…)	Mandated action	Forbidden action
11. § 26: 302–17	**Event:** Usurper seizes throne from Esarhaddon		(1) **Attitude:** Rejoice over his kingship (i. e. that of the usurper).
		(2) **Action:** Seize him and put him to death.	
			(3) **Action:** Submit to his kingship or swear and oath of servitude to him.
		(4) **Action:** Revolt against him, make other lands inimical to him, take plunder from him, defeat him and help Ashurbanipal to take his father's throne.	
12. §§ 27–28: 318–35	**Event:** Involved in plot to turn Esarhaddon against Ashurbanipal.		(1) **Action:** Cause a fight between Ashurbanipal and Esarhaddon by stirring up mutual hatred between them.
		(2) **Action** (speech): Refuse to carry out the orders of the plotter and argue against him (verbatim).	
13. §§ 29–30: 336–59	**Event:** Involved in a plot to turn Ashurbanipal against his brothers.		(1) **Action:** Obey or speak evil about Ashurbanipal's brothers in his presence, divide him from them. Let others who do this go free.
		(2) Come and **report** to Ashurbanipal as follows 'your father imposed a covenant on us and made us swear an oath' (verbatim).	
			(3) **Action/Attitude:** Look at Ashurbanipal or his brothers without submission.
		(4) **Action:** Contest them as you would on your own behalf (**Attitude**), saying 'your father set this in a covenant and made us swear it' (verbatim).	

Table 1: Structure of the covenant stipulations. *(Continued)*

	Scenario (When/If...)	Mandated action	Forbidden action
14. §§ 31–34: 360–96	**Event:** Ascension of Ashurbanipal upon the death of Esarhaddon.		(1) **Action** (speech): Say any evil word about Ashurbanipal's brothers or make them accursed, (saying) 'bring your hand against them for an evil deed'. Alienate him from Ashurbanipal, or say any evil word about them in the presence of their brother. Attempt to persuade Ashurbanipal to remove them from the positions that Esarhaddon assigned them. (2) **Action:** Try to revoke or undo the oath [...] think of or perform a ritual to revoke or undo the oath (**Attitude**).
		(3) You and your sons will be bound by the oath from this day on until what(ever) comes after this covenant.	
			(4) **Attitude:** Swear this oath with your lips only.
		(5) **Attitude:** Swear the oath wholeheartedly. (6) **Action:** Teach the oath to your sons.	
			(7) **Action:** Feign unclean illness.
		(8) **Action:** Take part in the covenant.	
15. § 35: 397–409	**Event:** (Someone transgresses or disregards the covenant) – Violation.	(1) **Action:** Guard the tablet like a god (**Attitude**).	
16. § 36: 410–13	**Action:** (Remove or destroy the tablet).	(Curse Section).	

3.1 Scenarios of potential danger in the covenant stipulations

The are, in total, sixteen distinct scenarios set out in the stipulation section of the covenant.[88] They begin with the following statement: 'When Esarhaddon, king of Assyria, passes away'.[89] This first scenario thus introduces the fundamental concern of the stipulations as a whole: the attempt to influence the aftermath of Esarhaddon's death. The use of the term *kīma* 'when' establishes this first scenario as an inevitability.[90] The event is framed in a general way, using the term *ana šimti ittalak* 'goes to his fate', thus indicating that the mandated and forbidden actions here apply regardless of the more precise circumstances of Esarhaddon's death.

The scenarios that follow, in contrast, are presented only as possibilities, generally introduced with the conjunction *šumma* 'if', and are described with a much greater degree of specificity. The second scenario, for instance, is presented as follows:

> If you hear any improper, unsuitable or unseemly word concerning the exercise of kingship which is unseemly and evil against Ashurbanipal, the great crown prince designate, either from the mouth of his brothers, his uncles, his cousins, his family (var. his people), members of his father's line; or from the mouth of magnates and governors, or from the mouth of the bearded and the eunuchs, or from the mouth of the scholars or from the mouth of any human being at all.[91]

Logically, this statement follows on from the previous scenario. Those bound by the covenant are to aid Ashurbanipal in coming to the throne after the death of Esarhaddon and, as such, a situation in which they learn of opposition to Ashurbanipal's *epēš šarrūte* 'exercise of kingship' is presumably relevant to this. From a temporal point of view, the scenario seems likely to pertain to the period around the death of Esarhaddon. Nevertheless, this is not stated explicitly, and therefore the stipulation can also be understood to extend beyond the period of succession: from the imposition of the covenant until the end of Ashurbanipal's reign. Unlike the first scenario, therefore, this second contingency is not a single event that will affect all people who swear to the covenant at the same time, but rather a recurrent potential scenario which requires continual vigilance.

---

88 Note that my delineation of these scenarios in my analysis differs from the segments into which the covenant composition is divided by the use of horizontal rulings. These rulings sometimes separate what I analyze here as a single scenario and its associated mandated and forbidden reactions into several sections.

89 SAA 2, no. 6, § 4: 46.

90 CAD K, 363–367 s.v. *kīma*.

91 SAA 2, no. 6, § 6: 73–80; Lauinger 2012, i 80'–ii 5.

The specification of particular parties in this section serves to highlight either those who are perhaps considered particularly likely to speak against Ashurbanipal's kingship or, possibly relatedly, those whose opposition to Ashurbanipal would be most dangerous. The male members of the royal family are the first group mentioned in this list: *ahhēšu ahhē–abbēšu mār–ahhē–abbēšu qinnīšu* (ND 4336: *nišēšu*) *zār bēt–abīšu* 'his (i. e. Ashurbanipal's) brothers, his uncles, his cousins, his family (ND 4336: his people), members of his father's line (lit. seed of his father's house)'.[92] These family members are followed by the highest officials in the Assyrian administration, *rabiūti pāhāti* 'magnates (and) governors'.[93] There then follows a catch-all phrase referring to all members of the palace's administrative sphere *ša–ziqni ša–rešāni* 'the bearded and the eunuchs'.[94] The list concludes with the particular mention of *ummânī* 'scholars', as well as *naphar ṣalmāt–qaqqadi mala bāšû* 'any human being (lit. black-headed one), as many as there are'. The term 'black-headed one' was used in the royal inscriptions of Esarhaddon to refer specifically to those over whom he ruled,[95] and thus it seems reasonable to assume that it references all subjects of the Assyrian crown, direct and indirect.

Drawing the attention of those entering the covenant to these 'likely suspects' serves to signpost the situations in which the entrants are to be most vigilant, namely in the presence of these classes of people. Of course, as is especially clear in the case of the governors, who would each have been the first-mentioned legal partner in their manuscript of the covenant, at least some of these suspected groups would themselves have taken part in covenant ceremonies.[96] Thus, the vigilance that the covenant composition aimed to instill, although it is depicted here as being directed towards others, also concerns the self in some instances. The composition also stresses the potential involvement in such scenarios of high-status groups that most people would normally have been expected to obey. In this way, the stipulations highlight the existence of certain scenarios in which it is necessary to disregard such authority. The message is clear: an Assyrian subject must be loyal to the king and his chosen successor, at the expense of all others.

92 SAA 2 no. 6, § 6: 76–77.

93 On the magnates, see Mattila 2000.

94 Note that the phrase *ša–ziqni* is only found when contrasted with *ša–rēši*. On this title, see Gross 2020, 264–269.

95 The term is generally used to describe Esarhaddon himself either as 'shepherd' or 'herdsman of the black-headed people'. See, for instance, RINAP 4, no. 48: obv. 34: *re-'u-ú tak-lum na-qid ṣal-mat* SAG.DU 'trusted shepherd, herdsman of the black-headed people'. On the phrase 'black-headed people', Karlsson 2020. 'Pastoral' epithets had a long history in Assyrian royal titulary (Cifola 1995, 4).

96 See discussion in Chapter 5.

From here, the scenarios become yet more specific, with the third scenario addressing the death of Esarhaddon *ina ṣahāri ša mar'ēšu* 'in the minority of his sons' (Table 1, no. 3). The scenario goes on to deal with the roles of Ashurbanipal and Šamaš-šumu-ukin after the death of Esarhaddon, as well as defining the roles after the death of Esarhaddon of all of his sons by Ashurbanipal's mother.[97] These sons of Esarhaddon are described in the text through their relationship with Ashurbanipal: *ahhēšu mār ummīšu* 'his (Ashurbanipal's) brothers, sons of his mother'. It seems unlikely that the phrase 'minority of his sons' refers to Ashurbanipal or Šamaš-šumu-ukin specifically, who were politically active at that time and thus probably not considered minors.[98] Rather, as these other sons of Esarhaddon are clearly relevant to this scenario, I consider that this phrase more likely applies to Esarhaddon's youngest sons, who may still have been children at the time of the covenant's composition.

This third scenario mirrors the first to a degree, as both deal explicitly with Esarhaddon's death. So too, the fourth scenario mirrors the second, as it returns to the possibility of the king's subjects hearing an *abutu lā ṭābtu lā de'iqtu lā banītu ina muhhi Aššur-bāni-apli* 'evil, improper, ugly word which is not seemly nor good to Ashurbanipal'.[99] Here the list of suspects is similar to that already set out in the second scenario, but subtly differs from it:

> Either from the mouth of his enemy or from the mouth of his ally, or from the mouth of his brothers or from the mouth of his uncles, his cousins, his family, members of his father's line, or from the mouth of your brothers, your sons, your daughters, or from the mouth of a prophet, an ecstatic, an inquirer of oracles, or from the mouth of any human being at all.[100]

The use here of the terms *nakru*, 'enemy' and *salmu* 'ally' could be interpreted as differentiating this scenario from the second in two main ways.[101] Firstly, the mention of a *nakru* 'enemy' seems to signal a shift in focus from the palace and administrative realm to a geographical or at least social zone that is further away from Ashurbanipal. Presumably the enemies of Ashurbanipal, at least those who were long established as such, were not given access to the Assyrian palace administra-

---

97 On Ashurbanipal's family, see PNA 1/1, 159–163 s.v. Aššur-bāni-apli and Novotny and Singletary 2009.

98 As noted, for instance, in Watanabe 1987, 180. Watanabe has since contradicted the assertion that these princes were not minors, however (Watanabe 2015, 186), illustrating that the precise meaning of this stipulation is still debated. See also Watanabe 2019, 180 for further discussion.

99 SAA 2, no. 6, § 10: 108–111.

100 SAA 2, no. 6, § 10: 111–119.

101 CAD N/1, 192–195 s.v. *nakru* 2; CAD S, 104–105 s.v. *salmu*. This pair of terms is also found in a prophecy text (Watanabe 1987, 180).

tive sphere in the way that the groups listed in the second scenario were. Nevertheless, the inclusion of *salmēšu* 'his allies' in the list possibly points to another intention behind this phrase's inclusion here. By making the assertion that his allies are also under suspicion, the fourth scenario emphasizes the need to anticipate seditious statements made by those of whom one might not necessarily expect such a thing.

The same list of Ashurbanipal's male relatives found in the second scenario follows this initial pair of suspects. The list then pivots from Ashurbanipal's relatives to the relatives of those entering the covenant: *ahhēkunu mar'ēkunu mar'ātikunu* 'your (pl.) brothers, your (pl.) sons, your (pl.) daughters'.[102] Thus, in social and spatial terms, once again the composition's focus moves away from Ashurbanipal and here moves explicitly towards those entering the covenant, wherever they may be and whatever their social status. This structure invites the swift transfer of attention from Ashurbanipal and his family to those entering the covenant and their own families, creating, perhaps, a sense of connection between the hearer and the crown prince. This list is shorter than that of the royal family members, as it includes only brothers, sons and, perhaps surprisingly, daughters. It seems likely that the composers were aiming here to echo the curse sections of the covenant, which reference several times the terrible consequences of breaking the covenant that will befall the hearers, as well as *ahhēkunu mārīkunu mārātikunu* 'your brothers, your sons and your daughters'.[103]

The list ends with another combination of new additions to the list, the *raggimu* 'shouter (i.e. prophet)' and *mahhû* 'estatic',[104] as well as the *mār šā'ili amat ili* 'dream interpreter'.[105] Here, the inclusion of groups possessed of special knowledge, sent by the gods, somewhat mirrors the mention of *ummânī* 'scholars' in the prior list.

The fifth scenario presents those entering the covenant with a new eventuality. In this case, they may not merely hear of someone speaking against Ashurbanipal, as in the previous scenarios; rather, it posits that someone may 'speak to you of rebellion and insurrection (with the purpose) of ki[lling], assassinating, and eliminating Ashurbanipal… or if you should hear it from the mouth of any-

102 SAA 2, no. 6, § 10: 115–116.

103 As in SAA 2, no. 6, § 69: 549. The formulation *mārīkunu mārātikunu* 'your sons and your daughters', is also common in the curse sections, as is *issātikunu mārīkunu mārātikunu* 'your women, your sons and your daughters'.

104 On these terms, see recently Stökl 2015, 58, as well as Watanabe 1987, 180.

105 CAD Š/1, 110–111 s.v. *šā'ilu*. The term is here defined as a 'diviner (interpreting dreams, practicing necromancy)'.

one'.[106] In this situation, an attempt is made to involve those entering into the covenant directly in a plot to kill Ashurbanipal. In contrast to the prior scenarios, the identity of the person who may do this is not explicitly stated. It seems possible that, in light of the lists of suspects presented in scenarios two and four, it is expected that those entering into the covenant will be able to infer this for themselves. The universal is emphasized here with the term *memmēni* 'anyone'.[107] Perhaps interestingly, the description of the scenario ends with the statement that, even if no one speaks to him directly and the subject only *šemû* 'hears' about such a plot, the mandated and forbidden actions remain the same.

The sixth scenario is similar to the fifth: 'If you should come into contact with perpetrators of insurrection, whether they are few or many, and hear (anything, be it) favourable or unfavourable'.[108] This can be seen as an escalation of the previous scenario, as those plotting against Ashurbanipal appear to have succeeded in forming a group, although it is emphasized that such alliances require a reaction regardless of their size: *lū ēṣūte lū ma'dūte* 'whether they are few or many'. In contrast to the fifth scenario, the sixth defines the plotters as *ēpišānūte ša bārte* 'instigators of an insurrection'.[109] In this way, the scenario develops the logic of the scenario that precedes it: whereas in the fifth scenario, it was necessary to state that 'anyone' talking of killing Ashurbanipal is relevant to the covenant, in the sixth it appears to be presumed that those bound by the covenant will themselves be able to identify an 'instigator of insurrection'. The statements that a plotter may make are described differently here: *dunqu lā dunqu tašammāni* '(and) you hear (anything, be it) favourable or unfavourable'. Thus, the idea is introduced that – depending on context, in this case the identity of the person speaking – it may be relevant to react even to 'favourable' statements. So too, it is not stated that these declarations need to pertain to Ashurbanipal specifically: the identity of those making the statements is sufficient to render them relevant. Thus, the sixth scenario represents not only an escalation of the situation described in the fifth, but also uses different terminology to describe similar circumstances, something which presumably would have served to develop the participant's understanding of their own task.

The seventh scenario addresses the next step in the escalation of the scenarios outlined in scenarios five and six, an outbreak of open rebellion:

106 SAA 2, no. 6, § 12: 130–134.

107 CAD M/2, 17–18 s.v. *memēni*; Watanabe 1987, 180.

108 SAA 2, no. 6, § 13: 147–149.

109 CAD B, 113–115 s.v. *bartu*.

> If an Assyrian or a client of Assyria, or a bearded (courtier) or a eunuch, or a citizen of Assyria or a citizen of any other country, or any living being at all besieges Ashurbanipal, the great crown prince designate, in country or in town, and carries out rebellion and insurrection.[110]

Here, for the first time, the clients appear to be mentioned explicitly, using the term *dāgil pāni ša Māt–Aššur* 'subject of the land of Aššur' as contrasted with *Aššurāya,* 'Assyrian'. Once more, the bearded courtiers and eunuchs are mentioned, as is the *mār Māt–Aššur* 'citizen (lit. son) of the land of Aššur' and the *mār māti šanītimma* 'citizen (lit. son) of another country'.[111] Here, then, the relevance of the stipulation not only to the provincial extent of the empire and its populace but also to the client states (and beyond) is highlighted as relevant. These various pairings serve to communicate the importance of vigilance towards both natives of Assyria and foreigners, as well as reinforcing the message that both high and low-ranking individuals are under suspicion. The statement *ina eqli ina libbi āli* 'in country or in town (lit. in the field or in the heart of the city)' draws attention to the necessity of vigilance in both categories of location. It is perhaps relevant to note that the 'country' here seems to refer to an agricultural zone (*eqlu* literally means 'field'),[112] as opposed to more remote, inhospitable regions. Thus, it is the settlements and the agricultural regions that are singled out as important.

The seventh scenario also explores the possibility that the covenant party may be captured by the rebels introduced in the previos scenarios: *šumma kî da'āni iṣṣabtūkunu* 'should they seize you by force', once again adding to the contingent possibilities laid out in the stipulations in the form of an escalation of the situation described.

The eighth scenario returns to the death of Esarhaddon:

> On the day that Esarhaddon, king of Assyria, your lord passes away, (on that day) Ashurbanipal, the great crown prince desi[gnate], son of Esarhaddon, your lord, shall be your king and your lord; he shall abase the mighty, raise up the lowly, put to death him who is worthy of death, and pardon him who deserves to be pardoned.[113]

This scenario is again presented not as possibility but as fact. The statement can be interpreted as a reiteration of the first scenario, but once more there are varia-

110 SAA 2, no. 6, § 14: 162–166.
111 CAD M/1, 315–316 s.v. *māru* 5. Here translated as 'citizen, native'.
112 CAD E, 249–252 s.v. *eqlu.*
113 SAA 2, no. 6, § 17: 188–194.

tions. The claim that Ashurbanipal will *ša duāki lā idukkūni* 'put to death him who is worthy of death' once he becomes king may be viewed as a subtle reference to the previous scenarios: Ashurbanipal will assume the throne when Esarhaddon dies and put to death anyone who revolts against him, taking on the role of judge.[114] This assertion therefore mirrors the first scenario while also serving as a logical conclusion to the narrative of the previous three scenarios. The statement that Ashurbanipal will *dannu lā ušappalūni šaplu lā imattahūni* 'abase the mighty, raise up the lowly' is also rather interesting, as it seems to imply substantial potential social upheaval upon Ashurbanipal's accession. This can perhaps be linked to the established technique in the stipulations of using lists of categories of persons to argue that a subject need not respect the established social hierarchy when it comes to vigilance. In fact, the stipulations made it clear that those at the upper echelons of Assyrian society are precisely those who are most dangerous to the person of Ashurbanipal. This scenario seems to imply that, once Ashurbanipal is king, the social position of his subjects will be made consistent with the status they deserve, with lowly but loyal subjects receiving their reward.

The ninth scenario concerns a palace revolt against Esarhaddon, thus initiating another narrative on the theme of rebellion. Here the people who may be involved in such a plot are not described by profession or category but rather by location, specifically their access to the *ēkallu* 'palace'.[115] This again appears to suggest the involvement of members of the royal family or high-ranking officials. It is specified that the scenario may take place during the day or at night, a first explicit statement that action may be necessary at any time. In contrast to the previous scenarios, which all refer to action or speech against Ashurbanipal, the insurrection described here is against Esarhaddon. Attention is also drawn to location: *ina hūli lū ina qabsi māti* 'whether on campaign or within the land',[116] perhaps suggesting that these are considered to be potential flashpoints.

The composition gives further details of what may happen in this scenario: 'If a messenger from within the palace at an unexpected time, whether by day or by night, comes to the prince saying: 'Your father has summoned you; let my lord come'.[117] While this scenario is certainly not completely consistent with the little that is known about Sennacherib's murder, one may wonder whether it was to

114 Watanabe 1987, 183.

115 CAD E, 52–61 s.v. *ekallu*. As discussed in Chapter 2.1, this term generally refers to the residence of the Assyrian king.

116 Watanabe 1987, 183.

117 SAA 2, no. 6, § 18: 201–204.

some extent inspired by that event.[118] Here again, the use of the phrase *ina lā simenīšu* 'at an unexpected time',[119] draws attention to the temporal aspect of the scenario. It is perhaps relevant that this phrase often refers to unusual astronomical and meteorological phenomena,[120] perhaps indicating that such an event may be expected to coincide with these things.

Scenario ten returns to the death of Esarhaddon, thus paralleling scenarios one, three and eight. In the same way that scenarios one and eight are closely linked, both thematically and linguistically, so too are scenarios three and ten. Scenario ten, like scenario three, addresses the possibility of Esarhaddon's death *ina ṣahāri ša mar'ēšu* 'in the minority of his sons'.[121] This time, it is suggested that a member of the palace administration, here described as *lū ša–rēši lū ša–ziqni* 'either a eunuch or a bearded (courtier)', may put Ashurbanipal to death and *šarrūtu ša Māt–Aššur ittiši* 'take over the kingship of Assyria'. Here, then, unlike scenario three, the suspicion is cast not primarily on the other members of the royal family, but rather on courtiers more widely – of course, members of the royal family were presumably also active at court.

In this way, the tenth scenario can be read as a progression of the narrative of scenarios eight and nine. In scenario nine, those in the palace, as well as messengers from the palace, are singled out as potentially dangerous. By mentioning bearded courtiers and eunuchs, the tenth scenario continues this focus on palace insiders. The use of the term *ina ṣahāri ša mar'ēšu* 'in the minority of his sons' here seems to be linked to the possibility of a courtier being able to seize the throne of Assyria after murdering Ashurbanipal. As an elaborate plot to kill Ashurbanipal, which does not seem to necessitate the minority of the crown prince, is set out in scenario nine, it seems most likely that this phrase again refers to Esarhaddon's other sons. In such a situation, it seems that the danger would exist that a courtier – even a eunuch – might seize the throne. While this may well be a reference to simple usurpation, one may also wonder if it refers to the possibility of an advisor ruling as regent in the case of a child king, namely one of Ashurbanipal's young brothers, coming to power.

The eleventh scenario is not a continuation of this narrative, and in this case the potential rebel is clearly a usurper: 'If anyone makes rebellion or insurrection

---

118 While not the same, the details of palace intrigue recounted in the description of the plot against Sennacherib by Esarhaddon's brother in the letter SAA 18, no. 100 perhaps bear some comparison to this scenario. See also Dalley and Siddall 2021, as well as the discussion of Sennacherib's murder in Chapter 1.1, 1.3.2 and Chapter 4.1.

119 CAD S, 268–271 s.v. *simanu.* The term is translated as 'season, proper time, time'.

120 Watanabe 1987, 183.

121 SAA 2, no. 6, § 22: 237.

against Esarhaddon, king of Assyria and seats himself on the royal throne'.[122] This reintroduces the death of Esarhaddon, the circumstances of which are not described in the tenth scenario, in the context of a rebellion. The theme is very closely connected to that of the eighth to tenth scenarios, and follows on from it in a logical manner. Here, the topic is advanced with the scenario of Esarhaddon's violent death and *memēni* 'anyone' sitting *ina kussie šarrūtīšu* 'on the royal throne' of Assyria.

The twelfth scenario comes back to the theme of speech:

> If (any) one of his brothers, his uncles, his cousins, his family,[123] (any) members of his father's line, or any descendant of former royalty or (any) one of the magnates, governors or eunuchs, (or any) one of the citizens of Assyria, (or) any foreigner, involves you in a plot, saying to you: 'Malign Ashurbanipal, the great crown prince designate, in the presence of his father. Speak evil and improper things about him.'[124]

This somewhat echoes the fifth scenario, as here once more the composition explores the possibility of an attempt to involve the covenant party in a plot. The list of Ashurbanipal's male relatives is initially identical to those in scenarios two and four, but it diverges with the inclusion of *zār šarri pāniūti* 'any descendant of former royalty'. This phrase seems to seek to highlight more distant members of the royal family, such as members of the bloodline of Sennacherib or Sargon II. Here, then, the composition places particular emphasis on such people as potential plotters. The decision to include these people in the list once more shows the use of repetition and slight variation to reinforce a point of interest while stressing different aspects of it at each iteration. So too, the mention of magnates, governors and eunuchs is familiar from previous scenarios, but they appear for the first time here as a group of three, rather than two groups of two: magnates and governors, along with bearded courtiers and eunuchs. This slight change was perhaps also intended to maintain and direct the attention of those listening to the stipulations.

Scenario twelve includes a hypothetical verbatim statement made by a plotter, as does scenario nine. While in situation nine, the plotter is a messenger lying to the listener, here he is ordering the listener himself to become involved in a scheme that requires telling untruths about Ashurbanipal to his father. The wording of this serves to parallel scenarios two and four, although this situation is more specific and the role of the covenant party is more involved. Once more, the danger of *lā ṭābtu lā de'iqtu* 'evil and improper' statements about Ashurbanipal is highlighted, serving to further emphasize their importance within the stipulations.

---

122 SAA 2, no. 6, § 26: 302–304.
123 The term *qinnīšu* 'his family' is preserved in this line on only one fragment: ND 4356.
124 SAA 2, no. 6, § 27: 318–325.

Scenario thirteen is very similar to scenario twelve, but here the focus is on potential strife between Ashurbanipal and his brothers:

> If someone involves you in a plot, be it one of his (i. e. Ashurbanipal's) brothers, his [unc]les, his relations, a member of his father's line, a e[unuch] or a bearded (courtier), an Assyrian or a foreigner, or any human being at all, saying: 'Slander his brothers, sons by his own mother, before him, make it come to a fight between them, and divide his brothers, sons of his own mother, from him'.[125]

One may perhaps presume that this scenario would take place later than the previous one, when Esarhaddon is already dead. In any case, the list is once more a slight variation on that which precedes it, perhaps with the intention of holding the attention of those listening to the stipulations.

The final scenario that involves Ashurbanipal's succession is scenario fourteen: it reiterates the inevitability of Esarhaddon's death and refers to Ashurbanipal's succession: 'When Esarhaddon, king of Assyria, your lord, passes away and Ashurbanipal, the great crown prince designate, ascends the royal throne'.[126] The cyclical structure of the scenarios, beginning and ending with the same event, serves to emphasize the centrality of Esarhaddon's death and Ashurbanipal's ascension to the throne of Assyria. Whereas in the first scenario, the necessity that the covenant parties participate in seating Ashurbanipal on the throne is highlighted, in the final one, Ashurbanipal's accession is portrayed as an inevitability. This structure serves to imply that the stipulations, themselves phrased as an inevitability, will ensure that Ashurbanipal takes the throne successfully.

Scenarios fifteen and sixteen turn the focus of the composition to the covenant tablets themselves, introducing first the notion that someone may break the covenant agreement and then that someone may destroy or damage the covenant tablet. In this way, they ensure that no one can invalidate its binding nature. These scenarios introduce the first curse section of the composition.[127]

3.2 Mandated reactions in the covenant stipulations

In the following, I separate the actions mandated in response to the scenarios of the stipulations into three categories. The first and broadest category mandates that the covenant party intercede on behalf of the crown. These actions, as framed

125 SAA 2, no. 6, § 29: 336–343; Lauinger 2012, v 1–8.
126 SAA 2, no. 6, § 31: 360–362; Lauinger 2012, v 24–27.
127 Discussed in Chapter 2.3.

in the composition, do not themselves stipulate prior communication with the king or the crown prince. The second group mandates that the covenant party report the details of the situation to the crown. The third category, meanwhile, mandates a particular attitude (see also Table 1). While these reactions are generally clearly conceptually connected to the scenarios described, this is not always the case.

In reaction to the first scenario, the death of Esarhaddon, actions are mandated: seating Ashurbanipal on the throne, protecting him and serving him.[128] The first of these reactions centres on the death of Esarhaddon, as seating Ashurbanipal on the throne is portrayed as an immediate effect of this event. The use of the second person gives immediacy to the stipulations, while use of plural verbal forms underlines their all-encompassing nature: the mandated actions apply to all subjects of the Assyrian king. In these instances, it is clear that a mandated action may require the participation not only of one covenant party, but of many, as it is unlikely that one person would be solely responsible for seating a new king on the throne. The phrasing of these demands also credits the covenant party with substantial influence, something that certainly would have been the case for some of the subjects entering into the covenant, but would not have applied universally. It seems possible that the framing serves to provide those who heard the stipulations with an impression of proximity to the crown and responsibility the stability of Assyrian rule. Such wording creates a collapsed impression of the hierarchy of the empire, in which there exists only the direct, bilateral relationship between subject and crown.

The impression of Ashurbanipal as easily accessible to the covenant parties is deepened by the required reaction to scenario two, in which the covenant party hears of opposition to the succession. Here, the covenant parties are mandated to *lā tallakāninni ana Aššur-bāni-apli mār–šarri rabiu ša bēt–rēdûti lā taqabbāni* 'come and report it to Ashurbanipal, great crown prince designate (lit. of the Succession Palace)'.[129] The use of the verb *qabû* 'to speak' again appears to refer to direct communication. One could interpret this as indicating that the people to whom these stipulations apply most are those who have physical access to the crown prince. Alternatively, it could once more be seen as a stylized means of creating an impression of intimacy between the subject and Ashurbanipal. It is also significant that it is Ashurbanipal, not Esarhaddon, to whom the covenant parties are mandated to report,[130] even though the scenario does not specify that Esarhad-

128 SAA 2, no. 6, § 4: 47–51.

129 SAA 2, no. 6, § 6: 81–82; Watanabe 1987, 179 f.

130 Compare with this the fragments of Sennacherib's succession covenant, which do not specify reporting to the crown prince. One fragment of the extended version of this covenant appears not

don has already died. The covenant therefore marks a shift in the balance of power between the king and his chosen successor, as it grants Ashurbanipal access to information that it does not afford Esarhaddon.

The mandated actions of scenarios three and four mirror those of one and two. Scenario three, the death of Esarhaddon during the minority of his sons, mandates that the covenant parties help Ashurbanipal to take the throne. The stipulation differs, however, in that the covenant parties are also required to assist Šamaš-šumu-ukin in taking the throne of Babylon:

> You will help Ashurbanipal, the great crown prince designate, to take the throne of Assyria, and you will help Šamaš-šumu-ukin, his equal brother, the crown prince designate of Babylon, to ascend the throne of Babylon. You will reserve for him the kingship over the whole of Sumer, Akkad and Karduniaš. He will take with him all the gifts that Esarhaddon, king of Assyria, his father, gave him; do not hold back even one.[131]

This can be interpreted both as safeguarding Šamaš-šumu-ukin and also as defining the limits of his role: he is entitled to the throne of Babylon and the *tidintu* 'gift' given to him by his father, and nothing more. The lack of other references to the crown prince of Babylon in the covenant composition reinforces this impression. Once more, the importance of communication is stressed through the requirement that the subject *kittu šalimtu kullu* 'keep absolute honesty' with Ashurbanipal and *ahhēšu mār ummīšu ša Aššur-bāni-apli* 'his brothers by the same mother as Ashurbanipal'. The allusion to Ashurbanipal's brothers by the same mother serves to define and restrict the number of children of Esarhaddon to be given special protection and loyalty by the subjects of the Assyrian crown.[132] The way in which the covenant parties are to serve them is also further specified in this scenario: 'Serve them in a true manner, speak with them with heartfelt truth, protect them in country and town'.[133] This statement further highlights a central theme of the stipulation section, with speech and honest communication being portrayed as key components of loyalty. The demanded reaction to scenario four is identical to that of scenario two and uses the same phrasing: the subject must report to Ashurbanipal.

---

to specify to whom the information should be conveyed (Frahm 2009a, no. 67: r. col. 7'), while the short version requires reporting to Sennacherib (Frahm 2009a, no. 69: obv. 3).

131 SAA 2, no. 6, § 7: 84–91.

132 Note that while Sennacherib's succession covenant does not make the same distinction, it seeks to protect the crown prince and the 'remaining younger sons' of Sennacherib (Frahm 2009a, no. 67: r. col. 7'). Eckart Frahm suggests that these younger sons would have been born to Esarhaddon's mother, Naqi'a (Frahm 2009a, 132). If this is correct, then both kings attempted to limit the number of legitimate sons they had through the identity of their mother.

133 SAA 2, no. 6, § 8: 97–100; Watanabe 1987, 180.

Scenario five does not only present one mandated action to the covenant parties, but rather sets out a hierarchy of preferred reactions to the potential situation of someone speaking to them about treason. The first mandated reaction states that 'you shall seize the perpetrators of insurrection, and bring them before Ashurbanipal, the great crown prince designate'.[134] There then follow two potential mandated reactions to the scenario. The first mandated reaction is: 'If you are able to seize them and put them to death, then you shall destroy (*lā tuhallaqāni*) their name (*šumšunu*) and their seed (*zar'ušunu*) from the land'.[135] This stipulation, then, takes into account the possibility that the covenant party may not be in a position to seize the plotter. The second option is presented thus: 'If, however, you are unable to seize them and put them to death, you shall inform Ashurbanipal, the great crown prince designate, and assist him in seizing and putting to death the perpetrators of rebellion'.[136] In this case, the covenant party is required to *uznē petû* 'inform (lit. open ears)' Ashurbanipal about the situation.[137] The covenant party will then act directly once more, assisting Ashurbanipal in seizing and putting them to death. The repetition of this aspect illustrates that this will be the outcome for the perpetrator regardless of the circumstances. Presenting the mandated reactions in this way makes explicit the necessity that the covenant parties use their own knowledge concerning the particular situation they are in in order to make a judgement about what the appropriate reaction might be. Perhaps interestingly, it is not stated as necessary to inform Ashurbanipal of the situation if the subject is able to put the perpetrators to death, nor must subjects do so prior to seizing them.

The mandated reaction to the sixth scenario (coming into contact with insurrectionists) is, once again, to report the guilty parties to Ashurbanipal. In the first instance, the subject is required to do this *libbakunu issīsu la gammurūni* 'being completely loyal to him'.[138] This can be interpreted as requiring a modulation of attitude as well as implying particular external behaviour. The subject is also mandated to report to Ashurbanipal and, after doing so, to 'seize and put to death the perpetrators of insurrection and the traitorous troops, and destroy their name and seed from the land'.[139] The repetition of these mandated actions, coupled with subtle variation, functions in a similar way to the repetition and variation in the descriptions of the scenarios. They would presumably have ingrained into the listen-

---

134 SAA 2, no. 6, § 12: 136–138.
135 SAA 2, no. 6, § 12: 138–141; Watanabe 1987, 181.
136 SAA 2, no. 6, § 12: 142–146.
137 CAD P, 352–353 s.v. *petû* 4 *uznu*. This phrase is also discussed in Watanabe 1987, 181.
138 SAA 2, no 6, § 13: 152.
139 SAA 2, no. 6, § 13: 159–161.

er a sense of familiarity with the actions required of them with respect to the crown prince. By varying the required response subtly for each scenario, the subject is perhaps even encouraged to think in incremental terms about unforeseen circumstances which may arise and the appropriate reaction to such a situation.[140] The attention of the subject is therefore continually redirected to a few recurring actions: being loyal, reporting disloyalty, seizing and putting to death plotters.

The mandated reactions to the seventh scenario require that, in the case of open rebellion, the subject *šumma… lā tazazzāni* 'take your stand' and *ina gammurti libbikunu* 'wholeheartedly (literally 'with the whole of your (pl.) heart')' protect Ashurbanipal. This latter demand is something that, while it would likely manifest in external behaviour, is an attitude best judged by the subject himself: the use of the term *libbu* 'interior; heart' implies that the action is largely internal and thus not necessarily discernible to others.[141]

In this case, the covenant partner's success in resisting the rebellion is also mandated, and he must 'rescue (*lā tušezzabāninni*) Ashurbanipal and his brothers by the same mother'.[142] This demand makes most sense when considering that the stipulations are written in second person plural. The covenant parties, all together, could presumably defeat any traitors. Once more, the term 'brothers by the same mother' is used, conferring a degree of protection onto a select group of members of the royal family. If the rebels capture the covenant parties, they are to 'flee (*lā tahalliqāni*) and come to Ashurbanipal'.[143] The manner in which the subjects are to achieve this result is apparently left up to them.

Despite this, scenario eight makes clear that, after Esarhaddon's death, the covenant party is to 'hearken to whatever he (Ashurbanipal) says and do whatever

140 Compare the arguments of Pamela Barmash in relation to the *Codex Hammurabi*. Barmash argues that, rather than functioning as authoritative law, the code served to set out 'paradigmatic sets of legal principles that illustrated legal principles by modifying variables, presenting a comprehensive treatment of a legal topic', from which scribes could learn (Barmash 2020, 279).

141 Note that, while it may be the case that such an attitude would have been expected to be expressed in certain actions, contemporary evidence does suggest that internal attitudes of the *libbu* can contrast to one's external actions. This sentiment is found in the covenant itself (SAA 2, no. 6, § 34: 385–387 and Lauinger 2012, v 49–52, discussed below). Another particularly interesting example is the statement in *Esarhaddon's Apology* (see Chapter 4.1) that, even though Esarhaddon lost his father's favour and was forced into exile, *šaplānu libbašu rēmu rāšišuma* 'deep down he (Sennacherib) was compassionate' and still wished Esarhaddon to succeed him (RINAP 4, no. 1: i 30). Beyond this, see also statements such as those found in the composition *Ludlul bēl nēmeqi* with regard to the god Marduk: 'as heavy as his hand is, his heart (*libbašu*) is merciful' (vv 33; see further discussion in Piccin and Worthington 2015, 116 f.).

142 SAA 2, no. 6, § 14: 170–172.

143 SAA 2, no. 6, § 15: 178–179.

he commands'.[144] The use of the verb *šemû* 'to listen' illustrates the importance of the faculty of hearing in the stipulations, as well as the equation of paying attention and being loyal. So too, the mandate that the parties obey Ashurbanipal's commands illustrates that it is not sufficient simply to follow the existing covenant stipulations: the subjects must follow the changing and developing orders of the king. In this way, the covenant parties cannot simply focus on their own mandate to protect Ashurbanipal, it is also part of their duty to pay attention to him and comply with his wishes.

When exploring the possibility of a palace revolt in scenario nine, perhaps the most detailed and specific worked example given in the entire stipulation section, the subject is required to act directly, while also regulating his attitude towards Ashurbanipal. He must guard the messenger strongly, 'until one of you who loves his lord and feels concern over the house',[145] has gone to the palace and confirmed the king's well-being (i.e. that it is not a plot). Only after this may the subject take the prince to the palace. In addition, the subject is expected to help resist any pretender to the throne by helping Ashurbanipal to take the Assyrian throne. This mandated action has already appeared multiple times in the stipulations, but here it is intensified by the addition of two demands. The first, that the subject *šumma attunu ina muhhi Aššur-bāni-apli ... lā tamaqqutāni lā tamuttāni* 'fall and die for Ashurbanipal' has also already been stated, although it here appears with the added demand that the subject *ša muhhīšu ṭābūni lā tuba''āni lā teppašāni* 'seek to do for him (Ashurbanipal) what is good'. Here again, the mandated action demands an external change in behaviour, but – in requesting the attempt to do him good – it seems to require a certain attitude not necessarily visible to other people. Secondly, the subject is required to *ina kēnāte ṭarṣāte lā tattanabbalāšunūni* 'continually serve him in a true and fitting manner', thus indicating that loyalty is demanded beyond the period of Ashurbanipal's immediate succession.

In the tenth scenario, which comes back to the possibility of Esarhaddon's death during the minority of his sons and the murder of Ashurbanipal, the subjects are again required to act. They are to break away from and resist Ashurbanipal's murderer, putting him to death. They are then to help one of Ashurbanipal's sons to the throne. In mandating this, it is specified, albeit somewhat implicitly, that the covenant parties are not to place either a non-royal or one of Ashurbanipal's brothers on the throne. This command is reinforced, if somewhat complicated, by the ones which follow it, which demand that:

144 SAA 2, no. 6, § 17: 194–196.
145 SAA 2, no. 6, § 18: 207.

> You shall wait for a woman pregnant by Esarhaddon, king of Assyria, (or) for the wife of Ashurbanipal, the great crown prince designate (to give birth), and after (a son) is born, bring him up and set him on the throne of Assyria, seize and slay the perpetrators of rebellion, destroy their name and their seed from the land, and by shedding blood for blood, avenge Ashurbanipal, the great crown prince designate.[146]

In this version of events, then, the subjects may place a son of Esarhaddon on the throne, but it must be a child who is not yet born. As the wording of the stipulation seems to imply that multiple of Esarhaddon's living sons are alive in this scenario, it seems likely that this is an attempt to bar any living sons of Esarhaddon from plotting against their brother by ensuring that the covenant parties will not support them. So too it may have been aimed at any members of the king's entourage who could have been tempted to select a rival prince to place on the throne of Assyria, with the belief that by doing so they might gain influence. Once more, the punishment for such seditious dealings is death, as the covenant parties are commanded to shed *dāmē kūm dāmē* 'blood for blood', avenging the death of Ashurbanipal.[147]

The mandated actions continue here, although they do not seem to be particularly closely associated with this scenario. Nonetheless, it is possibly of significance that they follow on from a scenario in which Ashurbanipal has been murdered. The subjects are to 'love Ashurbanipal, the great crown prince designate, son of Esarhaddon, king of Assyria, your lord, like your own lives'.[148] This stipulation, as noted by Kazuko Watanabe, marks a significant innovation in the covenant genre as well as in the relationship in Assyria between subject and monarch.[149] Here, the subjects are required to adopt a certain attitude, that of love towards their future king. Once again, the relationship between subject and crown prince is presented as close and direct. While such a love for one's ruler might show itself in action, particularly in laying down one's beloved life for Ashurbanipal. Nonetheless, it is again the case that other people would not necessarily be able to check whether or not a particular subject was obeying this stipulation.[150] As such, the subject himself was presumably required to exercise vigilance over his own emotional life, measuring his love for Ashurbanipal against his love for himself. The subjects are also required to speak well of Ashurbanipal's brothers in his

146 SAA 2, no. 6, § 22: 249–258; compare Lauinger 2012, iv 1–2.
147 As stated in Watanabe 1987, 185.
148 SAA 2, no. 6, § 24: 266–268, compare Lauinger 2012, iv 8–11. See also Watanabe 1987, 185.
149 Watanabe 2019, 250–51. For a discussion of the term *râmu* 'to love', see recently Fales 2023a, esp. 669 f.
150 As in instances of doing something 'wholeheartedly', see discussion above.

presence, again illustrating the importance of speech in the stipulations, as well as connecting it to the issue of the relationship between Ashurbanipal and his brothers, as set out in the tenth scenario.

The composition continues, again drawing a parallel between the family of Ashurbanipal and the families of the subjects entering into the covenant, as a new section begins and they are commanded to:

> Speak to your sons and grandsons, your seed and your seed's seed which shall be born in the future, and give them orders as follows: 'Guard this covenant. Do not sin against your covenant and annihilate yourselves, do not turn your land over to destruction and your people to deportation. May this matter which is acceptable to god and mankind, be acceptable to you too, may it be good to you. May Ashurbanipal, the great crown prince designate, be protected for (his) lordship over the land and the people, (and) may his name later be proclaimed for the kingship. Do not place any other king or any other lord over you'.[151]

In this section, the speech of the covenant parties is dictated to such an extent that it is stated verbatim. The covenant parties are given the responsibility not only to guard Ashurbanipal and selected members of his family, but also to ensure that their own family become loyal subjects of the Assyrian crown. This is to be achieved by demanding that their descendants observe the covenant in perpetuity. The temporal aspect of their duty is thus expanded far beyond the confines of Ashurbanipal's immediate succession.

In scenario eleven, the focus reverts to the immediate actions necessary in the case of a usurper seizing the throne from Esarhaddon. The subjects here are required again to seize the perpetrator and put him to death, again reinforcing the notion that death is the inevitable outcome for anyone who rebels against Ashurbanipal's kingship. They are also required to:

> Revolt against him and wholeheartedly do battle with him, make other lands inimical to him, take plunder from him, defeat him, destroy his name and his seed from the land, and help Ashurbanipal, the great crown prince designate, to take his father's throne.[152]

Again, the inevitable outcome is the ascension of Ashurbanipal to the throne. The term *šumma... ina gammurti libbikunu qarābu issīšu lā tuppašāni* 'you shall wholeheartedly do battle with him' also indicates that the subjects are required to modulate their attitude towards Ashurbanipal.

The twelfth and thirteenth scenarios, in which the subject is included in a plot to create tension between members of the royal family by *karṣu akālu* 'ma-

151 SAA 2, no. 6, § 25: 288–301.
152 SAA 2, no. 6, § 26: 310–317.

ligning' them to each other, both require precise responses from the covenant parties. In each instance, the subjects are to refuse to carry out the orders of the plotters and argue against him, using verbatim statements:

> (Instead) say to the envious person[153] who commands you and would make you become accursed: 'Where are his brothers or the servants who made themselves accursed to his father by slandering him in the presence of his father? Has not what Aššur, Šamaš and [Adad] said about him proved to be true? Did your father ... without (the consent of) Aššur and Šamaš? Let your brother be honoured, and stay alive'.[154]

The quotation appears to assume that the plotter will be a brother of Ashurbanipal, as it uses the terms 'your father' and 'your brother'. This is consistent with the scenario description, as the first mentioned group of suspects are Ashurbanipal's brothers. Nevertheless, the scenario description goes on to list many other groups, such as eunuchs, Assyrians and foreigners, as well as other male members of the royal family, such as Ashurbanipal's uncles. The wording of the verbatim statement could be seen as a reinforcement of the notion already conveyed by the prominent position of Ashurbanipal's brothers in the scenario description: they are the most likely potential perpetrators of such a scheme. The covenant party is apparently expected in his speech to allude to the fate of Esarhaddon's own brothers and servants, who were unsuccessful in opposing their father's choice of successor.[155]

In the thirteenth scenario, the covenant party is mandated to report if an individual, possibly a son of Esarhaddon, attempts to foment strife between Ashurbanipal and his brothers. The party is instructed to *šumma lā tallakāninni ana Aššur-bāni-apli mār–šarri rabiu ša bēt–rēdûte lā taqabbāni mā abūka adê ina muhhīka issīni issakan utammannāši* 'come and report to Ashurbanipal, the great crown prince designate as follows: 'Your father imposed a covenant on us and makes us swear an oath concerning it.''[156] This description, then, gives a template of how to approach Ashurbanipal about a plot. Specifically, the subject is required to cite the covenant and one of its constituent parts, the oath, using phras-

153 Note that the meaning of this term, *bēl qi'i* is uncertain (Watanabe 1987, 186), although the CAD translates *qi'i* as 'envious, jealous person' (CAD Q, 285 s.v. *qi'u*). Compare also the use of the term *qinû* in *Esarhaddon's Apology* to describe Esarhaddon's brothers when he was chosen as crown prince (RINAP 4, no. 1: i 23; Frahm 2009b, 35–39).

154 SAA 2, no. 6, § 28: 328–335.

155 Note that while the people being addressed in the scenario appear to be Ashurbanipal's brothers, the question 'Where are his brothers', probably refers to the brothers of Esarhaddon who plotted against him (Frahm 2009b, 36).

156 SAA 2, no. 6, § 29: 349–352; Lauinger 2012, v 5–8. On the tense, see Lauinger 2012, 115 and the discussion in Chapter 5.3.

ing that refers both to Esarhaddon and the second person plural group of subjects. Thus, the covenant text offers a novel mandated reaction to a scenario that is not dissimilar from the one that precedes it. It seems possible to interpret this less as an arbitrary variation and more as an implicit statement that it is the responsibility of the covenant party to choose the most expedient of these various possible reactions according to the specific context in which they find themselves.

The mandated statement to be made to the plotter also differs from that in scenario twelve, possibly for similar reasons: 'If someone does not protect him, you will fight them as if fighting for yourselves. You will bring frightful terror into their hearts, saying: 'Your (pl.) father wrote (this) in the covenant, he established it, and he makes us swear it.''[157] Thus, the text outlines the attitude with which the covenant party is to resist the plotter. Once more, the comparison is made between acting on Ashurbanipal's behalf and acting for oneself, inviting association on the part of the subject between himself and the crown. The subject is also commanded to alter the internal state of the plotter, affecting his *libbu* 'heart'. In scenario twelve, therefore, the subject reasons with the plotter by reminding him that the imposition of Sennacherib's succession covenant was successful, and that the gods were consulted when selecting Ashurbanipal as crown prince. In scenario thirteen, meanwhile, he is to remind the plotter that they are both made to swear an oath of loyalty to Ashurbanipal.

Scenario fourteen, the death of Esarhaddon, mandates that: 'you and your sons to be born in the future will be bound by this oath concerning Ashurbanipal, the great crown prince designate, son of Esarhaddon, king of Assyria, your lord, from this day on until what(ever) comes after this covenant.'[158] This is phrased as an inevitability, and does not clearly demand active participation on the part of the subjects. However, the earlier command that those entering the covenant teach their descendants about it means that the statement can also be interpreted as alluding to this demand. An injunction against attempting to undo the oath accompanies this mandated action. Thus, one can interpret it as stating that the parties must accept and submit to the covenant, and help in perpetuating it. This impression is reinforced by the mandate that they swear *ina gummurti libbikunu* 'wholeheartedly'. The repetition of this phrase reinforces the importance of self-monitoring. The statement that 'you shall teach it to your sons to be born after this covenant'[159] is then made explicitly once more, as is the demand that they

157 SAA 2, no. 6, § 30: 355–359; Lauinger 2012, v 11–15 and 18–23. Note that § 30 is duplicated in the Kullania manuscript. For further discussion of this section, see Chapter 5.3.
158 SAA 2, no. 6, § 33: 380–384; compare Lauinger 2012, v 44–48. Watanabe 1987, 189.
159 SAA 2, no. 6, § 34, 387–388; Lauinger 2012, v 52–54. Watanabe 1987, 189.

shall 'take part in this covenant of Esarhaddon, king of Assyria, concerning Ashurbanipal, the great crown prince designate'.[160]

There are no mandated actions of the final scenario (destroying the covenant tablet), presumably because such an action on the part of a covenant party is forbidden and its consequence therefore takes the form of divine curses, rather than human reaction. The fifteenth and penultimate scenario, meanwhile, i. e. that anyone may change, disregard or transgress the covenant, includes one mandated action: 'you will guard like your god this sealed tablet of the great ruler on which is written the covenant of Ashurbanipal, the great crown prince designate, the son of Esarhaddon, king of Assyria, your lord, which is sealed with the seal of Aššur, king of the gods, and which is set up before you'.[161] This language parallels the commands that the covenant parties treat Ashurbanipal as they would themselves, in that it employs a simile linking something local and personal with the crown, in this case the covenant tablet itself is compared to subject's own (presumably local) god.

3.3 Forbidden reactions in the covenant stipulations

As well as outlining the appropriate reactions to the scenarios, the covenant composers placed just as much emphasis on what the covenant parties were not permitted to do. The forbidden reactions to the scenarios are presented alongside the mandated ones and can be divided into similar categories. Particular actions are not permitted, nor are certain attitudes. Reporting, meanwhile, is not a forbidden category of reaction. Rather, its opposite, *pazāru* (D stem *puzzuru*) 'concealing',[162] is a common category of prohibited response.

The forbidden reactions to the first scenario include deposing Ashurbanipal or seating one of his brothers, *rabiūti ṣehrūti* 'elder (or) younger', on the throne. The first can be seen as reinforcing the mandated action, seating Ashurbanipal on the throne, by expressing the opposite action in the negative. This can therefore be interpreted as a way to repeat the statement of what the covenant party must do, but rephrasing it, thus increasing the attention of the listener. Nevertheless, the framing of the forbidden deed also provides the opportunity to deepen and specify the desired result of the stipulation. In this case, the subject is explicitly instructed not to place one of Ashurbanipal's brothers on the throne. This serves to communicate

160 SAA 2, no. 6, § 34, 390 – 392; Lauinger 2012, v 56 – 57.
161 SAA 2, no. 6, § 35, 405 – 409; Lauinger 2012, v 68 – 72. Watanabe 1987, 190 and Lauinger 2012, 117.
162 CAD P, 310 – 313 s.v. *pazāru*. Note that the verb is in the D stem.

the danger that the covenant seeks to protect against more explicitly, increasing the awareness of the subject regarding his task. The covenant party is also forbidden from changing or altering the *abutu ša Aššur-ahu-iddina* 'the word of Esarhaddon'. This statement serves a different purpose to the other forbidden reaction to this scenario. It introduces a theme that is more thoroughly explored in the final scenarios of the stipulation section, namely the importance of following and obeying the covenant and, by extension, Esarhaddon's wishes. This serves to reinforce the circular structure of the stipulations.

The following forbidden actions are then set out:

> You shall not sin against him, nor bring your hand against him with evil intent, nor revolt or do anything to him which is not good and proper; You shall not oust him from the kingship of Assyria by helping one of his brothers, elder or younger, to seize the throne of Assyria in his stead, nor set any other king or any other lord over you, nor swear an oath to any other king or any other lord.[163]

This section's initial portion serves to reiterate the danger posed to Ashurbanipal by his brothers. It is perhaps relevant to note that these brothers are not referred to as 'by the same mother', and in this way are differentiated from those whose protection is mandated. This repetition in the very first scenario is presumably intended to impress upon the listeners the particular danger posed by Esarhaddon's other sons, probably including but not limited to Šamaš-šumu-ukin, encouraging particular vigilance with regard to them. Another repeatedly prohibited action is also introduced in this section: the taking of an oath not imposed by Esarhaddon or Ashurbanipal. By introducing this stipulation, the covenant composers highlight the impossibility of keeping the present oath while swearing to another one. Thus, the composers seek to establish a monopoly on the use of oaths for the Assyrian crown, thus also conveying the message that anyone imposing a rival oath is committing treason.

In the second scenario, the subjects are forbidden from concealing the treasonous statements they have heard. Again, by denying the opposite of the mandated action, the composers of the covenant effectively reiterate the statement, strengthening it. The coupling of these two concepts, speaking and concealing, serves also to establish a connection between the former and loyalty and the latter and treason. Thus, full and honest speech is placed firmly at the core of a subject's appropriate behaviour, while concealment is its antithesis.

The prohibited responses to scenario three are actions. In this way, the forbidden reactions to a particular scenario often belong to the same category as the

163 SAA 2, no. 6, § 5: 66–72; compare Lauinger 2012, i 68'–79'.

mandated ones. While the subject is commanded to set Šamaš-šumu-ukin on the throne of Babylon, he must not hold back even one *tidintu* 'gift' given to Šamaš-šumu-ukin by Esarhaddon. The section continues as follows:

> You shall not sin against Ashurbanipal, the great crown prince designate, whom Esarhaddon, king of Assyria, has ordered for you, nor against his brothers, sons by the same mother as Ashurbanipal, the great crown prince designate, concerning whom he has concluded (this) covenant with you; you shall not bring your hands to (do) evil against them nor make insurrection or do anything which is not good to them.[164]

These forbidden actions therefore make more explicit the importance to this scenario of protecting – and limiting – the position of the other legitimate sons of Esarhaddon after his death. The statement that Šamaš-šumu-ukin must be given precisely those gifts allotted to him by Esarhaddon also serves to reinforce the notion that Esarhaddon's exact wishes must be respected after his death.

As is the case with the mandated reactions to scenarios two and four, the forbidden responses to scenario four mirror those of number two: the subject is instructed not to conceal from Ashurbanipal his knowledge of a plot. Nevertheless, the composition here further develops the prohibited response in such a scenario, stating that the subject must not do anthing *lā ṭābtu lā de'iqtu* 'evil or improper' to Ashurbanipal. It continues 'you shall not seize him and put him to death, nor hand him over to his enemy, nor oust him from the kingship of Assyria, nor sw[ear an oa]th to any other king or any other lord.'[165] While these statements are not necessarily directly related to the fourth scenario, there does seem to be a logical link between them, as here the subject is prohibited from joining or participating in a plot against Ashurbanipal. Nevertheless, the focus on handing Ashurbanipal over to an enemy here stresses the notion of a coordinated plot against the crown prince. These statements also serve to create a conceptual link between the fourth scenario and those which follow, as these form a narrative that describes a concerted attempt to overthrow Ashurbanipal.

The fifth scenario does not include any negative commands, while the sixth, coming into contact with insurrectionists, contains only one: participation in a covenant:

> You shall not take a mutually binding oath with (any)one who produces (statues of) gods in order to conclude a covenant before gods, (be it) by sett[ing] a table, by drinking from a cup, by kindling a fire, by water, by oil, or by holding breasts.[166]

164 SAA 2, no. 6, § 9: 101–107.
165 SAA 2, no. 6, § 11: 126–129.
166 SAA 2, no. 6, § 13: 153–156.

This therefore reinforces the importance of not participating in any covenants imposed by others, as well as providing further information about the kinds of ritual oath-taking activities that are excluded.

The subsequent injunctions are primarily inversions of mandated actions: making common cause with those revolting against Ashurbanipal is forbidden, as is revolting against him, making rebellion or doing anything to him *lā ṭābtu* 'which is not good'. This final statement serves to define disloyalty very broadly. In addition, the covenant parties are forbidden from having particular thoughts:

> You shall not, whether while on a guard duty [...] or on a [day] of rest, while resid[ing] within the land or while entering a tax-collection point, set in your mind an unfavourable thought against Ashurbanipal, the great crown prince designate.[167]

This mandates that the covenant parties exercise vigilance, watchfulness with the aim of protecting Ashurbanipal, over their own attitudes, not allowing themselves to have particular feelings. This statement, however, also serves to provide more information about the scenario that the covenant seeks to avoid. Particular locations and moments are here presented as especially dangerous, nevertheless, the inclusion of *ina ūme rāqi* 'a day of rest'[168] seems to seek to highlight the need for vigilance no matter the time or location.

Scenario eight again repeats the order not to seek any other king or lord against him. The need to reject anyone seeking to act against Ashurbanipal or Esarhaddon is expanded in scenario nine, in which a suspect messenger may come from the palace. In this instance, attention is to be directed away from the suspect individual, just as in other stipulations it is mandated that it be directed towards Ashurbanipal. There is also an injunction against letting the messenger leave, a negative reiteration of the command that the subjects *maṣṣartušu dunnunu* 'guard him strongly', although again this statement is used also to further define and specify the anticipated reaction. The negative stipulations then move on from the specific scenario, stating that the subjects cannot 'hold an assembly to adjure one another and give the kingship to one of you'.[169] This reiterates the denial that other oaths or covenants can be legitimate. The wording is different here, however, which serves to place a new emphasis on the situation: whereas in other cases, the subjects may not participate in an oath imposed by another king or lord, here they are portrayed as playing a collaborative role with the other conspirators.

167 SAA 2, no. 6, § 16: 180–185.

168 CAD R, 176 s.v. *rāqu*. The phrase *ūmu rāqu* is translated as 'work-free day'.

169 SAA 2, no. 6, § 19: 212 f.; Watanabe 1987, 183.

The composition continues, forbidding the subject from aiding anyone in seizing the Assyrian throne in Ashurbanipal's stead:

> You shall not help (anyone) from among his brothers, his uncles, his cousins, his family, or members of his father's line, whether those who are in Assyria or those who have fled to another country, or (anyone) in the closer palace groups or in the more remote palace groups or (any) groups great or small, or (any) of the old or young, of the rich or the poor, whether a bearded (courtier) or a eunuch, or (one) of the servants, or (one) of the bought (slaves) or any citizen of Assyria or any foreigner or any human being at all, any one of you, to seize the throne of Assyria, nor shall you hand over to him the kingship and lordship of Assyria.[170]

Here again, the phrase 'any one of you' allows explicitly for the fact that the subjects entering into the covenant may belong to one of the groups listed in this clause. The list echoes those used in the scenario descriptions but once again subtly deviates from them, introducing new possibilities or reformulating old ones: 'the closer palace groups' or 'the more remote palace groups', for instance, and '(one) of the servants' or 'the bought (slaves)'. These additions serve to reinforce the message that groups at the top of the imperial hierarchy are worthy of suspicion and must be monitored closely, while here also particularly stressing the need to monitor those further down the hierarchy as well. In particular, the groups stationed at a greater social distance from the king, such as the 'remote palace groups' must be subject to scrutiny. The inclusion of the members of the king's line 'who have fled to another country' notes the possibility of these events taking place at a substantial physical remove from the king. The inclusion of this statement is likely another oblique reference to the circumstances of Esarhaddon's own succession, as the plotters – Esarhaddon's brothers – did flee the country.[171] It seems possible that this is intended to preclude the possibility of a member of that particular branch of the royal family launching a coup.

The forbidden actions continue with the statement that the covenant parties shall not 'do for him what is not good, nor give him an improper counsel or direct him in an unwholesome course'.[172] Here, truthfulness as a tenet of loyalty appears again, although in this instance the subject is not merely suspected of withholding information – it seems that he is under suspicion of deliberately giving Ashurbanipal false advice in order to scupper either his succession or his reign. This can be interpreted as an indication that the statement particularly targets those in a po-

170 SAA 2, no. 6, § 20: 214–225.

171 As Watanabe notes (1987, 184), and Steymans also highlights (2006, 341), the statement 'those who are in Assyria and or those who have fled to another country' refers specifically to members of Ashurbanipal's family.

172 SAA 2, no. 6, § 21: 233–236.

sition to advise the royal family. Alternatively, it can be seen as once more rhetorically collapsing the distance between crown and subject, implying a universal availability of the Assyrian monarch and crown prince to their subjects.

The negative commands continue in scenario ten, which describes the death of Esarhaddon during the minority of his sons and the murder of Ashurbanipal. The subject is forbidden from making common cause with the murderer and becoming his servant. In this way, the composition denies the covenant parties the ability ever to transfer their loyalty to a ruler who has not been approved by Esarhaddon, even in the event of the death of his chosen successor. This statement serves as a reiteration, while also further specifying the intended consequences of the situation. The subjects are prohibited from participating in the murder of Ashurbanipal, the description of which functions to expatiate on the manner in which such an event could occur:

> You shall not give Ashurbanipal, the great crown prince designate, son of Esarhaddon, king of Assyria, your lord, a deadly drug to eat or to drink, nor anoint him with it, nor practice witchcraft against him, nor make gods and goddesses angry with him.[173]

Perhaps interestingly, a violent murder is not included in this description, instead more clandestine methods such as poison and activities which involve recourse to the supernatural are referenced. Again, it is relevant that here it is the covenant party himself who is under suspicion, and he is therefore encouraged to monitor his own actions.

Perhaps in anticipation of the scenarios which follow, the subject is also forbidden from slandering Ashurbanipal's brothers, the sons of his mother, in his presence. Again, the forbidden actions here provide a thematic link between this present scenario and those that follow it. The 'gifts' given to Ashurbanipal's brothers are mentioned once again. In the case of a usurper seizing the throne from Esarhaddon, a contingency advanced in scenario eleven, an attitude is forbidden: 'you shall not rejoice over his (i. e. the usurper's) kingship'.[174] Here, the stipulations allow for the possibility that the subjects, even in the plural, may not be able to do what is commanded of them, in this case putting the usurper to death. Nevertheless, they are told, if they are unable to do this: 'you shall not submit to his kingship nor swear an oath of servitude to him'.[175] As such, the initial inability to crush a usurper is no excuse for capitulation. Once again, the importance of not swearing an oath is stressed.

173 SAA 2, no. 6: § 23, 259–265; Lauinger 2012, iv 3–7.
174 SAA 2, no. 6, § 26: 304–306.
175 SAA 2, no. 6, § 26: 307–309; Watanabe 1987, 186.

In scenario twelve, the forbidden action reiterates the mandated one using different language, while also contributing to the description of the scenario: 'you shall not make it come to a fight between him and his father by stirring up mutual hatred between them'.[176] The same also applies to the thirteenth scenario, 'you shall not obey nor speak evil about his brothers in his presence, nor divide him from his brothers; you shall not let those who speak such things go free'. In this latter instance, the injunction against allowing those who say such things to go free serves to develop the desired consequences of such scenarios further. It is not enough simply not to participate: one must also act against those who do such things. By including this demand only in the second scenario that deals with the stirring up of strife between members of the royal family, this negative command can be viewed as a development of the logic of the stipulations, encouraging the covenant parties to think incrementally in such situations. A final prohibited reaction to this scenario is as follows: 'You will not look at Ashurbanipal, the great crown prince designate, or his brothers without reverence or submission'.[177] In this case, the subjects are required to regulate the manner in which they look at their monarch and his brothers. This language further highlights the importance of the use of the sense of sight in service of the king.

In the fourteenth stipulation, in which the scenarios come full circle, the importance of the maintenance of good relations between Ashurbanipal and his brothers is stressed:

> You shall not say any evil word about his brothers, sons of his own mother, before their brother nor try to make them accursed (saying): 'Bring your hand against them for an evil deed.' You shall not alienate them from Ashurbanipal, the great crown prince designate, nor shall you say any evil word about them in the presence of their brother. (As for) the positions which Esarhaddon, king of Assyria, their father, assigned them, you shall not speak in the presence of Ashurbanipal, the great crown prince designate, (trying to make him) remove them from their positions.[178]

In this situation, the covenant parties, and not the royal brothers, are presented as the drivers of a plot against Ashurbanipal. One could interpret this as an attempt on the part of the composers to build on the previous scenarios in casting strife between the legitimate members of the royal family as the fault not of the crown prince and his brothers, but rather of their subjects. In doing this, the covenant composition underlines the message that it is the responsibility of the cov-

176 SAA 2, no. 6, § 27: 326–327.

177 SAA 2, no. 6, § 30: 353–354; Lauinger 2012, v 9–11, see also Lauinger 2012, 112 and 115 f. § 30 is duplicated in the Kullania manuscript.

178 SAA 2, no. 6, § 31: 363–372; Lauinger 2012, v 24–36.

enant parties to maintain harmony among the Assyrian royal family and to ensure Ashurbanipal's smooth succession.

Finally, the subjects are also prohibited from attempting to invalidate the oath:

> You shall not smear your face, your hands, and your throat with ... against the gods of the assembly, nor tie it in your lap, nor do anything to undo the oath. You shall not try to revoke or undo (this) oath ... [...]; you shall neither think of nor perform a ritual to revoke or undo this oath.[179]

The reiteration of the statement highlights its importance while also casting light on different aspects of this projected contingency. In the first instance, particular methods of invalidating an oath are enumerated. In the second case, the stipulation touches on a forbidden attitude. Not only may the subject not attempt to undo the oath, he is not permitted to think about a ritual which will do this. The next denied act also concerns the attitude of the subject: *šumma... tamītu ša dabābti šapti tatammāni* 'you shall not swear this oath with your lips only'. In this case, the covenant parties are forbidden a lack of feeling: they must not swear without meaning it. By including such demands, the covenant composers raise the standard of loyalty, demanding that it be not only shown in action but also thought and felt. The last negative statement precludes non-participation in the oath by means of feigning *murṣu lā ellu* 'unclean illness'.[180] This is perhaps a logical conclusion, as the demand that the subject participate in the oath ensures his forced adherence to all other stipulations. The last two scenarios do not include forbidden actions, as they are themselves prohibited contingencies.

3.4 Directing vigilance in the covenant oath

That the covenant composition contains an oath section is somewhat unusual in comparison to prior *adê*-covenants. Of the extant covenant compositions that predate Esarhaddon's succession covenant, only SAA 2, no. 4, Esarhaddon's accession covenant, contains a first-person plural vow section.[181] As such, it seems clear that the oath was not a necessary written component of a covenant composition, al-

---

179 SAA 2, no. 6, § 32: 373–376; compare Lauinger 2012, v 37–40.

180 Lauinger 2012, 116.

181 SAA 2, no. 9, imposed by King Ashurbanipal on his Babylonian allies, also contains such an oath.

though it surely was mandatory to swear an oath in order to conclude a covenant.[182] The oath portion of the composition is as follows:

> May these gods be our witnesses: we will not make rebellion or insurrection against Esarhaddon, king of Assyria, against Ashurbanipal, the great crown prince designate, against his brothers, sons by the same mother as Ashurbanipal, the great crown prince designate, and the rest of the sons of Esarhaddon, king of Assyria, our lord, or make common cause with his enemy.
>
> Should we hear of instigation to armed rebellion, agitation or malicious whispers, evil, unseemly things, or treacherous, disloyal talk against Ashurbanipal, the great crown prince designate, and against his brothers by the same mother as Ashurbanipal, the great crown prince designate, we will not conceal it but will report it to Ashurbanipal, the great crown prince designate, our lord.
>
> As long as we, our sons (and) our grandsons are alive, Ashurbanipal, the great crown prince designate, shall be our king and our lord, and we will not set any other king or prince over us, our sons or our grandsons. May all the gods mentioned by name (in this covenant) hold us, our seed and our seed's seed accountable (for this vow).[183]

The scenario, mandated actions and forbidden actions found in the oath section of the covenant composition are in many ways consistent with those of the stipulations. Indeed, Parpola and Watanabe characterize the vow section as 'recapitulating the central points of the treaty'.[184] Despite the various points of similarity between the sections, however, this description is not entirely accurate, and there multiple ways in which the oath differs from the stipulations.

The initial portion of the oath is significant in that there is no explicitly stated scenario. Unlike the stipulations, which are clearly framed by the death of Esarhaddon, there is no mention of that event. Indeed, the first forbidden action is to 'make rebellion or insurrection against Esarhaddon', thus situating the relevance of the oath in the present, rather than, as in the case of the stipulations, primarily focusing on the period of succession which will take place in the future, upon the death of Esarhaddon.

The list of people against whom *sīhu* 'rebellion' and *bārtu* 'insurrection' are forbidden is short, encompassing Esarhaddon, Ashurbanipal, his maternal brothers and all other sons of Esarhaddon. This list is unique in the covenant com-

---

182 That the oath was a constituent part of an *adê*-bond is made clear by the appositional formulation found in Sargon II and Esarhaddon's royal inscriptions: *adê māmīt ilāni rabûti* 'covenant (sworn by) the oath of the great gods' (e.g. RINAP 2, no. 1: 264 and RINAP 4, no. 1: i 50). In Sennacherib's royal inscriptions, meanwhile, a similar but distinct formulation is found: *adê u māmīt ša Māt–Aššur* 'covenant and oath of Assyria' (RINAP 3/1, no. 16: iii 42).

183 SAA 2, no. 6, § 57, 494–512; Lauinger 2012, vi 77–93.

184 SAA 2, xlii.

position. In contrast to the stipulations, which guard against seditious activity performed against Esarhaddon to a degree, as in scenario nine, but which overwhelmingly direct focus towards Ashurbanipal, in the oath the king is mentioned first and his chosen crown prince is mentioned second. Furthermore, the remaining sons of Esarhaddon are then split into two categories. The members of the first category are defined through their relationship to Ashurbanipal, and to Ashurbanipal's mother. This category is mentioned several times in the stipulations as a protected group, and contrasted with the group framed simply as *ahhēšu* 'his (i. e. Ashurbanipal's) brothers', who are presented as a threat to Ashurbanipal. It is perhaps surprising, then, that a new formulation is used here, *rēhti mar'i ṣīt-libbi ša Aššur-ahu-iddina* 'the rest of the sons of Esarhaddon',[185] in a context that confers the same degree of protection to them as to Ashurbanipal and the sons of his mother. Here, this group is defined not by their relationship to Ashurbanipal but by their status as the offspring of Esarhaddon. In the oath, then, the hierarchy of the Assyrian Empire, which in the stipulations is subverted to a significant degree, particularly through mandated suspicion on the part of the empire's subjects directed towards the sons of the king, appears to be re-established to an extent, even if the position of these sons as last in the list suggests that they are less important that those who come before them.[186] This initial portion of the oath ends with the assertion that the oath-taker must not *pîni issi nakrīšu nišakkanūni* 'make common cause with his enemy' (lit. 'place our mouths with his enemy'). Here, then, the long lists of high-status groups found in the scenarios of the stipulations are replaced by a simple term found in the third scenario, *nakru* 'enemy'.

The next portion of the oath, meanwhile, does contain a scenario, and it is one that closely parallels those in the stipulations that relate to hearing statements against Ashurbanipal. Once again, however, the wording of the oath differs from the stipulations: *mušamhiṣūtu mušadbibūtu lihšu ša amat lemutti lā ṭābtu lā banītu dabāb surrāti u lā kēnāte... nišammûni* 'Should we hear of instigation to armed rebellion, agitation or malicious whispers, evil, unseemly things, or treacherous, disloyal talk'.[187] Here, the general subject is similar to that found in several of the stipulation scenarios, but several of the terms are different, such as *mušamhiṣūtu* 'sedition', *mušadbibūtu* 'instigation' and *lihšu* 'whisper(ing)'. These words are not found elsewhere in the covenant composition, and thus appear deliberately not

185 SAA 2, no. 6, § 57: 497; Lauinger 2012, vi 81. Watanabe 1987, 197. On the term *ṣīt libbi* 'offspring', see CAD Ṣ, 218 *s.v. ṣītu* 3b 2'.
186 On the identity of Ashurbanipal's birth mother, as well as his brothers, see PNA 1/1, 159–163 s.v. Aššur-bāni-apli. See also Novotny and Singletary 2009.
187 SAA 2, no. 6: § 57, 500–505; Lauinger 2012, vi 84–89.

to echo the scenario descriptions. The use of different terminology could perhaps be taken as an indication that the oath section had separate aims to the stipulations. In particular, it is noticeable here that, while the actions and words that must be reported form the focus of the scenario description, the identity of the possible perpetrator is not stated. One may wonder if this omission is indicative of the fact that, while the stipulations are willing to highlight particular high-ranking groups, it was not considered desirable for the empire's subjects themselves to repeat this when reciting the oath.

Another factor that distinguishes the oath from the stipulations is the identity of the people about whom the inappropriate words might be spoken. Whereas in the stipulations, this is always only Ashurbanipal, in the oath his brothers by the same mother are also included in this provision. Again, Ashurbanipal's brothers are featured and protected in the oath more prominently than in the stipulations. While the stipulations safeguard this group, they do not refer to reporting statements made against them. Thus, both the degree of emphasis and the specific circumstances of the protection afforded this group differ between the stipulations and the oath. The mandated action in such a scenario, however, is identical in the oath to that found in the stipulations: *nupazzarūni ana Aššur-bāni-apli mār–šarri rabiu ša bēt–rēdûti belīni lā niqabbûni* 'we will not conceal it but will report it to Ashurbanipal, the great crown prince designate, our lord'.[188] Thus, action and speech against Ashurbanipal and his brothers are to be reported to Ashurbanipal. Perhaps interestingly, the oath dispenses with the mandate to come to Ashurbanipal, stating only that it must be reported.

The final portion of the oath projects the focus forward in time by stating that it applies to the descendants of those entering it. No explicit action on the part of the oath-swearers is mandated in order to ensure this, but it should perhaps be understood as implicit. Here, in contrast to the other portions of the oath, Ashurbanipal's status is particularly highlighted, as is the period of time in which he will become king. So too, the oath-takers are forbidden to place *šarru šanûmma* 'another king' or *mār–šarri šanûmma* 'another prince (lit. another son of the king)' above Ashurbanipal. The use of the term *mār–šarri* 'son of the king' is particularly interesting here, and can be regarded as the closest that the oath comes to openly casting suspicion on Esarhaddon's own sons, something that occurs much more clearly in the stipulations. The oath ends as it begins, with the invocation of the deities of the covenant. It is thus framed by references to divine involvement and attention.

---

188 SAA 2, no. 6, § 57: 506f.; Lauinger 2012, vi 90–91.

3.5 Conclusions

The stipulations of the covenant provide those entering into it with a set of scenarios to which it is necessary to respond. These scenarios centre on the period of Esarhaddon's death and Ashurbanipal's accession, but they are not limited to it. Various frameworks are used when defining and describing the circumstances under which particular actions are required and forbidden. Particularly common is the use of lists of groups of people, in proximity to whom those entering into the covenant were required to be alert to possible disloyal conduct or speech. The stipulations subtly cycle through various constellations of these groups, frequently repeating key ones. In this way, the listener's attention is continually directed towards these people, as well as being frequently and surprisingly redirected towards new sets of potentially dangerous actors. Thus, the structure of the covenant stipulations somewhat mirrors the intended direction of the attention of the covenant parties after its imposition: they are to scale their vigilance based on their proximity to particular categories of people, repeatedly increasing their alertness when in direct or indirect contact with them. The high status of several of these groups and with it the potentially subversive character of the stipulations is not to be underestimated, and the fact that the oath section downplays this element may perhaps be evidence of its radical nature. The stipulations highlight the necessity of complete loyalty to Esarhaddon's chosen crown prince, at the expense of all others. In contrast to the description of likely perpetrators of dangerous acts in the stipulations, the phrasing of the acts themselves, such as speaking any *abutu lā ṭābtu lā banītu lā tarissu* 'improper, unsuitable or unseemly word' is kept fairly broad and fairly non-specific. In this instance, any word concerning Ashurbanipal that is not specifically good, could conceivably be considered to fall within this category. The stipulations, as is the case too with the oath, thus cast the net wide when describing the scenarios that require reaction.

The mandated and forbidden actions found in the covenant stipulations frequently work to reinforce and develop one another. It is possible to divide them into direct action, which does not necessarily require communication with the crown, reporting, and a particular attitude or emotion. Several of these mandated and forbidden actions are repeated for various scenarios, such as the demands that those entering the covenant do not conceal suspicious statements, but rather come and report it. I consider it probable that the repetition of such statements sought to allow the hearers to understand the rationale behind the selection of a particular response to a scenario. As such, it would have been possible for those bound by the covenant to know how to react even in scenarios subtly different to those laid out in the stipulations themselves. The stipulations, then, are not necessarily to be understood as hard and fast rules, to be followed to their letter, but rather as a set of worked examples, revealing the imagined functioning of the empire after the covenant's imposition.

It is interesting that, in the oath, the key forbidden and mandated actions are, on the one hand, making rebellion and insurrection or supporting an enemy of the crown, and, on the other hand, reporting various words and acts to Ashurbanipal. These two demands successfully capture two of the key messages of the stipulations, and therefore can be stated to function as reflecting them faithfully. They exclude, however, the important and prevalent mandating of an emotional bond directed towards Ashurbanipal, which reaches its zenith in the stipulations with the requirement that those bound by the covenant *râmu* 'love' Ashurbanipal like they love their own lives. The demand that those taking the oath report specifically to Ashurbanipal, however, does illustrate the message found throughout the stipulations that there exists a direct duty of loyalty to the crown prince on the part of the subject.

Chapter 4: Laying the ideological groundwork for enacting the covenant

As has been established in the previous chapters, the text of the covenant itself conveys a clear, if multifaceted, duty of vigilance. Close reading of the composition certainly suggests that Esarhaddon considered it more vital that some regions of the empire, specifically its provincial extent, be familiar with the stipulations of the covenant than others, namely the client states. Nevertheless, the covenant tablets themselves create a general impression of uniformity across the various administrative units of the empire, with each individual client state and province being sworn to identical stipulations using the same oath and curses.

Analysis of other royally-commissioned narratives written around this time tells a rather different story, however. This chapter explores the role of covenant in two of Esarhaddon's royal texts, both of which were drafted and disseminated shortly before the succession covenant's implementation. These compositions were aimed at elite groups in the royal and religious capitals of the empire, the cities of Nineveh and Ashur. I argue that both compositions were intended to reverse the narrative around *adê* 'covenant' which had previously existed in royally-commissioned narratives. In their royal inscriptions, previous Assyrian kings had presented covenants as a means of controlling Assyria's client rulers.[189] Esarhaddon's new narratives engage with this tradition and in order to convey a different message, firmly presenting covenant as something that is just as relevant, and probably more so, to the inhabitants of Assyria proper as it is to client kings.

The two royal texts discussed in this chapter were both disseminated, and likely also commissioned, in the year 673 BC, mere months before the imposition of Esarhaddon's succession covenant. It has been broadly accepted by modern scholars that both of these compositions, termed in the secondary literature *Esarhaddon's Apology* and *Esarhaddon's Letter to the God Aššur*, were composed at least in part with the intention of laying the ideological groundwork for the announcement of Esarhaddon's succession arrangements.[190] Despite this, the two narratives are substantially different to one another in their subject matter.[191]

The first composition deals with Esarhaddon's own succession, stressing that he had been the chosen successor of his father Sennacherib and accusing his brothers of patricide. Esarhaddon's claim that his father indeed intended him to

189 See recently Radner 2019, 310–312.
190 See Tadmor 1983, as well as the introductions to the texts themselves in RINAP 4, nos. 1 and 33.
191 For a general discussion of these royal narratives, see Chapter 1.3.2.

https://doi.org/10.1515/9783111323435-007

become king hinges on the statement made in the composition that Sennacherib had sworn the *nišī Māt–Aššur* 'people of the land of Aššur' to a succession covenant that elevated Esarhaddon to the position of crown prince.[192] After Sennacherib's murder, Esarhaddon's successful assumption of the kingship is presented in part as the result of the succession covenant, as he claims that the people of Assyria followed its demands and thus refused to support his brothers. This narrative serves as the prologue to a longer composition, a building inscription inscribed on clay prisms, which refers to covenants and oaths imposed on client kings in a manner that can be considered typical of the royal inscriptions of earlier Assyrian monarchs.

Esarhaddon's Letter to the God Aššur, meanwhile, is on its face a narrative that concerns the annexation of the northern client state of Šubria. The client king of this location is explicitly stated to have failed to keep the oath that he swore to Esarhaddon.[193] While this composition seemingly focuses on a client state, however, the manner in which the king of Šubria has disobeyed Esarhaddon primarily concerns the domestic situation in Assyria: he has refused to extradite Assyrian fugitives in Šubria back to the central region, despite Esarhaddon's repeated demands that he do so. The narrative details Esarhaddon's campaign in Šubria, where he takes the fugitives back by force and annexes the state, dividing it into two Assyrian provinces.[194] The *Letter to the God Aššur* is probably heavily indebted to a composition of the same genre written under Sargon II, and thus also attempts implicitly to draw a connection between the two kings.

This chapter is structured around two key innovations made in these texts. Firstly, I argue that these *adê* 'covenant' narratives, particularly that of *Esarhaddon's Apology*, are designed to recast the concept of legitimacy in the Assyrian context, elevating the status of covenant and ruling out usurpation, which previously had been a common and broadly accepted path to legitimate rule in Assyria. Secondly, I note that both royal compositions are of particular interest for their spatial inversions of traditional royal narratives around covenant. Despite the differences between the two texts, both seek to stress the importance of the covenant and loyalty of the residents of Assyria to the crown. The contemporary audience who would have come into contact with these compositions would themselves have belonged to this group, and therefore it seems reasonable to suppose that the narratives were intended in part to provoke among them some degree of self-reflection on this subject.

192 RINAP 4, no. 1: i 50–52.
193 RINAP 4, no. 33: i 23.
194 RlA 11, 63–64 s.v. Provinz. C. Assyrien.

4.1 Promoting covenant and delegitimizing usurpation in *Esarhaddon's Apology*

The composition known in modern scholarship as *Esarhaddon's Apology* constitutes the prologue to Esarhaddon's longest extant inscription, dubbed Nineveh A. While some exemplars have been found at other locations (Ashur and Susa), it seems clear that the preserved manuscripts were intended for use as foundation deposits at the *ēkal–māšarti* 'Review Palace' at the secondary citadel of Nebi Yunus in Nineveh.[195] The clay prisms upon which the *Apology* was written would have been hidden away from view, buried in the foundations of the building, and would not have been visible even to visitors to the palace after they were deposited.[196] The high-level scribes and scholars who were involved in the narrative's composition would, of course, have encountered the narrative, as would those who copied it in order to generate more manuscripts. Beyond this, one may imagine that palace personnel, as well as perhaps some high-ranking inhabitants of the city, or even visitors to it, would have been aware of the inscription. As such, it is likely that the composition would have been known primarily to those operating in the educated, scholarly milieu of Nineveh, and possibly not very widely beyond it, at least at the time of its initial circulation.[197] Colophons, short subscripts, are found on several exemplars, with the earliest dating to month Du'ūzu (IV) of 673, and the latest dating to month Nisannu (I) of 672 BC.[198] As such, those who did come into contact with the inscription would likely have done so in the year before Esarhaddon's succession arrangements were finalized.[199] As in the case

195 The text is published as RINAP 4, no. 1. Note that the *Apology* was also used as a prologue to two later prisms bearing a very similar inscription to Nineveh A, known as Nineveh F/S and Nineveh D/S (published as RINAP 4, nos. 5 and 6, see also Tadmor 2004, 273–276). These texts also describe building works on the Review Palace in Nineveh. On this palace, see most recently Maul and Miglus 2020. See also Kertai 2015, 150–153.

196 The use of royal inscriptions as foundation deposits in Assyria is discussed further in Chapter 1.3.2.

197 Barbara Nevling Porter has shown that Esarhaddon's foundation deposits were altered based on where they were to be deposited (Porter 1993, 94–105). While this may serve to some degree as evidence that some citizens of the respective location were expected to come into contact with them, it does also seem to imply that these inscriptions were not intended for circulation beyond their immediate geographical context.

198 RINAP 4, no. 1: vi 75. Exemplars 1, 16 and 26 bear date lines stating that they were drawn up in month Addāru (XII) of 673 BC, while ex. 29 was written in Nisannu (I) of 672 BC. Note that the text version of RINAP 4, no. 1 confuses ex. 6 for ex. 16, as noted in Novotny 2018, 206, fn. 13.

199 The version of Esarhaddon's annals that preceded Nineveh A – Nineveh C – can possibly be dated to 674 BC (Novotny 2018, esp. 208), providing a *terminus post quem* for the first recension of Nineveh A.

of most Assyrian royal inscriptions, the explicitly intended target audience of the inscription is future Assyrian royalty, a claim that is to be taken seriously.[200] Esarhaddon therefore sought to convey his message not only to those people who would have come into contact with the *Apology* at the time of its initial composition, but also did so in anticipation of it being read by future kings, as well perhaps as in anticipation of divine scrutiny.[201]

In his seminal work on the subject, Hayim Tadmor argued for the interpretation of *Esarhaddon's Apology* as 'an ideological *praeparatio*' for his own succession arrangements and their associated covenant.[202] In Tadmor's view, Esarhaddon's decision to nominate one of his younger sons, Ashurbanipal, as crown prince, along with his novel scheme to simultaneously designate Ashurbanipal's elder brother, Šamaš-šumu-ukin, as crown prince of Babylon, would have been controversial.[203] The *Apology* was therefore commissioned in order lay the groundwork for the succession covenant, partly by drawing parallels between the circumstances of Esarhaddon's own succession and those planned for Ashurbanipal. In Tadmor's telling, then, Esarhaddon's own succession is presumably taken to be legitimate and is presented as 'paradigmatic' for the future.[204]

Tadmor's compelling thesis has been widely embraced in subsequent years, although it has recently been critiqued by Andrew Knapp.[205] Knapp rejects the notion that Esarhaddon's decision to promote Ashurbanipal and Šamaš-šumu-ukin would have required extensive ideological preparation, arguing instead that the *Apology* was intended to bolster not the position of Ashurbanipal, but rather that of Esarhaddon himself.[206] It is surely correct that *Esarhaddon's Apology* was commissioned in part to boost Esarhaddon's own position, however, Knapp's argument relies in large part on a false dichotomy: in attempting to secure his chosen successor's status, Esarhaddon sought in no small part to safeguard his own position. Ashurbanipal's elevation eliminated, at least in theory, any motivations on the part of other royal men or their supporters to attempt to usurp the throne and, with it, any possible need to assassinate Esarhaddon. Knapp has elsewhere

---

200 Radner 2005, esp. 203.

201 The *Letter to the God Aššur*, in particular, is clearly written with a divine audience in mind (on the genre, see Pongratz-Leisten 1999, 227 f.).

202 Tadmor 1983, 45.

203 Tadmor 1983, 43–45.

204 Tadmor 1983, 45: 'The procedure undertaken for his own succession to the throne, as described in the *Apology* – became paradigmatic for his own acts. In both cases the first-born was by-passed: it was the will of the king and the gods that the younger prince – apparently better suited for the august office – should be preferred.'

205 These arguments are laid out in Knapp 2016 and are also discussed in Knapp 2015, 326–35.

206 Knapp 2016, 195.

also added his voice to those who interpret the *Apology* as evidence that Esarhaddon needed to stress his own legitimacy because it was actually he, rather than his brothers, who orchestrated his father's murder.[207]

Although Hayim Tadmor and those who have followed his argument, on the one hand, and others such as Andrew Knapp and those who have questioned Esarhaddon's role in Sennacherib's assassination, on the other, disagree about the reasons for which the *Apology* was commissioned, their theories are fundamentally compatible in that they consider the text in essence to be an exercise in boosting the legitimacy of Esarhaddon and his decisions. While this general premise is not necessarily wrong, I would like to suggest that the *Apology* can also be framed differently: in my view, it is an attempt to delegitimize an established means by which many Assyrian kings, including Esarhaddon himself, had previously come to the throne.

The *a priori* notion that it would have been necessary to recast Esarhaddon's own succession in order to stress his legitimacy at a difficult point in his reign has been put forward most recently by Andrew Knapp.[208] This suggestion taps into the assumption on the part of other scholars who have argued that the *Apology* would have been a necessary cover for Esarhaddon if he was indeed involved in his father's murder. In contrast, Hayim Tadmor focused instead on Ashurbanipal's elevation, but he made the related assumption that controversy would have been caused by the nomination of a younger son as crown prince and that the *Apology* sought to quell this in some way.

Is there sufficient evidence to assume that Esarhaddon and his advisors would have considered the composition of such a text a necessary response to any of these situations? The only Assyrian royal inscription that could perhaps be considered an 'apology' and predates *Esarhaddon's Apology* was composed during the reign of Šamši-Adad V (r. 824–811 BC).[209] He came to the throne after a bitter civil war with his brother, Aššur-da"in-aplu, about which he gives the following account:

> When Aššur-da"in-aplu, at the time of Shalmaneser (III), his father, acted treacherously by inciting insurrection, uprising, and criminal acts, caused the land to rebel and prepared for battle; (at that time) the people of Assyria, above and below, he won over to his side, and made them take binding oaths. He caused the cities to revolt and made ready to wage battle and war. The cities Nineveh, Adia, Šibaniba, Imgur-Enlil, Iššabri, Bit-Šaširia, Šimu, Šibhiniš, Tam-

207 Knapp 2020. Another important recent publication along these lines is Dalley and Siddall 2021.
208 Knapp 2015, 327: 'It is difficult to envision what sort of rhetorical situation might have elicited the *Apology* other than a need to bolster Esarhaddon's legitimacy.'
209 See RIMA 3, A.0.103.1: esp. i 39–53. For a brief recent discussion see Knapp 2015, 61; see also Tadmor 1983, 53 f.

nuna, Kipšuna, Kurbail, Tidu, Nabulu, Kahat, Ashur, Urakka, Raqmat, Huzirina, Dur-balaṭi, Dariga, Zaban, Lubdu, Arrapha, (and) Arbela, together with the cities Amedu, Til-abni, (and) Hindanu – altogether twenty-seven towns with their fortresses which had rebelled against Shalmaneser (III), king of the four quarters, my father, sided with Aššur-da"in-aplu. By the command of the great gods, my lords, I subdued (them).[210]

This narrative serves to introduce Šamši-Adad V's annals, which list his accomplishments by regnal year. The apologetic portion of the inscription is a mere fifteen lines long, six of which are dedicated to enumerating the twenty-seven settlements that sided with his brother in the struggle, many of which were important political, religious or economic centres. There are some parallels between the conditions under which Šamši-Adad V and Esarhaddon came to the throne, and the manner in which they chose to represent themselves in their royal inscriptions. Nevertheless, it seems clear that the circumstances of Šamši-Adad V's accession were far more turbulent than those of Esarhaddon. The latter was able to quell the opposition to him in less than two months, with his rival brothers fleeing the country. In contrast, huge swathes of Assyria supported Šamši-Adad V's brother upon their father's death, in a war that lasted some six years and constituted a period of prolonged crisis for the Assyrian crown.[211] In addressing the situation, Šamši-Adad V does not acknowledge his status as a younger son of the king, nor does he explicitly state that he had been his father's crown prince. *Esarhaddon's Apology*, meanwhile, spans ninety-one lines and thus offers a vastly more detailed account, including on the subject of the succession arrangements made by his father, something about which Šamši-Adad V is silent. Interestingly, Esarhaddon does not name any locations that supported his brothers, choosing instead to stress the loyalty of the Assyrian people to Sennacherib's succession covenant:

The people of Assyria, who swore by oil and water to the treaty, an oath bound by the great gods, to protect my (right to exercise) kingship, did not come to their aid.[212]

The people of Assyria, who had sworn by the treaty, an oath bound by the great gods, concerning me, came before me and kissed my feet.[213]

It is remarkable that the long tradition of Assyrian royal inscriptions, of which almost two thousand different compositions survive, yields only one text that pre-

210 RIMA 3, A.0.103.1: i 39–53.
211 Radner 2016, 47 f.; Fuchs 2008, 65–68.
212 RINAP 4, 1: i 50–52.
213 RINAP 4, 1: i 80–81.

cedes that of Esarhaddon which could be described as an 'apology'.[214] The inscription in question provides a brief note on the subject of a succession war between Šamši-Adad V and his brothers that lasted six years. The contents of this inscription therefore overlap with those of *Esarhaddon's Apology* only in a cursory fashion. The commissioning of apologies, then, does not seem to have been typical of Assyrian royalty under any circumstances. As such, it is difficult to justify the position held by some modern scholars that the dissemination of an apology would have been considered politically necessary in the Assyrian context, even in the case of patricide.

Whereas apologies were uncommon to the point of being almost unheard of in the Assyrian tradition, usurpation was itself a frequent occurrence. While Esarhaddon's own father, Sennacherib, is well attested as the crown prince of Esarhaddon's grandfather, Sargon II, such a scenario was the exception rather than the rule in the Assyria of this period. Sargon II himself spearheaded the overthrow of the former monarch, his own brother and their father's chosen successor, Shalmaneser V, while that same father, Tiglath-pileser III, also took the throne as the result of a coup against his predecessor.[215] Both of these men were certainly themselves royal princes, but neither had been selected for the throne by his father. While Tiglath-pileser seems to have assumed power fairly easily, execution of adversaries notwithstanding, Sargon did face some considerable resistance. In the core region of Assyria, the supporters of the different factions of the royal family vied for dominance, while Babylonia to the south and several western provinces took the upheaval as an opportunity to claim their independence.[216] Despite this, Sargon did not seek to assert that he had been his father's chosen successor. Instead, the inscription known today as the 'Aššur Charter', probably written at the very beginning of his reign, does not describe the circumstances of his accession, but rather uses the claim that he is restoring privileges to the inhabitants of Ashur to state that Shalmaneser V had mistreated them and that, as a result, the god Aššur *palâšu iškip* 'overthrew his (i. e. Shalmaneser's) reign'.[217] In this way, Sargon is very open about the fact that he had deposed his predecessor. In Sargon's view, it seems, Shalmaneser V's supposed loss of both popular and divine

214 Tadmor 1983, 53 f. There are 1891 known Assyrian inscriptions from the late third millennium until the reign of Sîn-šarru-iškun (r. 627/626 – 612 BC), see 'ORACC: The Royal Inscriptions of Assyria Online'.

215 On Tiglath-pileser III, see Garelli 1991, Zawadzki 1994 and RINAP 1, 1 – 18. On Sargon II, see Thomas 1993, Vera Chamaza 1992 and RINAP 2, 1 – 41. On succession and usurpation in Assyria, see Mayer 1998, as well as Radner 2016.

216 Radner 2016, 49 – 52.

217 RINAP 2, no. 89.

support, the crowning proof of which was presumably Sargon's own ability to seize the throne from him, was evidence enough that his brother was no longer a legitimate ruler. This almost meritocratic approach, which hinges primarily on the charisma of the monarch, differs strongly from that of Esarhaddon in his *Apology*, who rests his legitimacy on his father's choice of him as crown prince. While he does present himself as enjoying popular and divine support, he depicts these things as flowing in no small part from his father's nomination of him, something that he stresses was secured by the imposition of a succession covenant on his behalf.

When considering Assyrian succession practices beyond their characterization in royal inscriptions, it seems probable that most Assyrian kings did generally have a son, or perhaps another close male relative, whom they considered to be their heir. Despite this, relatively little evidence survives of formal attempts to ensure that they would successfully ascend to the throne after their predecessor's death.[218] In the case of Sennacherib himself, one of the best-attested Assyrian crown princes, Sargon gave his son considerable power while he was still alive, probably in order to prepare him for the challenge of kingship.[219] Despite their evidently close relationship, however, it is not clear that Sargon took any legal precautions to ensure that his chosen successor would be able to take the throne upon his death. This is particularly interesting given that Sargon's own path to the throne, usurpation, means that he can hardly have been unaware of the potential challenges his crown prince would face in asserting his claim. While some private legal arrangements may have been made, which were perhaps not widely publicized and do not survive, it is also possible that Sargon simply believed that Sennacherib would be capable of overcoming any resistance to his accession if it came. Indeed, as the Assyrian king was considered to be divinely chosen, the member of the royal family who had sufficient divine support to seize the throne was likely considered, by definition, the legitimate ruler. Sargon provided his favoured son with plenty of practical experience in overseeing royal projects, and perhaps considered this more or less sufficient to ensure Sennacherib's success.

---

218 I here take the view put forward by Karen Radner that 'the rules of succession were fluid' in Assyria (Radner 2010a, 26). It is perhaps of interest to note that, in Neo-Assyrian law generally, the eldest son does not seem to have been given preference above his brothers in questions of inheritance. Rather, an estate was generally split evenly between the sons (Radner 2003a, 900).

219 The surviving letters from Sennacherib as crown prince are published as SAA 1, nos. 29–40 and SAA 5, no. 281 and SAA 5, no. 281 and SAA 19, nos. 157–158. Note too that Tiglath-pileser III's crown prince, Shalmaneser V (who used the name Ululayu while crown prince), wrote similar letters to his father: SAA 19, nos. 8–11, see also Radner 2003b. Note too that Sargon may have depicted himself with Sennacherib in his reliefs (Botta 1849, 1: pl. 12), which would have served to underline his status.

The prevalence of usurpation and apparent relative lack of clear rules around succession also serve to cast some doubt on Tadmor's assertion that the relative ages of Esarhaddon's crown princes would have been so problematic as to merit the commissioning of such as text. While Esarhaddon's exact succession strategy had not been attempted before, the succession of younger sons was not unprecedented, and Babylonia had also been previously ruled by members of the Assyrian royal family.[220]

The picture of succession in the Assyrian context prior to 673 BC, therefore, is one in which the fact of belonging to the royal family was important, but, beyond this necessary qualification, it seems that it was often difficult to predict which male member of the family would end up on the throne. While Assyrian kings do seem to have chosen a favoured successor from among their sons, there does not seem in general to have been a strong ideological objection to a different son taking the throne if he was able to do so. To describe this situation using the conceptual framework of Max Weber, the Assyrian monarchy relied on familial charisma: a combination of traditional and charismatic qualities.[221] The king was part of a ruling family, itself imbued with charisma, but he also needed the individual qualities necessary to set himself up as king in favour of his male relatives, including popular support. Divine backing was a necessary criterion of Assyrian legitimacy, something that was derived more from charismatic sources than traditional ones: as in Sargon's description of his brother Shalmaneser V's demise, the king could also lose divine support if he failed to conduct himself properly.[222]

Beyond his royal inscriptions, Esarhaddon, who had successfully crushed opposition against him and seized the throne, fitted the profile of the charismatic Assyrian monarch. The narrative of *Esarhaddon's Apology*, however, rejects this conventional model of Assyrian legitimacy, presenting his succession differently. The *Apology* presents the popular and divine support enjoyed by Esarhaddon as linked to Sennacherib's succession covenant:

> I am my older brothers' youngest brother (and) by the command of the gods Aššur, Sîn, Šamaš, Bel, and Nabû, Ištar of Nineveh, (and) Ištar of Arbela, (my) father, who engendered me, elevated me firmly in the assembly of my brothers, saying: 'This is the son who will suc-

---

220 For an overview, see Frame 2008.

221 See Weber 1923, 51–53. Weber also uses the term in Weber 1980, 783; see also his broader discussion of the 'Veralltäglichung des Charisma', Weber 1980, 142–148. The translation of 'Gentilcharisma' as 'familial charisma' follows Reinhard Bendix (1960, 146, fn. 9).

222 Weber notes that this is a danger associated with charismatic rule: 'Bleibt die Bewährung dauernd aus, zeigt sich der charismatische Begnadete von seinem Gott oder seiner magischen oder Heldenkraft verlassen... so hat seine charismatische Autorität die Chance, zu schwinden' (Weber 1980, 140).

> ceed me.' He questioned the gods Šamaš and Adad by divination, and they answered him with a firm 'yes,' saying: 'He is your replacement.' He heeded their important word(s) and gathered together the people of Assyria, young (and) old, (and) my brothers, the seed of the house of my father. Before the gods Aššur, Sîn, Šamaš, Nabû, (and) Marduk, the gods of Assyria, the gods who live in heaven and netherworld, he made them swear their solemn oath(s) concerning the safe-guarding of my succession. In a favorable month, on a propitious day, in accordance with their sublime command, I joyfully entered the House of Succession, an awe-inspiring place within which the appointing to kingship (takes place).[223]

Esarhaddon presents Sennacherib's decision to elevate him as the result of divine will, nonetheless, his emphasis on the use of a covenant to ensure his succession serves to downplay the charismatic aspects of his legitimacy. Rather, by stressing that he was chosen by his father, he presents himself as having inherited a high degree of traditional legitimacy. Even his claim of widespread popular support fits this description:

> The people of Assyria, who swore by oil and water to the treaty, an oath bound by the great gods, to protect my (right to exercise) kingship, did not come to their aid.[224]

> The people of Assyria, who had sworn by the treaty, an oath bound by the great gods, concerning me, came before me and kissed my feet.[225]

This popular support, which could be construed as a sign of charismatic authority, and is used in this way in Sargon's Aššur Charter,[226] for instance, is here depicted as emanating from Sennacherib's succession covenant. Thus, the *Apology* frames Esarhaddon's rise as more reliant on tradition, his father's choice of him as successor, than his own charisma. Furthermore, the stress placed on covenant and oath in the composition tie the traditional, inherited nature of Esarhaddon's legitimacy with legal, bureaucratic authority. Weber himself presents legal domination as existing to some degree in opposition to traditional and charismatic domination, on the grounds that the latter two are based on notions of sanctity and sacredness and the former is not.[227] In the Assyrian case, however, all three modes of legitimate domination are anchored in divine will, which renders them compatible.

Whether or not one accepts the notion that Esarhaddon may have orchestrated his father's murder – the argument that he did not is more parsimonious and

223 RINAP 4, no. 1: i 8–22.
224 RINAP 4, no. 1: i 50–52.
225 RINAP 4, no. 1: i 80–81.
226 RINAP 2, no. 89.
227 Breuer 2019, 239f.

thus has more merit – it is clear that he did not come to the throne under easy circumstances:

> They started evil rumors, calumnies, (and) slander about me against the will of the gods, and they were constantly telling insincere lies, hostile things, behind my back. They alienated the well-meaning heart of my father from me, against the will of the gods, (but) deep down he was compassionate and his eyes were permanently fixed on my exercising kingship.[228]

Despite Esarhaddon's claim that his father still wished him to succeed *šaplānu* 'deep down', it seems clear from this passage that Esarhaddon fell out of favour with Sennacherib towards the end of the latter's reign, and that this forced Esarhaddon into exile. While he may technically still have been crown prince upon his father's death, then, it seems unlikely that he was still his father's *de facto* first choice as successor. As such, it is not actually necessary to attempt to show that Esarhaddon may have been implicated in the murder, or even that he was not crown prince, to characterize him to some degree as a usurper. Upon his father's death, Esarhaddon, like so many Assyrian monarchs before him, managed to use his charismatic authority to take the throne, thus besting what presumably were multiple factions supporting various of his brothers.[229] Beyond 'exterminating' the families of his brothers' supporters,[230] Esarhaddon does not seem to have needed to counter much resistance to his rule early in his reign, suggesting that his victory was decisive and his taking up the throne uncontroversial.

Why, then, did Esarhaddon decide much later, in 673 BC, to commission a royal inscription in which he disavows the charismatic authority to which he owed his success as the legitimate grounds for royal power? I consider it probable that Esarhaddon decided to re-litigate his own rise to power, providing a view that downplayed his personal qualities and highlighted the hereditary and legal aspects of his succession, but not because, as some scholars have assumed, these were necessary preconditions of legitimate rule in Assyria at the time. Elevation to the status of crown prince by the means of a covenant imposed by one's predecessor had not previously been the defining quality of royal legitimacy: Esarhaddon was seeking to make it such. This framing thus attempted to make the crown prince that Esarhaddon chose and about whom he concluded a covenant his only conceivably legitimate successor to the Assyrian throne, thus creating a situation that had not previously been the case in Assyria.

---

228 RINAP 4, no. 1: i 26–31.

229 Esarhaddon's claim that his brothers 'butted each other like kids for (the right to) exercise kingship' seems to suggest that there were rival factions (RINAP 4, no. 1: i 44).

230 'I exterminated their offspring' (RINAP 4, no. 1: ii 11).

4.2 Reframing the role of provincial subjects in royal narratives

In spatial terms, *Esarhaddon's Apology* tells the story of his forced flight from Nineveh and his victorious return to the city to be crowned as king. Although some parts of the narrative involve other locations, namely the place that Esarhaddon spent his exile and the land to which his brothers escaped upon his return to Nineveh, they remain unnamed and are described respectively as *ašar niṣirti* 'a secret place' and *māt lā idû* 'an unknown land'.[231] While the latter term may seek to reflect that Esarhaddon and his allies did not know where his brothers had gone,[232] this certainly was not the case of Esarhaddon's exile. The phrasing can thus be considered as calculated to render Nineveh, presence in it and absence from it, the main focus of attention in the composition.

The clear local focus of the *Apology* is further illustrated by the description of Esarhaddon's return from exile:

> With difficulty and haste, I followed the road to Nineveh and before my (arrival) in the territory of the land Hanigalbat all of their crack troops blocked my advance; they were sharpening their weapons.[233]

Esarhaddon describes his return from the unnamed place of his exile in terms of the royal capital of Nineveh, as well as using the antiquated term 'Hanigalbat' to refer to the provincial regions directly to the west of the core region, which had been under Assyrian control since the twelfth century BC.[234] Thus, the narrative is told in terms of locations associated with a traditional definition of Assyria. So too, those who had entered Sennacherib's succession covenant are consistently described in the narrative as the *nišī Māt–Aššur* 'people of Assyria', something which is perhaps to be understood in this context. In this way, the domestic in-

231 RINAP 4, no. 1: i 39 and i 84.

232 Note that Knapp and others have argued that the narrative found in the Hebrew Bible, that Esarhaddon's brothers escaped to Urartu, is reflective of a version of events put about by Esarhaddon and his supporters (Knapp 2020, 166, fn. 3). If this is indeed the case, then we must look elsewhere for an explanation of the decision to use this phrase than a genuine lack of knowledge about where the brothers had gone by the year 673 BC. One could imagine that the composers wish to stress that Esarhaddon did not know where his brothers were at the time that they initially ran away, thus excusing him from not having tracked them down immediately, or perhaps the phrase is used primarily for the stylistic reasons discussed here.

233 RINAP 4, no. 1: i 69–71.

234 On Hanigalbat in the Middle Assyrian period, see Reculeau 2022.

habitants of Assyria are presented as the only subjects of the empire whose participation in the succession arrangements was relevant. Whether this group encompassed the entire provincial extent of the empire, or merely – as is perhaps implied by the reference to Hanigalbat – a smaller grouping of provinces that were considered traditionally 'Assyrian', is not entirely clear.[235]

Esarhaddon's Apology inverts the typical covenant narrative of the royal inscriptions, in which the client ruler is treaty partner and has either broken his oath, necessitating retribution, or is in need of Assyrian support. In the *Apology*, it is the people of Assyria who are sworn to the covenant, they do not break their oaths, and it is Esarhaddon who receives their support. The contrast between the former trope and the latter innovation is underlined by the rest of the account in Nineveh A. The remainder of the text draws in part on previous royal inscriptions of Esarhaddon, but also includes sections composed as part of this new recension.[236] Immediately following the *Apology*, for instance, the scribes inserted a new section detailing Esarhaddon's punishment of Nabû-zer-kitti-lišir the 'governor of the Sealand', who initiated a rebellion and thus broke his covenant.[237] Hayim Tadmor views the inclusion of this episode as a 'new and pertinent message' that 'any transgressor of the loyalty oaths [of 672 BC] will be similarly punished by the great gods'.[238]

Although the description of Nabû-zer-kitti-lišir's death at the hands of the gods may have been intended as a deterrent to those who came into contact with the inscription, it is also relevant that it differs starkly from the narrative that proceeds it. From a spatial perspective, this juxtaposition draws a clear line between the loyal people of Assyria, some of whom presumably were expected to come into contact with this text, and the treacherous foreign rebel located far-away. While covenants unite these groups as important instruments of imperial control for both of them, their different status is clear. In this way, in addition to perhaps serving as a veiled threat, the Nabû-zer-kitti-lišir episode actually underlines the message of the *Apology* that the important, privileged actors in succession covenants are not foreigners, but rather Assyria's domestic inhabitants.

Esarhaddon's Letter to the God Aššur, meanwhile, presents a very different narrative to the *Apology*, despite having probably been composed at roughly the

---

235 The narrative itself makes clear that Esarhaddon was sheltering further west, but this is downplayed in the text. For the theory that Esarhaddon spent this period in Harran, see Leichty 2007.

236 On the recension history of the text, see Eph'al and Tadmor 2006, as well as Jamie Novotny's discussion of the recension history of Esarhaddon's annals (Novotny 2018).

237 RINAP 4, no. 1: ii 40–64.

238 Tadmor 1983, 47.

same time. The genre of text, a royal narrative framed as a letter to a deity, is attested throughout the Neo-Assyrian period,[239] but the most complete *Letter to the God Aššur* that has survived until the modern day is that of Sargon II.[240] Esarhaddon's own letter very clearly exists in conversation with that of Sargon: both relay the details of a campaign that took place in the ruler's eighth regnal year, both of which took place in the mountain regions adjoining the Assyrian heartland to its north and northeast. While some have suggested that Sennacherib also wrote a *Letter to the God Aššur* describing a campaign in Judah, several scholars have argued that this fragmentary text is to be attributed to Sargon, not Sennacherib.[241] In either case, the various connections between Sargon's composition and that of Esarhaddon provide clear evidence that the latter king wished to make allusions to the exploits of his grandfather.

The decision to make an implicit link between Esarhaddon and Sargon in the form of a *Letter to the God Aššur* is an interesting one. Sargon's sudden death while on campaign, and in particular the failure to recover his body, were considered to be evidence of a grave negative judgement of that king made by the gods.[242] As a result of this, Sennacherib, Sargon's crown prince and successor, played down his close relationship with his father in various ways, such as not citing him in his inscriptions and changing the royal capital from the newly built Dur-Šarruken 'Fortress of Sargon' to Nineveh.[243] As discussed at greater length in Chapter 6, the reckoning with Sargon's mistakes continued well into the latter portion of Esarhaddon's reign. Sargon's difficult legacy renders it all the more noteworthy that Esarhaddon commissioned a narrative that seems deliberately to allude to, and in ways subvert, a text composed in the name of that monarch. It seems possible that Esarhaddon and his advisors may have considered this text, and the campaign that it describes, a high point in Sargon's reign, dating to the period when he still

239 Pongratz-Leisten 1999, 227 f.

240 RINAP 2, no. 65, with bibliography. Sargon's letter details that monarch's eighth campaign, recounting his military defeat of King Rusâ I of Urartu. Throughout the text, Sargon makes several references to oaths and covenants, accusing his enemies of not keeping them, and professing that he himself is the guard of the *samnu* 'oath' sworn by Enlil and Marduk. See discussion in Pongratz-Leisten 1999, 134–138. Note that Beate Pongratz-Leisten terms both this text and that of Esarhaddon a 'Königsbericht an Aššur'. On the political ideology of the text, see Fales 1991.

241 The text in question is published as RINAP 3/2, no. 1015, with commentary and bibliography. For the view that this *Letter to the God Aššur* dating to the reign of Sennacherib is such a text, see Na'aman 1974. For the argument that it should be attributed to Sargon, see Fuchs 1994, 314 and Frahm 1997, 230–232.

242 On the circumstances of Sargon's death, see recently RINAP 2, 37. On the aftermath of Sargon's death, see Frahm 1999.

243 Frahm 2017b, 183.

enjoyed divine support.[244] Beyond this, the decision to commission such an composition shortly before the imposition of the succession covenant is consistent with Esarhaddon's apparent desire to reframe his relationship to his predecessors, something which is also clearly evident in the *Apology*. In the *Letter to the God Aššur*, Esarhaddon presents himself to a certain degree as a new Sargon. One may wonder if Esarhaddon and his advisors made this choice in part due to Sargon's own success in selecting a crown prince who came to the throne seemingly without opposition: an unusual feat by the standards of the time. So too, it may be relevant that Sargon's eighth campaign, recounted in his *Letter to the God Aššur*, represents an instance of brotherly co-operation within the royal family. It features Sargon's *ahu talīmu* 'equal/favourite brother', Sîn-ahu-uṣur, who took part in the campaign in his role as *sukkalmahhu* 'mighty vizier'.[245] The former phrase is used in Esarhaddon's succession covenant to refer to Ashurbanipal and Šamaš-šumu-ukin, perhaps indicating that the monarch took some inspiration for the pre-existing notion of the *ahu talīmu* when devising, or at least communicating, his succession plan.[246]

Esarhaddon's Letter to the God Aššur differs from that of Sargon in various ways. Sargon's narrative is full of action: Sargon conducted his campaign primarily against Assyria's historic rival, the kingdom of Urartu, the political centre of which was located in the vicinity of Lake Van in what is now eastern Turkey. Sargon's royal missive to Aššur describes the passage of his army through the mountainous terrain between the Assyrian core region and Urartu, emphasizing the difficulty of the journey. The narrative enumerates the local client rulers of the Assyrian Empire who presented Sargon with tribute as part of the campaign. Sargon then inflicts, according to the text, a decisive defeat upon the king of Urartu, Rusâ I, and his allies. The members of this group are variously – albeit briefly – described as breaking Assyrian oaths and covenants.[247] Sargon's campaign also includes the invasion of the buffer state of Muṣaṣir, which was a holy centre in the region on

244 As Sargon II seems to have been charged with being insufficiently pious towards the gods of Assyria (see Chapter 6.2.1.), a composition emphasizing his relationship to the god Aššur may have been seen as worth highlighting.

245 Note that Sîn-ahu-uṣur refers to himself as the *talīmu* brother of Sargon, while Sargon does not appear to use the term with reference to Sîn-ahu-uṣur (RINAP 2, no. 2002, see also Bartelmus 2007, 293). The precise implications of the term *talīmu* in this context is the subject of discussion, see for instance Bartelmus 2007. The term *sukkalmahhu* is known only in reference to Sîn-ahu-uṣur. As he is also attested as *sukkallu rabiu* 'great vizier', it seems probable that the titles were synonymous (PNA 4/1, 135 s.v. *sukkalmaḫḫu*). On Sîn-ahu-uṣur, see PNA 3/1, 1128 s.v. Sīn-aḫu-uṣur.

246 SAA 2, no. 6, § 7: 86.

247 On Urartu, see Fuchs 2012.

account of its temple to the state god of Urartu, Haldi, which Sargon sacked.[248] In contrast to Sargon's composition, which revolved around not only Assyria's greatest historical rival but also a complex network of client states located to the east and northeast of Assyria, the extant portions of Esarhaddon's narrative ostensibly focus on just one small client state, Šubria, located in the headwaters of the Tigris.[249]

Like Muṣaṣir, Šubria was one of a string of buffer states located between Urartu and Assyria. Beyond the clear parallels between the timing, setting and genre of these texts, and in particular the shared emphasis on the devotion of the monarch to the god Aššur, the particulars of their narratives suggest very different priorities. In the case of *Sargon's Letter to the God Aššur*, the narrative underscores Sargon's external struggles, presenting him as quelling the client states and neutralising an external threat with the aid of the god Aššur. Although Esarhaddon's narrative nominally also describes the defeat of a disobedient client king, the problem that Esarhaddon is addressing has an internal source. Ik-Teššub, the king of Šubria, has broken his covenant with Assyria by refusing to extradite Assyrian fugitives who had taken refuge in Šubria. The unusual practice of offering fugitives a safe haven was traditional in Šubria, and was likely religiously motivated.[250] There has been significant speculation among modern scholars about the identities of the refugees Esarhaddon wanted back, with some suggesting that Esarhaddon's brothers, those who had escaped *ana māt lā idû* 'to an unknown land', may have been among their number.[251] While this is certainly a possibility, the composition does not focus on only one group. Rather, the fugitives are described as:

> ..., robbers, thieves, or those who had sinned, those who had shed blood, [... eu]nuchs(?), governors, overseers, leaders, (and) soldiers who fled to the land Šubria.[252]

The list thus moves from generic criminals to members of the Assyrian imperial administration, seeming to suggest an overlap between these categories, and thus directing suspicion towards those belonging to the state apparatus. While some of these epithets may have applied to the murderers of Sennacherib and their co-con-

248 RINAP 2, no. 65.
249 On the buffer states to the north of Assyria that play a central role in both texts, see Radner 2012.
250 As argued in Dezső 2006 and Radner 2012, 263 f. See also Hipp 2015, 50.
251 A theory put forward by Erle Leichty (1991). Note also that Na'aman (2006), in an attempt to connect this text and the biblical evidence claiming that the brothers escaped to Urartu, suggests that Esarhaddon was seeking to exchange the Urartian fugitives he captured in Šubria for his brothers. Thus also Fuchs 2012.
252 RINAP 4, no. 33: obv.? ii 2–3.

spirators, the description casts a wider net than this, presenting every *halqu munnabtu mār Māt–Aššur* 'runaway Assyrian fugitive' as the subject of the dispute between Esarhaddon and Ik-Teššub.[253] Thus, all inhabitants of the empire's provincial extent who attempt to escape Assyrian justice are included by this description. In this way, the text uses the traditional Assyrian royal narrative of the punishment of a disobedient client ruler to present an inverted image. Esarhaddon's revenge against Ik-Teššub, who has broken his covenant, is primarily a means by which to ensure that disloyal Assyrians receive retribution. In addition, the covenant imposed upon Ik-Teššub is depicted as impacting not only the subjects of the client king but also those directly under the rule of Assyria, as this *adê*–covenant eliminates the ability of the latter group to escape the wrath of the Assyrian monarch. In stark contrast to *Sargon's Letter to the God Aššur*, in which Urartu is presented as the main antagonist, Esarhaddon's text presents the Urartian monarch as an ally to whom Esarhaddon returns the Urartian fugitives he finds in Šubria:

> [In] order to keep the *adê*-treaty and because of the truth and justice the great gods gave to me, I inquired, questioned, investigated, (and) denounced those people. I did not hold back a single Urartian fugitive (and) not one escaped. I returned them to their land.[254]

The final extant statement concerning the Assyrian fugitives lays out their gruesome fate:

> (As for) all of the [runaway] fugitives who had abandoned their owners and fled to the land Šubria, … […] I cut off [th]eir [hands] (and) removed their noses, eyes, (and) ears. [(As for) …] who had not run away to another country, I punished (them). I returned every […] … to their (text: 'his') land and to their owners. […] and they celebrated, rejoiced (and) blessed my kingship.[255]

This section of the narrative makes the point that all fugitives from Assyrian justice face the same punishment regardless of their location. Thus, in a manner similar to the covenant's stipulations, which seek to collapse the geographical expanse of the empire and emphasize the direct relationship between crown and subject, this royal composition denies that a change in physical distance from the crown will alter the ability of the monarch to punish his people. To this end, it is perhaps interesting that, while Sargon's text highlights the challenges of traversing the stretch between the core of Assyria and Urartu, this aspect of the campaign is less emphasized in the extant portions of Esarhaddon's composition. Instead, much of the text

253 RINAP 4, no. 33: i 16.
254 RINAP 4, no. 33: rev. iii 32'–34'.
255 RINAP 4, no. 33: rev. iii 23'–27'.

is taken up with a lengthy exchange of letters between Esarhaddon and Ik-Teššub. These missives are probably simply literary devices, as opposed to quoting from real letters sent by the Šubrian king.[256] Nevertheless, the decision to include what is framed as their correspondence in the text is interesting. *Esarhaddon's Letter to the God Aššur* presents long-distance communication between the Assyrian king and his clients as quick and simple.[257] So too, failure to comply with direct commands from the king is met with prompt action: in this case, perhaps significantly, the annexation of the client state and its integration into the provincial system.

A similar dynamic with regard to long-distance communication can be found in the *Apology*, which presents information as travelling to Esarhaddon *urruhiš* 'quickly', despite his location in a *ašar niṣirti* 'secret place': 'I, Esarhaddon, who with the help of the great gods, his lords, does not turn back in the heat of battle, quickly heard of their evil deeds.'[258] This statement is echoed again later in the Nineveh A inscription, in relation to the foreign rebel leader, Nabû-zer-kitti-lišir.[259] The impression given by both of these texts is that of an Assyrian king able to traverse his dominion not only physically, but also by proxy: the monarch has access to the full extent of his empire by means of information supplied by others. This situation thus parallels to some degree the manner in which the stipulations of the succession covenant are intended to function.

In contrast to the *Apology*, *Esarhaddon's Letter to the God Aššur* was likely intended to be presented publicly: read aloud in the city of Ashur in the same way Sargon's letter had been four decades earlier.[260] Despite this difference between the manner in which a contemporary audience would have accessed these compositions, however, they would likely have been known in particular to a small group of elite residents of Assyrian cities. Both compositions were likely known in Esarhaddon's immediate social vicinity, his entourage, some members of which would have been involved in their composition. Again, these people are to be seen not merely as the originators of these royal narratives but also as a

256 Note that client rulers did write to the Assyrian king, and indeed some letters from a previous Šubrian king, Hu-Teššub, to Sargon II have survived until the present day (SAA 5, nos. 44–45). As such, it seems probable that Ik-Teššub did at least write some letters to Esarhaddon. Nonetheless, the letters quoted in the text seem likely to be literary compositions.

257 See discussion of the Assyrian Empire's communication network in Chapter 1, as well as Radner 2014a.

258 RINAP 4, no. 1: i 53–55.

259 RINAP 4, no. 1: ii 50–51: 'I heard of his (i.e. Nabû-zer-kitti-lišir) evil deeds (while) in Nineveh'.

260 As set out by A. Leo Oppenheim in 1960. For more recent discussion, see Frahm 2019, 141, as well as Pongratz-Leisten 1999, 273 f. and 2013.

key part of their audience. The *Apology*, meanwhile, would quite probably have been known beyond this in a Ninevite context, specifically among those associated with Esarhaddon's Review Palace and the building works taking place on it at this time. The *Letter to the God Aššur* would likely have been known to the people who were associated with the Aššur temple in the city of Ashur, a significant proportion of the elite and specialist contingents of that settlement's population.[261]

The fact that these texts appear to have been composed for these narrow audiences serves to reinforce and sharpen their geographical message. Beyond simply commissioning two royal narratives in which covenants function in a manner that is more pertinent to the domestic, provincial, portion of the empire than they are to the client states, Esarhaddon appears to have been most interested in conveying this particular message to his entourage, as well as two urban population groups located at close geographical and social range. Both texts are themselves devoid of any explicit calls to vigilance of the kind found in the covenant itself. Nonetheless, the innovative focus of the narratives on the very people who made up their contemporary audience may have served as a rather jarring message that Esarhaddon considered the greatest potential threats to the Assyrian crown to be among their number. Such an apprehension may have prompted a vigilant response in itself, but even more so would have been likely to sensitize these groups to the significance of the stipulations of Esarhaddon's succession covenant when they heard them a few short months later.

4.3 Conclusions

The analysis of *Esarhaddon's Apology* and *Esarhaddon's Letter to the God Aššur* as narratives that aim to achieve immediate political results has revealed several common threads. While it has been observed by previous scholarship that these texts were intended in no small part to prepare their audiences for Esarhaddon's imminent announcement of his succession arrangements, it is worth stressing that the concept of *adê* 'covenant' was at the heart of both of these narratives. Considering *Esarhaddon's Apology* in the light of the reality of Neo-Assyrian royal succession, it seems clear that the narrative constituted to some degree a disavowal of established Assyrian succession practices, even though it was not explicitly framed as such. By professing that it was in large part the power of Sennacherib's succes-

261 Note too that Ashur does not receive booty from the campaign (Oppenheim 1979, 130 and 132). This could be taken as a warning directed specifically towards the people of Ashur. Leichty connects this with the theory that Esarhaddon's brothers' rebellion was based in Ashur (Leichty 1991, 57, fn. 19).

sion covenant that had ensured his accession, Esarhaddon sought to minimize the traditionally large role played by charisma in Assyrian conceptions of legitimacy and stressed the existence of a legal, administrative component to Assyrian kingship, something that had not previously been particularly prominent. In projecting the centrality of this component back to the circumstances of his own succession, Esarhaddon and the composers of his inscription portrayed the imposition of his own covenant, and the importance he afforded it, as reflecting a practice of succession that predated his own rule.

Similarly, *Esarhaddon's Letter to the God Aššur* can perhaps be viewed as referring, obliquely, to the seditious deeds of Esarhaddon's brothers, and members of the Assyrian state administration more generally, around the time of Sennacherib's murder. Whether or not the text is to be interpreted as directly concerning this incident, it seems unlikely that the parallels between the characterization of Esarhaddon's brothers as having fled Assyria in the *Apology* and the fugitives described in *Esarhaddon's Letter to the God Aššur* as having 'sinned' and 'shed blood' were accidental. Assuming that the flight of Esarhaddon's brothers was common knowledge among the elite milieu in Ashur, the reference would have been difficult to miss. The narrative thus implies that those who broke their succession covenant were punished for doing so, again emphasizing the role of legal measures in securing Esarhaddon's present status as king. Beyond this implied reference to his father's succession arrangements, however, Esarhaddon also draws parallels between himself and his grandfather, Sargon II. Despite the shocking consequences of his death on the battlefield when the king's body was lost to the enemy, Sargon II's elevation of Sennacherib to the position of crown prince was unusually successful by Neo-Assyrian standards. Esarhaddon's desire to associate himself with that monarch may have been motivated in part by the desire to draw parallels between this aspect of his own agenda and that of Sargon, as well as the peaceful transition of power from that king to Sennacherib, despite the extremely difficult circumstances of the former's death.

Another common thread in the texts, and one that has perhaps been less appreciated until now, is the strong geographical focus of both narratives. They are focused not on elevating the status of succession covenants in general conceptions of Neo-Assyrian rule, rather, they explicitly stress the relevance of these legal bonds to their respective contemporary audiences. In both cases, they are levelled at small groups of high-status individuals based in the urban centres of Nineveh and Ashur. Both compositions, in different ways, make use of the traditional narratives referencing client treaties that would have been familiar to this milieu. In each case, the narrative inverts the conventional view that covenants pertain principally to the empire's client states, instead focusing on the empire's domestic affairs. These narratives both portray the enemy within, therefore, as far more dangerous and pressing than any potential foreign adversary. Despite this, the compositions take care not to simply portray the people of Assyria as having the same status as clients. *Esarhaddon's Apology* takes pains to stress the loyalty of the Assyrian people, contrasting them with the disobedient client kings and foreign leaders who are vanquished in the remainder of the inscription Nineveh A. In *Esarhaddon's Letter to the God Aššur,* meanwhile, the punishment of Assyrian fugitives is embedded in a classic tale of a client ruler who has broken his oath. Thus, both compositions work to differentiate Assyrians from non-Assyrians, placing the former firmly above the latter, while also casting veiled suspicion on the loyalty of the inhabitants of Assyria proper.

The decision to commission and disseminate these texts ahead of the imposition of Esarhaddon's succession covenant is telling. In particular, it belies the language of the preamble of the covenant manuscripts, which present the administrative zones of the empire – client state and province alike – as equally bound by the terms of the covenant. While this may have been true from a legal and administrative point of view, Esarhaddon and his advisors seem to have considered the covenant more germane to some than to others. From a spatial point of view, it is clear that the provinces were considered more relevant than the client states. More specifically than this, however, it seems that a small selection of key target groups were prioritized as the recipients of ideological groundwork in the form of these compositions. That these groups appear to be close, both socially and physically, to the person of the king, and specifically those who spend time in the cities of Nineveh and Ashur, is to some degree consistent with various aspects of the covenant composition, such as the stipulations and the list of divine witnesses. Nonetheless, the decision specifically to single out these groups further distinguishes them from the rest of the subjects of the Assyrian Empire. Thus, while Esarhaddon seems to have gone to some significant lengths to impose his succession covenant across the geographical extent of his empire, it was nonetheless those at close range to him whom he seems to have considered particularly dangerous. This belief may have been rooted in part in memories of his own succession, as Ashur may have been a stronghold of his brothers' supporters. Beyond this, it may also have reflected a sense of the kinds of people who would have had a realistic chance of launching a successful coup against Esarhaddon or his crown prince. In either case, by propagating such royal narratives among these groups, it seems likely that Esarhaddon sought to increase the perception that his succession covenant was important and pertinent to them, improving the chances that they would heed its call to vigilance once it was imposed, shortly after the dissemination of these compositions.

Chapter 5: Putting the covenant into practice

The previous chapters of this study have shown that Esarhaddon's succession covenant coveys to its audience a clear duty of vigilance. So too, a few select groups of Assyrian subjects would have been exposed in the months preceding the announcement of Esarhaddon's succession arrangements to narratives that complemented and echoed the covenant's message. This chapter, meanwhile, seeks to answer the question of how – on whom and under what circumstances – the covenant was implemented and its duty of vigilance communicated. The moment of the covenant's imposition was the key opportunity for the Assyrian crown to responsibilize its subjects, transferring to them the duty to become vigilant on its behalf. Despite its great importance to the covenant's real-world impact, the practical and logistical aspects of its imposition across the empire have been explored by modern scholars only in a fairly cursory way. Those who have analzed the imposition of the covenant have often done so through the prism of whether or not, and in what way, it was likely enacted on the client kingdom of Judah. These arguments have then been used as a way to argue for or against a direct connection between the succession covenant and various close parallels found in the Old Testament. This avenue of investigation is, of course, of great interest considering the vast influence of the Hebrew Bible, but does not reflect the relatively low contemporary status of Judah within the Assyrian Empire at the time. Those who have looked beyond Judah, meanwhile, have generally focused on the religious context of the covenant's imposition, exploring its apparent connection with the New Year ceremony (*akītu*) of the god Nabû in the temple at Kalhu. Again, such an approach is valuable in its own right, but does not answer questions regarding the practicalities of communicating the message of the succession covenant.

The following investigation seeks to supply a narrative of the covenant's implementation in which the logistical and practical aspects of communicating the covenant's contents form the primary focus. This is not to assume that the covenant was implemented uniformly across the entire empire, or that Esarhaddon and his advisors aimed to communicate its contents to all social groups. Rather, it is to place such questions, which previously have been treated as peripheral concerns, at the heart of the analysis. Following this approach, I firstly stress that the covenant ceremony, which would have involved the communal swearing of an oath, a clear public display of loyalty to the crown, constituted in itself a situation of heightened vigilance. In this sense, the covenant served not only as a measure that was intended to protect the Assyrian crown in the future, but also as an immediate means of testing allegiance to Esarhaddon and his decisions. In the second section of the chapter, I argue that – in contrast to its characterization in the sev-

 https://doi.org/10.1515/9783111323435-008

enth century royal narrative – the covenant ceremony itself as it actually took place is best conceptualized as a 'staggered event', held over the period of some several months and in many locations across the empire. Finally, I posit, following arguments made by Jacob Lauinger, that the covenant was not intended to be enacted only once, in 672 BC, but rather to be implemented in an iterative fashion, something that would have served to embed regular events of heightened vigilance into Assyrian public life. Considering the conclusions of the previous three chapters, it is perhaps unsurprising that it appears that, in contrast to the interests of many modern scholars studying this topic, Esarhaddon and his advisors took much greater pains to implement the covenant thoroughly and effectively within the provincial portion of the empire than in the client states.

5.1 Scaling vigilance at the covenant ceremony

Before embarking on a detailed exploration of the logistics of the covenant ceremony, it is worth dwelling on the experience of entering and concluding a binding agreement of this kind, in particular because swearing to the succession covenant was in itself probably associated with a high degree of vigilance. Firstly, as will be discussed further below, it seems clear that at least some participants in the covenant ceremony would have had to travel from their place of residence to an Assyrian administrative centre, either in the Assyrian heartland or the provinces. These locations would in themselves have borne clear signs of Assyrian imperial power.[262] The ceremonies would most commonly have taken place in temples or other sacred spaces, before the gods, and were possibly integrated into other religious celebrations. As such, one would have been acutely aware of divine scrutiny during the process. The ceremonies seem likely to have involved the ritual slaughter of more than one sheep, probably a ewe and her young, along with the dissection and manipulation of the carcasses.[263] The participants would have listened to the covenant text, which would have been read aloud, and would have been required to repeat the oath section verbatim. They would also have been required to participate in ritual actions of self-cursing, many of which seem to have been designed to internalize the covenant: smearing oil on oneself, drinking water, consuming bread and wine, eating honey.[264] The practice and logic of at least some of

262 The royal citadel at Tell Tayinat, ancient Kullania, where one of the exemplars of the covenant tablet was found, is a good example of the expression of late Neo-Assyrian imperial power within the provincial system (Harrison 2014, esp. 92–93).
263 SAA 2, no. 6, §§ 69–70: 547–554; Lauinger 2012, vii 40–47.
264 See Chapter 2.3.

these acts would have been familiar to those doing them, as they were also used in Neo-Assyrian legal practice more widely.[265] Such repeated self-cursing would have constituted instances of heightened vigilance directed towards the self, in which one's own intentions towards the crown and degree of loyalty towards it would have been the focus of attention.

The act of coming to and participating in the covenant ceremony would also in itself have been a signal of loyalty to the Assyrian crown. Convening these ceremonies would have been an opportunity for top-down surveillance, with non-attendance a sign of potential deviance. Although, as discussed below, some scholars consider the findspot of the Kalhu covenant tablets to be evidence of absenteeism on the part of these client states, there is no definite evidence that anyone refused to enter the succession covenant. Nevertheless, the possibility of such an eventuality should not be dismissed out of hand. Neo-Assyrian legal documents show that refusal to take an oath as part of legal proceedings was common.[266] This was presumably due to the real fear of incurring divine wrath if a participant broke their oath or swore falsely. In addition to promising to be loyal to the Assyrian crown, it was required of those swearing to the covenant that 'while you stand on the place of this oath, you shall not swear this oath with your lips only but shall swear it wholeheartedly'.[267] In this way the stipulations referred not only to future possible scenarios but to the immediate situation in which the participants in the covenant ceremony found themselves. This gives the oath and the self-curses that they would have had to perform an immediate relevance and, presumably, a direct sense of the danger of not participating with the correct attitude. The distress caused by the belief in proximate divine punishment, not only of oneself but of one's family, would have been very real to the inhabitants of the Assyrian Empire, and must not be underestimated.[268] That participation in the covenant ceremony was considered indicative of loyalty to the crown – and that failure to take part was suspicious – is further illustrated by the statement in the composition that 'you shall not feign unclean illness but take part in this covenant'.[269]

In addition to the top-down surveillance possible at a covenant ceremony, the language of the composition, which uses the second-person plural and the first-per-

265 The practice of swearing by 'oil and water' existed in Neo-Assyrian law more generally (Faist 2020, 159 f.).

266 Faist 2020, 166.

267 SAA 2, no. 6, § 34: 385–387; see also Lauinger 2012, v 49–52.

268 Faist 2020, 166: 'Die Angst vor der göttlichen Strafe muss eine für uns heute kaum vorstellbare psychologische Wirkung gehabt haben.'

269 SAA 2, no. 6, § 34: 389–392; see also Lauinger 2012, v 54–57. On the reading of the adjective modifying *murṣu* 'illness', see Lauinger 2012, 116.

son plural, is an accurate reflection of the communal nature of these events. Participants would therefore have been aware of the scrutiny not only of the gods and the crown, but would also have been in the company of their peers. This, in turn, would likely have prompted a high degree of vigilance directed towards the self, and with it an anticipatory adaption of one's own behaviour. The communal nature of covenant ceremonies and the lateral vigilance they would have provoked thus model the sustained lateral vigilance that the covenant sought to instill in those bound to it. In this way, the covenant ceremony served as a microcosm – and also a key instance – of the culture of vigilance that it was designed to provoke.

The ritual actions undertaken by those swearing to the covenant would potentially have provided an opportunity for the vigilance directed towards the self and others to be heightened even further at various points during the ceremony. By ritually internalizing the covenant, it became possible for the subject to be destroyed from the inside out by breaking its terms. In this way, the forbidden actions, thoughts and feelings enumerated in the covenant composition needed no witnesses, they could be sensed and punished from within the body: 'May the great gods of heaven and earth turn water (and) oil into a curse for you.'[270] Beyond this, these acts were also intended to prompt the memories of those who entered into the covenant. On this subject, an excerpt from a prophecy collection written up in final years of Esarhaddon's reign after the covenant's implementation is perhaps instructive:

> In your hearts you say, 'Ištar is slight,' and you will go to your cities and districts, eat (your) bread and forget this covenant. (But when) you drink from this water, you will remember me (i.e. the goddess Ištar) and keep this covenant which I have made on behalf of Esarhaddon.[271]

This text is unusual in many ways and should not necessarily be taken as representative of popular understanding.[272] Nevertheless, the connection drawn here between the act of drinking and memory of a covenant seems to imply, in this interpretation at least, that these acts are also seen as scaling vigilance from within: prompting a participant to remember their covenant and to act according to its demands. In this telling, then, these ritual acts overcome the limitations of human attention, ensuring effective vigilance.

The covenant ceremony thus seems to have functioned to draw the attention of participants not only to the supra-individual goal of the protection of the Assyr-

270 SAA 2, no. 6; § 61: 523; see also Lauinger 2012, 104 f, vii 13–14. See also discussion of swearing by 'oil and water' in Neo-Assyrian law more generally (Faist 2020, 159 f.).
271 SAA 9, no. 3: iii 7–15.
272 See discussion in Chapter 6.2.2.

ian crown, but also to direct this attention towards various actors in order to modulate it both during and after the covenant's initial imposition. The ceremony drew attention towards the crown, and in particular its power and ability to surveil its population, something that was made possible in large part by the participation of the gods, to whom the subjects' attention was also directed. Moreover, the participants' vigilance was orientated variously towards their peers and themselves, something that was modulated throughout the ceremony. As already discussed in Chapter 3, the stipulations of the covenant composition can themselves be interpreted as working with the ebb and flow of human attention, strategically employing repetition and variation to guide the listeners' awareness towards particular aspects of the scenarios and reactions that they set out. While it is the variations that presumably caught the listeners' immediate attention, with the repeated formulations fading into the background of their awareness, repetition facilitates memorization, and as such its use can perhaps be interpreted as an instance of the covenant's composers taking into account the nature of human memory. This latter point finds some support in the prophecy quoted above, which seems to suggest that the ritual acts of the covenant ceremony served to direct attention towards the conditions of the covenant long after the conclusion of the ceremony. Beyond this, if the suggestion laid out below that the covenant was itself imposed iteratively is correct, then the subjects of the Assyrian Empire would have gone through this experience several times. In this way, the regular scaling of vigilance towards oneself and each other on behalf of the crown would have become a feature of Assyrian public life. So too, it would surely have served as a powerful tool for instilling the duty of vigilance in those subjects compelled to undergo it, as well as constituting an opportunity for regular and immediate top-down measuring of individual and group loyalty towards the crown.

5.2 The covenant ceremony as a staggered event

Some three decades after the imposition of Esarhaddon's succession covenant, in 645 BC, King Ashurbanipal, by now the ruler of the Assyrian Empire, circulated a royal inscription that characterized its imposition as follows:

> In the month Ayyaru (II), the month of the god Ea – the lord of humankind – on the twelfth day – an auspicious day, (the day of) the bread donation(s)(?) to the goddess Gula – he (i. e. Esarhaddon) assembled the people of Assyria – great and small (i. e. of high and low social status) – (and) of the Upper and Lower Sea(s). In order to protect my (position as) heir des-

ignate and afterwards (my) exercising the kingship of Assyria, he made them swear to a covenant (*adê*), an oath bound by the gods; he made the agreements strong.[273]

This description of the covenant ceremony is instructive in various ways. It is interesting that, rather than specifying the location in which the ceremony took place, Ashurbanipal simply characterizes his father as having *pahhuru* 'gathered together' the people of Assyria and the 'Upper and Lower Sea(s)' (i. e. the Mediterranean and the Persian Gulf), with the latter referring to the client states and the former the provinces. While the inscription itself goes on to mention the *bīt ridûti* 'House of Succession', it does not explicitly link the covenant ceremony with this location.[274] Instead of stressing place, the inscription specifies and describes the date of the covenant's imposition. The day in question is placed within the context of the Assyrian religious calendar: the festival associated with the date is cited, as is the status of this day as *ūm magāri* 'auspicious day'. The Neo-Assyrian calendar was full of such festivals, as well as auspicious and non-auspicious days for undertaking particular activities, a concept that was closely tied to notions of divine approval.[275] By including the date of the covenant ceremony, as well as its hemerological information, therefore, the composers of Ashurbanipal's royal inscription stress divine approval for the event, and portray Esarhaddon as a ruler who specifically takes this into account in his actions.

5.2.1 Holding multiple ceremonies within the core region

Despite the insights that this royal inscription provides into those aspects of the covenant ceremony that Ashurbanipal and his advisors much later considered worth highlighting, there is good reason to believe that the description does not accurately represent the covenant's implementation process. As already argued by Simo Parpola in 1983, it is clear to modern scholars that the covenant was not enacted over the course of just one day.[276] This is not merely a logistical assessment: the contemporary evidence also suggests that, even within the core region, the covenant was imposed over a protracted period. One indication of this is given by the short informational subscripts ('colophons') on the covenant manuscripts themselves, as well as some date lines on royal inscriptions that were apparently written at around the same time. The royal inscriptions are dated to the

273 RINAP 5/1, no. 9: i 9–16.
274 See also discussion of this term in Chapter 2.1.
275 See A. Livingstone 2017, as well as Ermidoro 2017.
276 Parpola 1983, 4; also quoted in Fales 2012, 149.

18th of Ayyaru 672 BC, 'when the covenant concerning Ashurbanipal, the senior son of the king, who (resides in) the House of Succession, was made'.[277] In contrast, the preserved colophons of the covenant tablets themselves, meanwhile, are dated variously to the 16th (one Kalhu exemplar) and the 18th of Ayyaru (two Kalhu exemplars).[278] Although the Kullania tablet preserves a colophon, the numeral designating the day is damaged: it is between 16 and 19.[279] There are various potential reasons for these discrepancies in dating. It is possible, for instance, that the date was intended to refer to the day upon which the tablet was inscribed and not that of the ceremony itself.[280] Nonetheless, even if this were the case, it is striking that scribes were still drawing up covenant tablets several days after the date of the ceremony according to Ashurbanipal's later account, cited above.

Of course, it is tempting simply to dismiss Ashurbanipal's claim that the ceremony took place on the 12th as a mistake on the part of the scribes some thirty years later. It is relevant to note here that the earliest manuscripts of the inscription put the date as the 18th of Ayyaru, 'the *huntu*-holiday of the god Šamaš, the hero'.[281] Some of the manuscripts were then altered to state that the ceremony took place on the feast day of Gula, not Šamaš. A few of the scribes making that change seem to have decided that, since the feast day of Gula was not on the 18th, but the 12th, they would alter this too.[282] I have not been able to detect any particular political incentive on the part of Ashurbanipal and these scribes for inaccurately portraying the ceremony as occurring on the latter day rather than the former. Therefore, it seems most likely that the primary motivation for this correction was a desire to render the date accurately, and in particular to reference the feast day correctly, perhaps for religious reasons. As such, while it is not impossible that the date given in this inscription is merely a later inaccuracy, it seems more

277 RINAP 4, no. 77 ex. 6: 63B–64B and no. 93: 40.

278 16th: ND 4336C; 18th: 4354D, 4354F. Note that fragment 4354F does not preserve the month date, and thus – while it is presumed in the secondary literature that it is dated to Ayyaru, this is not certain.

279 Lauinger 2012, 122; Lauinger 2013, 122, fn. 50.

280 Jacob Lauinger notes this, as well as the fact that these dates may not be different (2015, 292, fn. 14). Lauinger contrasts his position to that of Mordechai Cogan (Cogan 1977, 99), who views the colophons as referring to dates on which the festivities took place. It is perhaps relevant that in Neo-Assyrian legal practice more generally legal documents were dated to the day on which they were drawn up (and thus the date on which the legal agreement was made), and not necessarily the day on which the transaction was executed, something which sometimes happened significantly later (Radner 1997, 91, esp. fn. 504; Faist 2012, 211–13).

281 Cogan 2005, 10.

282 Jeffers 2018; RINAP 5/1, 182.

likely that this date reflects the fact that those being sworn to the covenant were adjured over the course of several days.[283]

This impression is strongly supported by probably the most effusive sources concerning the logistical planning for the covenant ceremony: three letters sent to Esarhaddon by his chief scribe, the probable overseer of the covenant's composition, Issar-šumu-ereš.[284] The correspondence appears to have taken place in the month Nisannu (I) of 672 BC, and concerns the decision-making around the correct moment for various groups, apparently from the core region, to *erēbu* 'enter' and *šakānu* 'conclude' the covenant. In one of the letters, Issar-šumu-ereš mentions 'the scribes, haruspices, exorcists, physicians and augurs staying in the palace and living in the city',[285] while in another he is planning the covenant ceremony for 'the scribes of the cities of Nin[eveh], Kilizu and Arbela', who have already arrived, while those of Ashur '[have] not (yet) [come].'[286] In the same letter, Issar-šumu-ereš is planning the adjuration of the 'citizens' of Nineveh and Kalhu. These letters, if the widely accepted view that they pertain to the succession covenant is correct, provide a fascinating insight into the organization of the covenant's imposition. Issar-šumu-ereš suggests several possible dates for the ceremony, variously mentioning the 8th, 15th and 16th of the month Nisannu, as well as the 20th, 22nd and the 25th of a month that is not specified. Of course, these letters provide an insight into the planning stage of the covenant's imposition and it is entirely possible that the king rejected some of these suggested dates, choosing others instead. Nonetheless, the letters themselves appear to be written temporally very close to the days that Issar-šumu-ereš is suggesting that the ceremony take place: in one case he states that a group should conclude the covenant *iššiāri* 'tomorrow'.[287] So too, the groups seem generally already to have arrived or to be coming imminently. This indicates that these plans were fairly advanced and, as such, it seems unlikely that they were completely cancelled in favour of holding a covenant ceremony for all members of the empire one month later. Instead, the letters seem to imply an *ad hoc* situation, in which the chief scribe was responsible for gathering together citizens and specialist groups from important cities in the core region for a ceremony.

---

283 Novotny has also argued that the mention of both the 12th and the 18th of Ayyaru potentially demonstrates that the ceremony lasted at least seven days (Novotny 2023, 386).

284 SAA 10, nos. 5–7, see also commentaries in Parpola 1983, 3–6. Parpola's interpretation, that these letters pertain to the succession covenant (Parpola 1983, 4) has generally been accepted. See also the discussion of Issar-šumu-ereš in Chapter 1.3.3.

285 SAA 10, no. 7: obv. 6–11; Parpola 1983, 6.

286 SAA 10, no. 6: obv. 6–11; Parpola 1983, 4.

287 SAA 10, no. 7: rev. 2; Parpola 1983, 6.

In this regard, it is particularly interesting that the dates suggested for the various groups do not seem to align. In one letter, Issar-šumu-ereš suggests the dates 'the 20th, the 22nd and the 25th' for a group that is not specified.[288] In another, he recommends that the scribes of Nineveh, Kilizi and Arbela, who have already arrived, enter the covenant immediately, with the still-absent Ashur scribes presumably doing so at a later date, although he does not specify this. The citizens of Nineveh and Kalhu, who are also not currently available but will 'be free soon',[289] meanwhile, should enter the covenant on the 8th day of the month. Alternatively, those who have already arrived should be dismissed and told to come back and enter the covenant on the 15th, concluding it on the 16th.[290] In another letter, meanwhile, he states that the specialists of the palace and the city will enter the covenant (rather than concluding it) on the 16th of Nisannu.[291]

Even accounting for the possibility that these dates were not all accepted,[292] it is clear that Issar-šumu-ereš is not expecting all of these groups to take part in the covenant ceremony at the same time. The impression given is one in which the chief scribe is overseeing the adjuration of several discrete groups with busy schedules, and that their simultaneous participation in the covenant is sometimes logistically useful, but not necessary. In addition to finding dates that worked for the human participants, Issar-šumu-ereš also had to choose divinely sanctioned dates: when suggesting that the participants enter into the covenant on the 15th of Nisannu, he notes that the hemerologies state that swearing an oath on that particular day is not sanctioned, and thus notes that the participants should only conclude the covenant on the 16th.[293] The letters also clearly show that the process of imposing the covenant had begun at least one month before the official date given in Ashurbanipal's royal inscription, or on the colophons of Esarhaddon's royal inscriptions.

The question that naturally flows from this conclusion is to what degree the groups mentioned in Issar-šumu-ereš's letters can be considered representative of the manner in which the empire's subjects were sworn to the covenant. Simo Parpola has put forward the suggestion that the, presumably large, number of scribes required to draw up the covenant tablets may have been adjured early.[294]

288 SAA 10, no. 5: obv. 8–10; Parpola 1983, 6.
289 SAA 10, no. 6: obv. 20–21; Parpola 1983, 4f.
290 SAA 10, no. 6: rev. 1–19; Parpola 1983, 5.
291 SAA 10, no. 7: obv. 12–14; Parpola 1983, 5f.
292 A possibility mentioned by Simo Parpola (1983, 4) and further discussed by Jacob Lauinger (2015, 291). Eckart Frahm takes the view that this is what happened (Frahm 2009a, 135).
293 SAA 10, no. 6: rev. 1–19.
294 Parpola 1983, 4.

Jacob Lauinger supports this view, stating that 'it seems only logical that the scribes who were entrusted with the task of drawing up what would become the destinies of the entire empire would need first to be bound to protect the crown prince themselves.'[295] It seems perfectly likely that some of the scribes mentioned in Issar-šumu-ereš's letters are the same scribes who were required to draw up the covenant tablets, and even if this were not the case, the letters strongly suggest that all of these scribes would have been sworn to the covenant. Nonetheless, it does not follow that this would have been a prerequisite of inscribing the tablets themselves. Indeed, if one accepts this notion, it becomes difficult to imagine how the first manuscript of the succession covenant could have been produced, as the covenant tablets themselves were necessary for the covenant's imposition.[296] Furthermore, while some of the people mentioned in addition to the scribes could conceivably have been involved in preparing the covenant tablets, it seems rather unlikely that some of them would have been, such as the *mar'ē* 'citizens' of Nineveh and Kalhu. As such, it is more likely that the letters represent the logistical reality of the covenant's imposition: that, even within the core region, it was imposed over a protracted period, rather than the groups mentioned in the letters representing an anomaly in this regard.

Why, then, did the subsequent royal narrative choose to portray the covenant ceremony as taking place on a specific day? Ashurbanipal's royal inscription takes pains to highlight the divine sanctity of the date upon which the covenant took place, something that Issar-šumu-ereš's letters show was indeed considered important at the time. Beyond this, several scholars have seen a link between the Ayyaru dates associated with the covenant's imposition in the royal narrative and the festival calendar of Assyria, linking the timing to the New Year ceremony (*akītu*) of the god Nabû.[297] The ceremony centred on the marriage of Nabû and his consort Tašmetu on 4th Ayyaru, which was followed by a programme of ritual activity lasting until 11th Ayyaru. Nabû was the son of Marduk, the head of the Babylonian pantheon, and thus himself a crown prince of sorts. Furthermore, the covenant composition explicitly cites him as the 'bearer of the tablet of fates of the gods',[298] which would have rendered him a fitting figure to oversee the imposition of a Tablet of Destinies, a document sealed by a god the contents of which are fated to occur.[299] That there existed a close association between Ashurbanipal and Nabû

---

295 Lauinger 2015, 291.

296 The composition itself includes references to the covenant tablet, see Chapter 3.

297 This position has been set out and expanded recently by Jacob Lauinger (2013; 2019). See also Fales's arguments on this subject (2012) and those of Barcina (2016).

298 SAA 2, no. 6, § 105: 660 f.; Lauinger 2012, viii 59 f.

299 On the covenant as a Tablet of Destinies, see discussion in Chapter 2.3.1.

at this time is clear from three letters, two of which are addressed to Esarhaddon and one of which is written to Ashurbanipal. These letters discuss Nabû's ritual marriage, linking the proceedings directly to the status and health of the crown prince.[300] The clear decision to emphasize the close relationship between the crown prince and Nabû in Kalhu should be understood in the context of Esarhaddon's wider cultic innovations, which were a reaction in large part to those of his father.

Sennacherib had sought to elevate the status of Aššur above Marduk and the Assyrian Nabû; Esarhaddon, in contrast, took steps to show his devotion to both.[301] One other such measure was his own decision to ascend to the throne on the 8th of the month Addaru (XII), Nabû's *eššēšu*-festival.[302] The fact that the Kalhu covenant tablets were found in the Throne Room of the *akītu*-complex at the Nabû temple in that city strengthens the association between this ceremony and the covenant's imposition.[303] Beyond this, it seems possible that the groups, or at least some of them, mentioned in the letters of Issar-šumu-ereš were also adjured in the Nabû temple at Kalhu, as he makes reference to the citizens of Nineveh and Kalhu entering the covenant *šapla Bēl Nabû* 'under (the statues of) Bel (= Marduk) and Nabû'.[304] Despite this, it is worth noting that Nabû had temples in other locations, such as Nineveh, so it would have been possible to enter the covenant in front of Nabû elsewhere.[305] So too, it is worth stressing that the divine seals of Aššur are closely associated with the city hall of Ashur, and were likely kept in that city.[306] As such, the covenant manuscripts may have been drawn up in Ashur, or the seals themselves may have been transported to the location(s) in which the covenant manuscripts were inscribed. It is also worth noting that, if different groups

---

300 SAA 13, nos. 56, 70 and 78.

301 Robson 2019, 77. See also the deities mentioned in the short version of Sennacherib's succession covenant (Frahm 2009a, no. 69). Note that Nabû and Marduk do not seem to be included (although the relevant portions of the text are very fragmentary). The text may mention the gods of the *bīt akīti*, and the very damaged colophon bears the legend [... *in*]*a bīt akīt* [...] '...in the *bīt akīti* (of)...'. It seems fairly likely, then, that this covenant was imposed in the *akītu*-house of Aššur in that city (Frahm 2009a, 135).

302 Robson 2019, 78.

303 See Lauinger 2013, 111 for discussion and bibliography (esp fns. 46 and 47).

304 SAA 10, no. 6: obv. 22–b.e. 23.

305 Radner 2006, 368. Tablets were also kept in the Nabû temple at Nineveh (Fincke 2004, 55).

306 Sennacherib's seal inscription dedicates the seal to Aššur, while the Old Assyrian seal mentions the *bīt ālim* 'city hall' of Ashur (Watanabe 2020, 75–81; on the latter seal see also Dercksen 2004, 90). Note also that the the city hall in Ashur came to be associated with the temple of Nabû (A. George 1986, 141; Dercksen 2004, 95; see also the relevant line of the Divine Directory of Ashur, SAA 20, no. 49: 68).

from the core region were adjured separately, then it is possible that some were required to travel to Kalhu to participate in covenant ceremonies and that others were required to swear to the covenant elsewhere.

In summary, the royal narrative associating the ceremony with mid-Ayyaru is a reflection of the importance to the Assyrian crown of aligning the covenant ceremony with the Assyrian religious calendar in general and Nabû's *akītu*-ceremony in particular. While characterizing the process of swearing Assyria's subjects to the covenant as taking place on one particular day is to some degree inaccurate, it is also likely that this period coincided with the ritual elevation of Ashurbanipal to the role of crown prince, something that may have been linked to Nabû's *akītu*-ceremony. This would render the emphasis on mid-Ayyaru in the narrative of Ashurbanipal's elevation accurate, even though the covenant was actually imposed over a much longer period. It also seems possible that some of the more prominent subjects of the Assyrian king, namely client kings and provincial governors, may have been required to participate in the covenant ceremony in central Assyria at around this time. The dating of the manuscripts to this period seems to support this theory.

This notwithstanding, it is not necessary to assume that all client kings and provincial governors would have been expected to participate in the covenant ceremony simultaneously. Indeed, while there are clear religious considerations for Esarhaddon and Ashurbanipal's inscriptions associating the covenant ceremony with dates in mid-Ayyaru, there may be other reasons that the extant covenant manuscripts of the eastern client kings and the governor of Kullania seem to have been drawn up during this period. There may for instance, have been logistical reasons that for them, or their emissaries, to convene in the core region, possibly Kalhu, at this time.

Hans Ulrich Steymans has suggested that the covenant tablets of the eastern client kings were kept in the Nabû temple at Kalhu because emissaries from these states may have made regular deliveries of tribute, specifically equids, to this location.[307] While this association is disputed,[308] a connection between the covenant ceremony's timing in the first months of the year and the delivery of equids to the core region has various pieces of circumstantial evidence to support it. The delivery of annual tribute was one of the key obligations of client states to Assyria. In his analysis of this system, J. N. Postgate has suggested that annual tribute was to be delivered around the new year (i.e. in April) or in the autumn after the cam-

307 Steymans 2003, 2004, 2006, and 2013. See also discussion in Radner 2006, 372–373, as well as Lauinger 2013 and 2019.
308 Fales 2012, 151, fn. 115.

paigns.[309] His theory is based in no small part on the fact that these would have been the most suitable times for such an undertaking on climatic and logistical grounds. That at least some client states did indeed send delegations around the new year is also confirmed by documentary evidence.[310] Furthermore, as Karen Radner has argued, there is reason to believe that the delivery of tribute was sometimes combined with participation in religious festivals.[311] In support of this observation, Radner cites a letter to Sargon II in which emissaries arrive in Kalhu (which at this time was probably still the capital) at around the same time as a festival is taking place:

> [The festi]val has been celebrated; the god [...] came out and returned in peace. May Nabû and Marduk bless the king! ... I have received 45 horses of the [pala]ce. The emissaries from Egypt, Gaza, Judah, Moab and Ammon entered Kalhu on the 12th with their tribute. The 24 horses of the (king) of Gaza are with him. The Edomite, [Ashdo]dite and Ekronite [...]. The emissary from Que [...] is departing and going [...] the Bow [River]; the [...] of the Commander-in-Chief is with him.[312]

In addition to this, Esarhaddon appears to have renovated both the Review Palace at Nineveh and the one at Kalhu around the dates associated with the covenant's imposition.[313] This is significant because deliveries of tribute were likely stored there.[314] Furthermore, in Nineveh at least, the king conducted annual inspections of the contents of this palace at the beginning of the year.[315] That some ceremonial activities took place in Nineveh on the occasion of Ashurbanipal's nomination as crown prince seems likely, as his succession palace was probably located there.[316] One may imagine that delegations from the client kingdoms, who would have arrived to deliver their tribute to newly renovated palaces, may have been required to stay until they had sworn to the new covenant. It is impor-

---

309 Postgate 1992, 255; Postgate 1974b, 121. On tax and tribute in the Neo-Assyrian context, see also Radner 2007.

310 See discussion in Postgate 1974b, 121. The two letters in question date to the reign of Sargon II: SAA 5, no. 52 and SAA 15, no. 60. Note that the former letter also references attendance of Šubrian emissaries at the *māšartu* 'review'.

311 Radner 2007, 219.

312 SAA 1, no. 110: obv. 4–7 and rev. 4–17.

313 *Esarhaddon's Apology*, discussed in Chapter 5, is part of a longer inscription commemorating building works on the Review Palace at Nineveh, on which see Maul and Miglus 2020. On the Review Palace at Kalhu under Esarhaddon, see Kertai 2015, 159–165.

314 A. Otto 2015, 482–84; Kertai 2015, 148.

315 As noted in Kertai 2015, 148. RINAP 4, no. 2: vi 25–43.

316 Jamie Novotny argues that the oath-swearing ceremony took place in Nineveh on these grounds (2023, 386).

tant to state, however, that tribute was typically delivered by emissaries, not by the client ruler himself. Thus, such a reconstruction could mean that – in contrast to the letter of the covenant text – these people would have entered the covenant on behalf of their rulers.[317]

While the provincial governors and officials mentioned in the covenant compositions were not required to deliver tribute on behalf of their provinces in the same way, they were sometimes involved in the delivery of tribute from client states.[318] As such, some could have been expected to have made similar deliveries at around this time too. It is also perhaps relevant that review palaces were used for the administration, equipment and training of the army and thus may have been points of assembly for provincial governors and members of the administration in their military capacities too.[319] Reports sent to the king by Nabû-šumu-iddina, the *hazannu* 'inspector' of the Nabû temple, show that the various provinces were required to send equids to the capital in the first months of the year (the reports are dated 6th Nisannu (I) to 4th Simanu (III)).[320] These horses were delivered to Kalhu, Nineveh and Dur-Šarruken, and the author reports on what is happening at the Review Palace in Nineveh. The dates neatly overlap with the time period in which the covenant was imposed. These deliveries likely coincided with annual preparations for war, as the yearly campaign would have likely begun in the fourth month of the year.[321] In this way, it is perhaps possible that the ceremony or ceremonies that took place in mid-Ayyaru (II) of 672 BC were organized in a manner similar to the earlier ones mentioned in Issar-šumu-ereš's letters: undertaken in a staggered fashion, and timed in no small part according to the availability of particular groups to travel to the core region.

5.2.2 Sending the covenant tablets out of the core region

As the discovery of a manuscript in the inner sanctum of a small temple in the provincial capital of Kullania has shown, the covenant tablets were not merely im-

317 See further discussion in Tushingham 2023, 46.

318 Radner 2007, 219.

319 Postgate 2007, 349. Note also Postgate's statement that 'we have to assume that province by province the individual governors were head of both civilian affairs and the military hierarchy' (2007, 334). This impression is, of course, strengthened by the language of the covenant composition itself (see Chapter 2.2).

320 Postgate 1974b, 18; SAA 13, nos. 81–123. See also a letter written during the reign of Ashurbanipal, which seems to refer to something similar (SAA 21, no. 79).

321 Postgate 1974b, 18.

posed in the core region and then forgotten. The covenant composition itself confirms this view, mandating that:

> You will guard like your god this sealed tablet of the great ruler on which is written the covenant of Ashurbanipal, the great crown prince designate, the son of Esarhaddon, king of Assyria, your lord, which is sealed with the seal of Aššur, king of the gods, and which is set up before you.[322]

Thus, it was expected that provincial governors, and if the covenant composition is to be relied upon, also client rulers, would *naṣāru* 'guard' the covenant *kî ilikunu* 'like your (pl.) god'. The provincial governor of Kullania seems to have done this, setting up the tablet for display in a small temple. It is worth noting that this findspot bears comparison with the Kalhu tablets in various ways. Like those manuscripts, it was found, for instance, close to a raised platform.[323] The temple was also part of a larger religious complex, and opened onto an expansive cobblestone plaza to its west and south.[324] Such an arrangement seems ideal for the adjuration of large groups, and the tablet could presumably have been taken out into the temple's portico and presented to a crowd.[325] It seems likely, therefore, that provincial governors were not required simply to set up their respective covenant tablets in their provincial capitals, but also to impose the covenant itself upon those members of the provincial administration, and perhaps the rest of the population, who had not taken part in a central ceremony. Indeed, it is also possible that some provincial governors did not themselves take part in the ceremony, instead sending a representative or perhaps having the covenant tablet sent to them for them to swear to in a local ceremony. That arrangements of both types took place in the case of the imposition of other covenants is well attested.[326]

Various elements of the covenant composition support this interpretation. Perhaps most tellingly, the oath to which participants were required to swear is itself included on the covenant tablet, something that comparison with other extant covenant compositions shows was not always the case.[327] The high degree of detail with which the provincial population groups expected to swear to the covenant

322 SAA 2, no. 6, § 35: 404–409, Lauinger 2012, v 68–72.

323 Harrison and Osborne 2012, esp. 137; see also Harrison 2014.

324 Harrison and Osborne 2012, 133.

325 Steymans 2013; Lauinger 2019.

326 For discussion of the former arrangement, see Radner 2006, 358 f. For an example of the latter, see SAA 21, no. 28: rev. 12–19; see also no. 75: rev. 4, which also mentions the sending of a covenant tablet, although the section is quite fragmentary.

327 Lauinger also notes this (2019, 97). See too discussion of the oath section of the composition, Chapter 3.4.

are enumerated in the composition could also be interpreted as a sign that the tablet was expected to be used outside the sphere of immediate royal oversight. The list potentially instructs those in the provinces concerning the groups of people that most needed to be adjured. In this way, the covenant composition itself acts as the means by which not only the legal and religious bond between crown and subject is transmitted, but also provides instructions regarding the manner in which its transmission was to be achieved.

As mentioned above, the client ruler manuscripts also demand that these rulers 'guard' their covenant tablet, which seems to imply that they too were expected to keep this divine object in a sacred place within the client state. Nonetheless, it is worth pointing out that the list of people to whom the covenant applies is far less extensive in the client state manuscripts than it is in the provincial version of the composition, simply including the ruler and his direct male descendants, as well as his subjects at large. This seems a relatively clear indication that the Assyrian crown did not consider itself in practice to have a similar level of control over the manner in which the covenant would be implemented in the client states, even if it did expect it to be imposed there.[328] Whether it was the case at all that client rulers were required to store the covenant tablets in local temples, however, has been the subject of debate among modern scholars. This has come about not least because the covenant tablets of several client rulers appear to have remained in Kalhu, rather than being taken to their respective capitals.

Various possible explanations have been put forward to explain the find location of the Kalhu covenant tablets. The simplest is that these client rulers, or their emissaries, never came to swear to the covenant or to pick up their tablets.[329] In this reading, the Assyrian crown intended that these tablets be taken back to the client states, but failed to impose the covenant on these rulers at all. This suggests a wider scenario in which those client states who did swear to the covenant received their tablet, with the request that they set it up, whereas other clients, potentially a significant percentage, neglected to take part in proceedings. It is perhaps interesting to note that the city-rulers whose tablets were found at Kalhu ruled over very small principalities in the Zagros. Therefore, if they did indeed fail to collect their covenant manuscripts, one wonders whether the Assyrian crown would have tolerated such actions from larger, more powerful client states.

Alternative explanations have also been put forward, including that of Hans Ulrich Steymans, who has argued that, in the Assyrian conception, these eastern clients would have been considered ill-equipped to guard a holy tablet on the

---

328 See also discussion in Chapter 2.
329 Fales 2012, 151; Radner 2019, 314.

grounds that they 'lived in tents and had no temple buildings'.[330] In this view, the status of the eastern city-rulers as mountain dwellers means that they form an exception among the client states, with those rulers who were accepted as more culturally similar to the Assyrians being given covenant tablets to display in temples. It is worth noting that the Zagros city-lords certainly did not live in tents,[331] which can probably be considered to undermine this theory. Finally, Jacob Lauinger has suggested that the practice of not giving client rulers their covenant tablets was applied universally, with the manuscripts kept within the provincial extent of the empire, to be imposed when subjects of the client state came to deliver their tribute.[332]

While there is not sufficient evidence to rule out any of these suggestions entirely, it seems to me that the argument that all client rulers were supposed to set up the tablet locally but some failed to do so relies on fewer unsubstantiated assumptions than the others. This particular debate is often framed around the question of the extent to which the inhabitants of the client state of Judah would have been exposed to the covenant composition, and thus whether it is likely that the remarkable parallels between the succession covenant and various passages in the Book of Deuteronomy are due to direct influence. Whether or not a tablet was set up in the temple at Jerusalem, however, these three possibilities all present a situation in which the Assyrian crown clearly intended that the covenant be imposed on all the client states. Beyond this and very crucially for our specific context, however, it does not seem that those planning the implementation of the covenant expected to have a similar level of control over the process, or penetration into the local population, as was envisioned for the provinces.

5.3 The covenant ceremony as an iterative event

The text of the covenant does not only suggest that it was designed to facilitate its initial imposition at a distance from the empire's central administration. Rather, as Jacob Lauinger has argued, there are various indications that it was designed to be imposed not once but many times, in an iterative fashion. To support his thesis, Lauinger has contended that the language of one of the covenant's stipulations suggests that Assyria's subjects are expected to swear to it in perpetuity:

---

330 Steymans 2013, 9. On Assyrian perception of and rule in the Zagros region, see Lanfranchi 2003.

331 Potts 2014 and Balatti 2017.

332 Lauinger 2019.

> You shall not look at Ashurbanipal, the great crown prince designate, or his brothers without reverence or submission. If someone does not protect him, you shall fight them as if fighting for yourselves. You shall bring frightful terror into their hearts, saying: 'Your father wrote (this) in the covenant, he established (the covenant), and he makes us swear (it)'.[333]

In contrast to the other verbs in the stipulation, this final statement, *utammanâši* 'he makes us swear' is in the present tense. Lauinger views this as evidence that those bound by the oath were required to swear to it repeatedly.[334] To this observation I would like to add that the form of the same verb used in the introduction to the divine adjuration section, *titammâ*,[335] can be interpreted in a similar vein. In this case, it is not the tense that is at issue, but rather the stem. The verb is an imperative in the Gtn stem. Kazuko Watanabe has interpreted this as a case in which this stem is used in its distributive meaning, 'swear each individually'.[336] While this is one possible interpretation, it is not the only one, as the Gtn stem more commonly has iterative, rather than distributive, force.[337] I consider these the more probable connotations of the verbal form here, and thus propose the translation 'swear, all of you, again and again!'. These two instances of the verb *tamû* 'to swear' both imply, then, that those entering the covenant were expected to swear to it more than once.

Further details of the stipulations support this view, in particular the requirement that those bound by the covenant teach its oath to their children and grandchildren: 'You shall teach it (i.e. the oath) to your [sons] to be born after this covenant'.[338] The duty to teach the oath to one's progeny, if it is taken seriously, can be taken to assume either that those bound by the covenant will have memorized the oath, or that they will have access to a covenant tablet in the future. As the latter option would presumably make the former more likely, these interpretations need not be taken as mutually exclusive. In addition, the possibility that the oath was designed for use beyond the covenant's immediate imposition in 672 BC would go some way towards explaining the discrepancies between its own broader focus and the more specific focus of the covenant's stipulations.[339]

What might this have looked like in practice? That the covenant tablets were indeed integrated into religious life, at least within the provincial extent of the em-

---

333 SAA 2, no. 6, § 30: 353–359; Lauinger 2012, v 9–15.

334 Lauinger 2019, 96.

335 SAA 2, no. 6, § 3: 25; Lauinger 2012, i 29.

336 Watanabe 1987, 178.

337 Gtn stems can express repeated, habitual, or continuous action: Huehnergard 2011, 411 f., see also von Soden 1995, 139.

338 SAA 2, no. 6, § 34: 387 f.; see also Lauinger 2012, v 52–54.

339 Chapter 3.4.

pire, is clear from the find context of the Kullania covenant manuscript. Jacob Lauinger, who remains the only modern scholar to have presented a detailed vision of how this might have worked, interprets the small temple at Kullania as another New Year Festival house of Nabû and thus suggests that the covenant tablet was incorporated into an annual New Year ceremony (*akītu*) held in Kullania, Kalhu and across the provincial extent of the empire. He suggests that local communities and representatives from client states alike would have gathered together at various Nabû shrines across the provinces for the New Year ceremony to pay tribute and reaffirm their oaths to the king.[340]

I accept Lauinger's broad thesis that the covenant was designed to be imposed repeatedly and that this was likely achieved in part by integrating its use into the religious fabric of provincial life. The suggestion that those emissaries who delivered annual tribute to Assyria would have also been required to swear to the covenant is an interesting one and, as discussed above, the timing of the covenant ceremonies in 672 BC is suggestive in this regard, even if there is no direct evidence to substantiate it. Nevertheless, I consider it probable that Lauinger's specific model of repeated imposition as part of a New Year Festival celebration associated with Nabû seeks too strongly to homogenize the religious and economic landscape of the Assyrian provinces.[341] On the economic side, it is necessary to stress that, while client rulers in the west sometimes delivered their tribute to stations in the provinces, and thus the provinces sometimes played a role in the collection of tribute, inhabitants of the provinces themselves were not required to deliver tribute. Rather, the inhabitants of a province were required to pay tax.[342] It is certainly true that the Assyrian taxation system evolved from the practice of extracting tribute from vassals, and thus the two are certainly to be regarded as similar in some respects,[343] however, there are a few differences. Taxes were levied on agricultural and livestock production, but also on travel and commercial activities.[344] While Jacob Lauinger notes that the *akītu*-ceremony was an amalgam of rites, some of which he argues marked the harvest,[345] the date of the New Year Festival of Nabû in the heartland does not neatly accord with the harvest in Northern Mes-

---

340 Lauinger 2019, 95.

341 That Esarhaddon promoted Nabû widely and associated him with his succession plans, is clear (Rubin 2021, 367–432). Nonetheless, evidence for a programme of the sort that Lauinger describes is still lacking.

342 On this system, see Radner 2007; see also Postgate 1974b and 1992.

343 Radner 2007, 226 f.

344 Radner 2007; Postgate 1974b. On taxation on agricultural production, see also Postgate 1989, 149 f.

345 Lauinger 2019, 94.

opotamia. While the celebration is held in early Ayyaru (April-May), the harvest would not have taken place until significantly later.[346] As such, while it seems clear that the New Year Festival of Nabû was associated to some degree with agricultural activity,[347] it does not seem likely that it coincided with the delivery of tax on agricultural yield, which would have been paid in kind as a percentage of the total yield.[348] It is also necessary to note that some individuals and groups were granted special tax status by the king, such as the people of the city of Ashur.[349] If entry into the covenant was contingent on tax payment, these groups would potentially not be required to participate.

Beyond this, it was the task of the provincial administration to collect the tax of that province.[350] Lauinger's theory has provincial officials and client rulers alike gathering together to deliver tribute at temples of Nabû 'whether custom built or repurposed' across the empire's provincial extent.[351] As stated above, it is correct that Esarhaddon's reign marked a period of increased recognition of Nabû, and that the god was associated with the well-being of the crown prince. Nonetheless, for Lauinger's theory to be persuasive, one would either have to accept that there were temples to Nabû – complete with New Year Festival complexes – in each of the approximately seventy provinces of the Assyrian Empire, or that some inhabitants of the empire were required to pay their tax in a different province to the one in which they lived. This latter option is not consistent with the available evidence on tax collection in the Assyrian Empire, which indicates that this was a key responsibility of each provincial administration. While the former possibility seems more plausible, it perhaps goes too far in assuming the imposition of religious practices from the core region on the farther-flung provinces. Most importantly, there is no solid evidence to suggest that the temple at Kullania was associated with Nabû, let alone that similar temples dedicated to Nabû were set up across the empire's provincial extent.

346 Grain loans were often made between the twelfth and second months of the year (March-May), when people were running low shortly before the harvest (CTN 6, 42). So too, several legal documents dating from the first to third months record employing harvesters (e.g. Günbati et al. 2020, nos. 11, 12, 13, 26 and 33), indicating that the harvest was not yet finished by this point. Note that the Assyrian calendar did not correspond perfectly to the solar year, and intercalary months were used several times during Esarhaddon's reign. As such, a particular Assyrian month does not always equate reliably to a specific Julian month.

347 The letter from Nabû-šumu-iddina to Ashurbanipal mentioned above (SAA 13, no. 78) describes Nabû visiting the palace threshing floor and a garden (obv. 15–17).

348 Postgate 1979, 205. On the details of this process, see also Postgate 1974b, 196 f.

349 Radner 2007, 223: RINAP 4, no. 57: iii 3–15. This inscription of Esarhaddon dates to ca. 679 BC.

350 Radner 2003a, 889.

351 Lauinger 2019, 95.

While it is perhaps narratively less satisfying than the detailed model put forth by Lauinger, I consider it more likely that the repeated enactment of the covenant took place, like its initial imposition, in a far more *ad hoc*, locally specific, manner. That the covenant was not intended to be entered and concluded in front of Nabû alone is abundantly clear from the several passages in the covenant composition in which the local gods of those entering a covenant are explicitly mentioned. In fact, this is clear evidence that it was considered important that a subject be held to account not only by the gods of the Assyrian heartland, but also by their own gods.[352] It seems therefore distinctly unlikely that the people of the provinces and the client states would universally have been required to take part in a covenant ceremony centred so entirely around the god Nabû. Instead, I contend that the covenant would have been re-imposed before local gods, and probably according to the local calendar. While there is no reason that this might not have been combined with other state tasks that would bring the community together, such as tax collection or local festivals, as well perhaps as other activities, such as building works, the local muster or perhaps even legal activities,[353] there is no need to postulate an identical practice applied universally across the empire's entire extent.

Although the crown probably considered it neither possible nor desirable to standardize entirely the covenant's repeated local imposition, it could reasonably have expected to be able to monitor it to some degree. The well-established Assyrian state communication network would have facilitated reporting on such arrangements within the provincial extent of the empire. In the client states, meanwhile, the Assyrian ambassador, *qēpu* 'trusted one', was tasked with the maintenance of diplomatic relations and reporting back to the Assyrian monarch.[354] In this way, the Assyrian crown may have anticipated some degree of oversight when it came to the continued relationship between the covenant and the local population. Regardless of the precise manner in which the subjects of Assyria were expected to engage with it, the general expectation that the covenant tablet be set up locally could itself be construed as a way of measuring the loyalty of these groups to the crown.

352 See discussion in Chapter 2.3.2.
353 On connections between the covenant and legal practice, see also discussion in Chapter 7.3.
354 On these officials, see Dubovský 2012.

5.4 Conclusions

The covenant ceremonies were held both in the central region and throughout the provinces, and possibly the client states. These events would have been exercises in the scaling of vigilance, producing situations in which subjects would have been highly aware of divine and state surveillance, and would have participated in the monitoring of their peers and of themselves. The ritual acts of internalization associated with taking the oath would have been particularly important in this regard. They served to ensure that any disloyal action or attitude would be easily detectable, at least from the Assyrian perspective, as well as seeking to guarantee that participants in the covenant would not forget their duty of vigilance when no longer participating in the ceremony.

While the practical implementation of Esarhaddon's succession covenant cannot be reconstructed with complete confidence, it is abundantly clear that the central administration made a substantial effort to bind Assyria's subjects to its terms. The early months of 672 BC were taken up with planning for and enacting Ashurbanipal's elevation to the status of crown prince. The covenant ceremonies held in the empire's central region took place over a protracted period and were timed around important cultic festivals. Nabû and probably also Bel were considered particularly important in this, and as such at least some of the ceremonies likely took place at the Nabû temple in Kalhu. Nonetheless, while the planners of the covenant's imposition clearly took ritual and hemerological considerations into account when deciding on its timing, these were not their only priorities. So too, they considered logistical factors, such as the availability of particular social and administrative groups to attend at particular times. In the case of those travelling to participate in the covenant ceremonies from farther afield, they may have sought to combine these events with other reasons to travel to the central region, such as the gathering and distribution of horses and military supplies ahead of the campaign season. Perhaps partly in order to prepare for these events, the monarch ordered renovations of the Review Palaces at Nineveh and Kalhu. While the notion that there was a large ceremony in mid-Ayyaru is probably correct to some extent, the overall implementation of the covenant in the Assyrian heartland is to be viewed as having taken place in a protracted and piecemeal fashion.

This statement also holds true beyond the core region. The covenant tablets of each province would either have been transported back to the provincial capital by members of the local administration or by delegates of Esarhaddon tasked with moving these divine objects around the realm. It seems probable that here the relevant members of the provincial administration, civil and military, would have been adjured to the covenant in local ceremonies. According to the covenant's stipulations, the tablets would then have been incorporated into religious life in these capitals, although I tentatively reject Jacob Lauinger's argument that they would have been used universally as Tablets of Destinies in New Year ceremonies of Nabû across the provinces. Rather, I propose a more *ad hoc* system in which they were associated in various ways, possibly with building works, military activity, local festivals, and legal decision-making: in short, with community acts that were sponsored or overseen by the state. Nonetheless, it is clear that the dissemination of covenant tablets throughout the provinces meant that they themselves functioned as crucial nodes in the state-wide communication network that transmitted the very duty of vigilance.

The extent to which a comparable system was enacted in the client states is unclear, and the find location of the Kalhu manuscripts appears to defy such a conclusion. However one reconstructs the covenant's imposition on the client states, the cultural penetration of the covenant would likely have been significantly shallower in the client states than the provincial system, although it seems probable that those who had direct contact with the Assyrian authorities would have been aware of their new duty to the crown. What is clear is that the imposition of the covenant appears to have been prioritized in the empire's provincial zone, and this is consistent with the conclusions of Chapter 4 regarding the ideological preparation undertaken on behalf of the succession arrangements. Furthermore, if the find location of the tablets of the Zagros city-lords can be interpreted as evidence that the Assyrian crown did not ensure that all client rulers actually attended covenant ceremonies, then it is possible to postulate that practical implementation was prioritized not only in the distinction between province and client state, but also within these two categories. Such a finding would accord with the conclusion of Chapter 4, which suggested that *Esarhaddon's Apology* and *Esarhaddon's Letter to the God Aššur* would only have been available to small groups in specific locations within the provincial system.

Part 2: **Responses to the call to vigilance**

Chapter 6: Responses to covenant at Esarhaddon's court

The Assyrian royal court can be considered the place in the empire where both social and physical distance from the king was shortest. The English term 'court' does not have an exact Akkadian equivalent, but the Assyrian monarch had a palace household and, within this, an entourage who would have advised him and attended to his needs.[1] The monarch's residence in Nineveh, or at times in Kalhu, can be seen as the primary location of this group. Nonetheless, these individuals – like the king himself – would have been highly mobile.[2] Some of Esarhaddon's learned advisors, who form the main focus of this chapter, would have accompanied him on his travels, going on campaign with the Assyrian army, as well as travelling independently to carry out tasks either on behalf of the monarch or in a private capacity.[3]

The royal advisors in Esarhaddon's entourage were some of the most powerful people in the Assyrian Empire. So too, several of them would have been aware of Esarhaddon's tactical and ideological deliberations when it came to drafting and implementing his succession covenant. Indeed, some of them, such as Issar-šumu-ereš, were likely more intimately involved in the minutiae of the practical organization of such things than Esarhaddon himself would have been. After the covenant's implementation, the royal advisors would also have been involved in reflecting on how well it had succeeded in its aims. In the period of upheaval following the discovery of the conspiracy against Esarhaddon that was put down in 670 BC,[4] these individuals would also have been privy to the monarch's own feelings on the subject and to his reaction to the situation. In this way, the position of Esarhaddon's advisors was one of a group of insiders, whose perspective was not the same as that of the crown itself but was probably based in large part on an intimate knowledge of the king's thought processes and mental state. So too, the surviving documents that reference covenants that these people produced in the final years of Esarhaddon's reign may well have been composed at the behest of the monarch, or at least with his encouragement or knowledge. Nonetheless, it

1 Gross 2020, 7 f. See also Barjamovic 2011.

2 Esarhaddon regularly resided in Nineveh and Kalhu (Radner 2003c, 168). Esarhaddon himself went on campaign, and indeed he died *en route* to Egypt with his army (RINAP 4, 6–8, see also Grayson 1975, no. 1: iv 30–33 and no. 14: 28'–30').

3 Note the degree of the various court advisors' mobility depended in part on their respective specialisms (Robson 2019, 105).

4 For further discussion, see Chapter 1.

 https://doi.org/10.1515/9783111323435-009

is possible to differentiate such texts from royally-commissioned documents, which speak entirely on behalf of the king himself. This chapter examines two letters sent by close advisors to the king, a literary composition known in modern scholarship as *The Sin of Sargon*, and a prophecy compilation. These documents are all revealing of various aspects of the reactions to the covenant's imposition among Esarhaddon's closest advisors. In particular, they shed light on the perceived interaction between the succession covenant and the plot against Esarhaddon of 670 BC.

The sources discussed in this chapter, as well as others from the Assyrian state archives, reveal a royal court that was rocked by the uncovering of a conspiracy against the king. The king's entourage was based on a system of royal patronage.[5] As such, despite their power, the monarch's advisors were at the best of times in a precarious position that was largely contingent on their good personal relationship with the monarch. As a result of this, Eleanor Robson has recently noted that membership in the king's entourage would have been 'constantly in flux'.[6] At a time when those who had lost the monarch's trust were being put to the sword, this would likely have been all the more the case.[7] This chapter attempts to show that the upheaval at the Assyrian court during this period was itself framed by several of those closest to Esarhaddon in terms of the succession covenant and its impact, and was even interpreted as proof of its efficacy.

6.1 The succession covenant in letters from royal advisors

The final years of Esarhaddon's reign, from 672 BC until his death in 669 BC, constitute a period in which the number of extant letters sent to the monarch from his close advisors increases sharply. Simo Parpola has argued that some 170 of the 247 scholarly letters that he published in SAA 10 can be dated to the years 671–669 BC. It is certainly possible that this is in large part an arbitrary accident of preservation, although the assumption that it does in some part reflect an increase in correspondence sent to the king as a result of the imposition of the covenant's duty of vigilance is also plausible. Beyond this, as many of the letters pertain to the health of the king, it can perhaps be taken as a sign that the monarch experienced a period of ill health at that time.[8] Parpola, at least, views the correspondence dat-

5 As argued in Radner 2011a.

6 Robson 2019, 79.

7 As stated in two Babylonian chronicles recounting the reign of Esarhaddon (RINAP 4, 6–8, see also Grayson 1975, no. 1: iv 29 and no. 14: 27').

8 See in particular the discussion in Radner 2003c.

ing from these years as 'exceptional', while acknowledging that many letters from both this period and others have been lost.[9]

It is perfectly possible that the high volume of letters that Esarhaddon received at this time is to be viewed to some degree as a response to the terms of the succession covenant, specifically the injunction to report to the crown. Nonetheless, it is worth noting that, at least when it came to Esarhaddon's closest advisors, only two surviving letters appear explicitly to refer to responses to or repercussions of the covenant (see Table 2). In this way, it appears that while Esarhaddon's advisors may have increased the volume of letters that they sent to Esarhaddon to some extent as a result of the covenant's imposition, they explicitly referenced it only seldom. The relative absence of any mention of the covenant could, of course, be taken as a sign that the covenant was not particularly instrumental in motivating this particular group to write to the monarch. The evidence is far from conclusive on this point, but it is certainly interesting to compare this to the situation among other social groups and in other places, particularly the provinces, for which far more references to the covenant are attested.[10] Despite their relative infrequency, however, the references to the succession covenant from the court that do survive provide insights into beliefs among the king's cohort concerning its effects.

6.1.1 The chief *asû*-healer, Urdu-Nanaya

The two extant letters to Esarhaddon from his closest advisors that refer to the succession covenant were probably written at around the same time. Simo Parpola suggests a date of early 670 BC, possibly in the month Ayyaru (II).[11] One of these missives was sent by Urdu-Nanaya, Esarhaddon's chief *asû*-healer.[12] Urdu-Nanaya had been promoted to this position only recently, in 671 BC, possibly replacing the courtier Ikkaru in this role.[13] In contrast with Esarhaddon's chief scribe, discussed in Chapter 1.3.3, and the king's *āšipu*-healer discussed below, little is known about Urdu-Nanaya's family background or his career trajectory. Despite this, it is possible to state that he was an Assyrian, and that he would have been a highly trained and likely highly experienced scholar.[14] One may wonder

9 SAA 10, xxix–xxx.
10 Compare Chapter 7.1 and 7.2.
11 As argued by Simo Parpola (1983, 121, no. 133 and 238, no. 247).
12 SAA 10, no. 316. See also Parpola 1983, no. 247.
13 SAA 10, xxvi; PNA 3/2, s.v. Urdu-Nanāia no. 2, 1411.
14 It is worth pointing out that not all members of the entourage were Assyrian. One member of the 'inner circle' as defined by Simo Parpola was Babylonian (SAA 10, xxvi). Beyond this, Anatolian

whether the timing of his elevation to the king's cohort indicates that he was brought in as a response to the unrest of the period.

In the Assyrian conception, the art of healing was divided into two distinct categories: *āšipūtu*, the lore of the *āšipu*-healer, and *asûtu*, the lore of the *asû*-healer.[15] The former category of healer is often referred to in the secondary literature as an 'exorcist', although this is such a loaded term that I will avoid it here, instead following Eleanor Robson's definition of *āšipūtu* as 'healing through reconciliation of human clients with the divine world'. The *asû*-healer, meanwhile, is often described in modern terms as a doctor or physician, and Robson characterizes this branch of healing as focused on 'the reduction of bodily discomfort through therapeutic means.'[16] These two forms of healing were closely aligned, as is illustrated in the case of Urdu-Nanaya by the fact that he and Adad-šumu-uṣur, the king's *āšipu*-healer, worked together closely at least some of the time.[17]

Table 2: References to the succession covenant in Esarhaddon's court correspondence.

Publication	Relevant extract	Sender	Sender's occupation	Date	Provenance
1. SAA 10, no. 316	Aššur and the great gods bound and handed over to the king these criminals who plotted against (the king's) goodness and who, having concluded the covenant of the king together with his servants before Aššur and the great gods, broke the covenant. The goodness of the king caught them up.	Urdu-Nanaya	Chief *asû*-healer	Ayyaru (II), 670 BC	Esarhaddon's Court
2. SAA 10, no. 199	Is it not said in the covenant as follows: 'Anyone who hears something (but) does not inform the king…'?	Adad-šumu-uṣur	The king's *āšipu*-healer	Ayyaru (II), 670 BC	Esarhaddon's Court

and Egyptian scholars also served the monarch during the final years of Esarhaddon's reign (Radner 2009).

15 Schwemer 2015, 27; Lenzi 2008, 70–71, with fn. 70.

16 Robson 2019, 52; on the role of the *asû*-healer, see also Robson 2008.

17 Robson 2019, 108.

In total, fourteen letters sent from Urdu-Nanaya to Esarhaddon are currently known to modern scholars.[18] While this is a significant number, the relative infrequency with which high-ranking *asû*-healers wrote to the king in comparison with other key advisors, along with some references in Urdu-Nanaya's letters, have led Robson to suggest that *asû*-healers tended to see the king for in-person consultations rather than communicating by letter.[19] Urdu-Nanaya's correspondence indicates that he was generally at the monarch's disposal, and thus in close proximity to him. As such, he was likely privy to the day-to-day mental and physical status of the king to a high extent, even by the standards of Esarhaddon's most trusted advisors. Nonetheless, it was not always possible for even Urdu-Nanaya to simply enter the king's chambers without requesting an audience with the monarch first.[20] As what we know of Urdu-Nanaya's quotidian tasks comes to us from letters that he wrote to Esarhaddon, it is unsurprising that his duty to minister to the king is the aspect of his work about which we know the most: he refers in his letters to meetings with the king, but also sends him medicines and instructions concerning their administration by royal attendants. However, he also repeatedly alludes to treating other members of the royal family: most frequently the king's sons, as well as a royal baby of unspecified sex, and Esarhaddon's mother, Naqi'a.[21] This suggests that Urdu-Nanaya would not only have had a personal relationship with the sitting monarch, he would also have interacted with the members of the royal family more broadly.

The one reference in his correspondence to what Urdu-Nanaya calls the *adê ša šarri* 'covenant of the king' is found in a letter dating to a period of illness on the part of Esarhaddon. This illness clearly also coincides with the discovery of the conspiracy against Esarhaddon, which in Urdu-Nanaya's telling is causing the king to lose faith in his loyal staff. Urdu-Nanaya begins the substance of the letter in this vein, recounting an incident that took place between Esarhaddon and *urdānīšu* 'his servants', in which the king chastised them for not attending to him in his illness with the same dedication shown by the attendants of his predecessors:

> The speech that the king, my lord, made to his servants about the former kings who had fallen ill: 'How did their servants sit up with them all nights and carry them on litters! How (well) did they keep watch over them!' – the king, my lord, made a speech about men, and all the

18 One letter, SAA 10, no. 327, is not explicitly from Urdu-Nanaya but is attributed to him by Simo Parpola on the basis of the handwriting (Parpola 1983, 258, nos. 265(+)266(+)267).

19 Robson 2019, 108.

20 Robson 2019, 108; Radner 2010b.

21 See Robson 2008, 473 on the gender of the patients of royal *asûs* and *āšipus*.

> vigil[ant servant]s who have remembered their orders are dead of throbbing heart because of this speech of the king.[22]

Urdu-Nanaya is presumably to be included as one of the *unzarhi hardūte* 'vigilant servants'[23] of the king who figuratively *ina tirik libbi mētu* 'are dead of throbbing heart' due to the Esarhaddon's speech.[24] Urdu-Nanaya seems to attribute the monarch's distrust and displeasure with his loyal servants not to their own behaviour, but rather to the recent conspiracy, which has unsettled Esarhaddon:

> Aššur and the great gods bound and handed over to the king these criminals who plotted against (the king's) goodness and who, having concluded the covenant of the king together with his servants before Aššur and the great gods, broke the covenant. The goodness of the king caught them up. However, they made all other people hateful in the eyes of the king, smearing them like a tanner with the oil of fish. The king, my lord, is one who fears the gods. Aššur, Šamaš, Bel and Nabû, who have given you confidence, will not abandon the king and the crown prince, but will secure the rule of the king and the crown prince until far-off days.[25]

Here, Urdu-Nanaya references the *adê ša šarri* 'covenant of the king' in connection with the *parrisūte* 'criminals' who, recently, have plotted against the crown. Nonetheless, this allusion is not an instance of Urdu-Nanaya fulfilling his own promise to report on malign activity. Indeed, in Urdu-Nanaya's telling, the plot is firmly situated in the past, and the conspirators have been punished. As such, Urdu-Nanaya argues, the monarch does not need to continue in his current state of suspicion. In this sense, Urdu-Nanaya can be construed as calling, if anything, for less vigilance on Esarhaddon's part, as opposed to participating in the dynamics of vigilance himself.

Nonetheless, it is also relevant to note that Urdu-Nanaya does associate the successful quashing of the conspiracy with the 'king's covenant'. It is perhaps particularly interesting that he stresses the fact that the traitors had themselves 'concluded the covenant of the king together with his servants before Aššur and the great gods'. Although Esarhaddon's succession covenant was not the only *adê* ever imposed that sought to protect Esarhaddon's life, it seems reasonable to suppose that Urdu-Nanaya mentions this partly in order to reflect the alarming speed with which these people broke their covenant oath after swearing it in 672 BC. So too, Urdu-Nanaya's reference to *urdānīšu* 'his (i.e. the king's) servants' highlights

22 SAA 10, no. 316: obv. 7–obv. 20.

23 SAA 10, no. 316: obv. 16: *un-⸢za⸣-[ar-hi har]-⸢du⸣-te.*

24 As stated by Simo Parpola (1983, 240, no. 247).

25 SAA 10, no. 316: obv. 20–rev. 14.

another, possibly disturbing, element of the incident: the perpetrators were among Esarhaddon's subordinates and, as we know from other sources, high-up in the hierarchy of the administration. As established in the first half of this study, it was precisely such people who were targeted by the covenant. This indicates that Esarhaddon had known to suspect possible treachery from this quarter. Nonetheless, the identities of the conspirators may well have been interpreted by some as evidence of the covenant having failed, as it did not prevent those who had sworn to it shortly beforehand from acting against the crown. Urdu-Nanaya's letter gives the impression of a monarch shaken by these events, and uncertain whom to trust. Perhaps interestingly, in another letter, likely written close in time to this one,[26] the *asû*-healer defends what Esarhaddon regards as Urdu-Nanaya's failure to properly diagnose and treat the king's illness. He also states that Esarhaddon can corroborate his medical opinion by ordering the haruspices to perform an extispicy.[27] While Urdu-Nanaya was not a casualty of Esarhaddon's purge of his officials in 670 BC, these letters seem to suggest that, in the political climate of the time and with his most important patient sick with a disease he was struggling to cure, the chief *asû*-healer was feeling the precarity of his situation.

It is probably in the context of Esarhaddon's illness, and of Urdu-Nanaya's own need to secure the trust of the king, that the *asû*-healer assures the monarch that his covenant has worked, ensuring the punishment of the traitors, and that it will be similarly successful in the future. In order to do this, Urdu-Nanaya focuses in particular on the relationship between the gods and the king, assuring him that he has divine support. The king is someone who 'fears the gods', and thus Aššur, Šamaš, Bel, and Nabû, all deities closely associated with the covenant,[28] will not 'abandon' the king or the crown prince. By mentioning the crown prince, he explicitly links divine support to Esarhaddon's succession plans and continues to assure the monarch that the gods will secure his rule and that of the crown prince *ana ṣâti ūmē* 'until far-off days'.

Interestingly, these statements on the part of Urdu-Nanaya are paralleled by both the literary composition and prophecy compilation discussed below. These texts were likely composed at around this time and explore the relationship between the gods and the king, as mediated by covenant. As such, Urdu-Nanaya's assurances should likely be seen in the context of a larger discourse on these topics that was taking place in Esarhaddon's cohort during this period. That Urdu-Nanaya

---

26 SAA 10, no. 315, see also commentary in Parpola 1983, 229 f., no. 246.

27 As indeed did happen, as illustrated by extant extispicy queries, noted in Parpola 1983, 237, no. 246. The medical queries are published as SAA 4, nos. 183–199, several of which pertain to the crown prince, Ashurbanipal.

28 See discussion in Chapter 2.3.

is addressing these comments to the king indicates that Esarhaddon was aware of this discourse, and one can speculate that he may have been one of its main promulgators.

Unlike the scholarly texts discussed below, Urdu-Nanaya's comments also make a clear connection between these reflections and the plot that was put down in 670 BC, as well as Esarhaddon's illness. Urdu-Nanaya is not himself participating in the vigilance mandated by the covenant, and indeed he does not attribute the thwarting of the conspiracy against Esarhaddon to vigilance on the part of Assyria's subjects. Instead, he highlights the role of the gods in punishing those who have broken the covenant. Urdu-Nanaya's letter seems designed to some degree to curb Esarhaddon's own vigilance towards his subjects. This seems to indicate that at least one of Esarhaddon's closest advisors was concerned about the monarch's distrustful attitude, perhaps on his own account but maybe also more generally. In a way, Urdu-Nanaya here frames the covenant as a reason that the king himself does not need to be so suspicious of those around him: it guarantees the failure of plots against Esarhaddon, so his wariness is unwarranted.

6.1.2 The king's *āšipu*-healer, Adad-šumu-uṣur

The other attested reference to the succession covenant in a letter from a member of the royal entourage was sent to the king by Adad-šumu-uṣur, Esarhaddon's personal *āšipu*-healer.[29] In contrast to Urdu-Nanaya, a lot is known about Adad-šumu-uṣur's background. He belonged to the same illustrious scholarly family as the king's chief scribe, his nephew Issar-šumu-ereš.[30] He can thus be considered a firm member of the traditional Assyrian intellectual establishment, and someone who would have grown up in close proximity to royalty. Adad-šumu-uṣur's father, Nabû-zuqup-kenu, had been an influential scholar under Sargon II, and was probably his chief scribe, but appears to have lost his court position under Sennacherib. Nabû-zuqup-kenu's primary base under both monarchs seems to have been Kalhu, a location that Sennacherib largely neglected.[31] Nabû-zuqup-kenu's trajectory would likely have made Adad-šumu-uṣur highly aware of the inherent instability of the patronage relationship between scholars and monarchs from a relatively young age.[32] Nonetheless, Esarhaddon's rise to power seems to have occasioned

---

29 SAA 10, no. 199. See also Parpola 1983, no. 133.

30 See discussion and bibliography on Issar-šumu-ereš in Chapter 1.3.3.

31 Robson 2019, 76; 256. See also Frahm 1999, Šašková 2010a and May 2018.

32 On the fickle nature of royal patronage for members of this family, see Radner 2015a, 53–55 and Radner 2017b, 221–223.

the elevation not only of Adad-šumu-uṣur but of his family as a whole, the descendants of Gabbu-ilani-ereš, to the highest echelons of the court. Adad-šumu-uṣur himself appears to have had a close relationship with his royal master, and – if the surviving letters are representative – was one of his most prolific correspondents.[33]

Adad-šumu-uṣur's close family ties to Esarhaddon's chief scribe in 672 BC, in addition to his status as a senior scholar, render it likely that he would have been at the very least well aware of, and possibly actively involved in, the drafting of the covenant composition and the process of its intellectual and practical implementation. As such, he was likely intimately acquainted with the provisions of the covenant, as well, like Urdu-Nanaya, as quite probably having encountered *Esarhaddon's Apology* and the *Letter to the God Aššur*, the compositions discussed in Chapter 4. Beyond this, Adad-šumu-uṣur would, like Urdu-Nanaya, certainly have himself been sworn to the covenant. Adad-šumu-uṣur is attested as a *ṭupšarru* 'scribe', but was also the king's personal *āšipu*-healer.[34] As discussed above, an *āšipu* performed incantations and rituals designed to dispel the evil forces that made a person sick,[35] and Adad-šumu-uṣur undertook and organized the performance of such works for the king and other members of the royal family. He also advised the monarch more generally. As such, he was intimately involved in attempts to aid Esarhaddon during his illness in the year 670 BC, as well as trying to coax him out of his seclusion. It seems certain that he, like Urdu-Nanaya, would have been privy to the plot against Esarhaddon and its fallout.[36] Despite his role as close advisor to Esarhaddon and *āšipu*-healer to the royal family, Adad-šumu-uṣur apparently also had time for various other activities, such as seeing private patients and writing scholarly manuscripts for the library of the Nabû temple at Kalhu.[37]

Some forty-eight letters from Adad-šumu-uṣur to the king are known to modern scholars.[38] This likely reflects his importance within the monarch's entourage,

33 SAA 10, xxv–xxvi.

34 May 2018, 111.

35 Illness, in the Mesopotamian conception, had various possible causes, including deities, demons, ghosts and witchcraft. For an overview, see Schwemer 2015.

36 Note that Adad-šumu-uṣur is named in SAA 16, no. 60: obv. 7', a letter about the conspiracy sent from the provinces (see Chapter 7.2.1.).

37 As discussed by Eleanor Robson, who suggests that he may well have had a residence in that city (2019, 108 f.).

38 Published most recently as SAA 10, nos. 1, 3, 24, 185–232, 256, 259 and 281. SAA 16, no. 167 may also be attributed to Adad-šumu-uṣur, although the name of the sender is not preserved. Adad-šumu-uṣur also wrote astrological reports to the king. Three such documents are signed by him (SAA 8,

but is probably also a side-effect of his various duties away from court. Despite the number of surviving letters from Adad-šumu-uṣur, however, he mentions an *adê*-covenant in just one letter. The communication was likely written roughly contemporaneously to the letter sent by Urdu-Nanaya, possibly slightly after it, as Adad-šumu-uṣur refers to the recent convalescence of the king. While this could be mere coincidence, it seems possible that this is an indication that the covenant was considered particularly relevant at Esarhaddon's court during these months in particular.

Although the timing of Adad-šumu-uṣur's letter, and to some degree the professions of the two men as healers of the king, are closely aligned, the manner in which the covenant is cited in his letter and that of Urdu-Nanaya differs substantially. After Adad-šumu-uṣur's greeting to the monarch, the letter is preserved only very fragmentarily, although a broken reference to one of Esarhaddon's sons, Sîn-per'u-ukin, implies that the first part of the letter is a report on that prince's health.[39] The broken portion of the text also introduces a man who has told Adad-šumu-uṣur some significant information, as the better preserved latter portion of the letter indicates:

> He said [as] follows: 'The god told me, 'If you do not tell, you will die; and if you tell it to somebody belonging to the entourage of the king, and he does not make it known in the palace, he will die.' My mother was charged to go, (but) she did not tell (anything) in the palace. (Instead) she spoke in the presence of Bi[...] and his wife and sister. None of them told anything, and she and the others died.' Now that (the illness of) the king is being taken away, he (finally) spoke out to me, and I wrote to the king, my lord. Is it not said in the covenant as follows: 'Anyone who hears something (but) does not inform the king ...'? Let them now summon him and question him![40]

The name of the man, as well as the details of his message, are lost, but this section of the letter provides various insights into conceptions of, and responses to, Esarhaddon's succession covenant. Unlike Urdu-Nanaya, Adad-šumu-uṣur cites the covenant in the context of his own act of reporting to the king, apparently in order to explain and justify his decision to do so. In this way, Adad-šumu-uṣur frames the covenant as causing him to act in accordance with its stipulations. Despite this, Adad-šumu-uṣur's characterization of the covenant's demands differs in interesting ways from the text of the succession covenant itself. As Simo Parpola has already observed, Adad-šumu-uṣur's supposed quotation of the composition is in

nos. 160–162), and Hermann Hunger attributes another unsigned one to him as well (SAA 8, no. 163).

39 PNA 3/1, 1139–1140 s.v. Sīn-pir'u-ukīn.

40 SAA 10, no. 199: rev. 5'–rev. 22'.

fact a loose paraphrasing of the various injunctions that: 'If you hear any evil, improper, ugly word ... you shall not conceal it but come and report it to Ashurbanipal, the great crown prince designate, son of Esarhaddon, king of Assyria'.[41] Adad-šumu-uṣur's version, perhaps significantly, transfers the statement from the second person to the third person, and uses the new subject *mannu ša* 'anyone that...'. The stipulation is also cut significantly: it includes two of the four verbs used in the stipulation, *šemû* 'to hear' and *qabû* 'to speak, report'. Those omitted are *alāku* (+ventive) 'to come' and *pazāru* (D stem *puzzuru*) 'to hide, conceal'. Perhaps most tellingly, Adad-šumu-uṣur misidentifies the person to whom those bound by the covenant are required to report: rather than Ashurbanipal, it is Esarhaddon. That Adad-šumu-uṣur considers this an acceptable representation of these stipulations is striking, and perhaps indicates that he does not consider it necessary to interpret the covenant's injunctions literally.

The dynamics of reporting to the crown illustrated in the letter also differ substantially from those described in the covenant. The person in question has reported his message not to its intended recipient directly, but rather to Adad-šumu-uṣur, one of the king's advisors. Beyond this, he tells a tale of several people being made aware of the information that needed reporting, speaking about it with each other, but not telling it *ana mazzassi pāni ša šarri* 'to somebody belonging to the entourage of the king'. Given that the covenant text is so focused on male actors, it is also noteworthy that three of the five people involved in this scenario are women. Adad-šumu-uṣur's informant seems to have waited until an appropriate moment to report, doing so only now that the king is apparently recovering from his illness. The result of this failure to relay their information to the palace, however, is consistent with the characterizations of such an eventuality in both the covenant curses and Urdu-Nanaya's letter: they have died. The manner in which they have died is not specified: are these casualties of Esarhaddon's purge? Or have they perhaps died of illness, interpreted by the speaker as divinely-wrought? Otherwise, it may be worth noting the parallel between the claim that these people are *mētu* 'dead', and Urdu-Nanaya's claim that Esarhaddon's servants *ina tirik libbi mētu* 'are dead of throbbing heart' when faced with the monarch's displeasure. That the gods, in the eyes of the man reporting to Adad-šumu-uṣur, have been intimately involved in the fate of those who did not speak out is clear from his claim that 'the god told me' *ilu iqṭebia* that failure to speak out would result in his death. Thus, as in Urdu-Nanaya's letter, the gods are portrayed as punishing those who fail to act in a manner consistent with the stipulations of the covenant. The speaker, however, does not himself explicitly reference the covenant, although this does not necessarily rule

41 Parpola 1983, 121, no. 133.

out an awareness on the part of the speaker of the similarities between the demands of the divinely enforced covenant and the statement addressed to him by 'the god'. Instead, it is Adad-šumu-uṣur who makes a connection between the scenario described in the letter and the covenant's stipulations.

As in the case of Urdu-Nanaya's letter, a top advisor draws a connection between recent events and the succession covenant. Adad-šumu-uṣur, meanwhile, makes the claim that the covenant is functioning as intended less explicitly than his colleague, but his portrayal of himself as following the terms of the covenant does imply that it is achieving the desired effect. So, too, does the claim that those who fail to report are dying. This is despite the various ways in which the act of reporting described in the letter is actually not consistent with the covenant's precise stipulations.

6.2 Scholarly musings on covenant and vigilance

Not all of the people involved in the process of deciding on Esarhaddon's succession arrangements and the subsequent composition and implementation of the covenant need necessarily have been based at court. Nonetheless, Esarhaddon's court was likely the nexus for such activities, as well as the drafting of royally-commissioned compositions more generally. The scribes involved in this process would have been accustomed, when composing or copying the first-person royal inscriptions, to speaking with the voice of the monarch, stressing his successes and glossing over his failures. Logistically speaking, it seems likely that Issar-šumu-ereš and perhaps some of his colleagues were the power behind the throne when it came to determining the wording of the succession covenant and the manner in which the covenant would be promoted in the royal inscriptions at this time. Esarhaddon would surely have had ultimate authority over and responsibility for such matters, but he would not have composed the texts himself.

Nevertheless, having themselves generated the official royal narrative around the covenant would not have precluded the scribes at Esarhaddon's court from having to swear to its oath. It seems probable that these scribes continued to reflect on the covenant, its duty of vigilance and its implications for the crown and for them. Such musings are hard to locate in the historical record, not least because it is generally difficult to distinguish between the sensibilities of a scribe and those of the king. This section makes the case, however, that two courtly texts, the literary composition *The Sin of Sargon* and a prophecy compilation, provide possible evidence of some aspects of the intellectual discussions that immediately followed the imposition of the succession covenant.

6.2.1 Vigilance in *The Sin of Sargon*

The literary composition given the modern title *The Sin of Sargon* is known from just one exemplar, found at Nineveh.[42] The lack of other manuscripts implies that it was not a widely copied text, and may suggest that it was not intended for circulation outside the milieu in which it was first composed. This milieu was likely that of the Assyrian court, not least because the text has clear affinities with the contents and style of Esarhaddon's royal inscriptions.[43] It is in part on the basis of these similarities that Simo Parpola has suggested that the text was first composed in 671 or 670, thus situating it temporally one or two years after the imposition of the succession covenant.[44] While this dating is by no means certain, it seems to me quite likely, for reasons discussed further below, that the text was indeed written in the aftermath of the plot to kill the king that culminated in Esarhaddon putting many of these officials to the sword. It is highly probable that the composer was an individual or part of a group with ready access to Assyrian royal inscriptions, and possibly even someone involved in their composition. So too, the content of *The Sin of Sargon* seems to relate to various issues that would have been familiar primarily to those with a relatively high degree of access to the royal family. Indeed, it seems possible that it was written for King Esarhaddon, possibly as part of a wider conversation that was taking place around his religious policy, his relationship to the gods and the resistance to his rule.

Regrettably, the manuscript of *The Sin of Sargon* is only fragmentarily preserved. While Simo Parpola has provided possible restorations for many sections of the text, these suggestions should be incorporated into a historical analysis such as this one only with caution. The first preserved section of *The Sin of Sargon* begins with the statement: '[I am Sennach]erib, the [circumspect] kin[g ...]'.[45] The section continues from the perspective of Sennacherib, who reflects on the deeds of the gods and also the fate of his father, Sargon II, who was killed on campaign and whose body was never recovered and thus never given the appropriate funerary rites of Assyrian tradition.[46] By means of extispicy, Sennacherib attempts to ascer-

---

42 See editions by Alasdair Livingstone in SAA 3, no. 33 and Tadmor, Landsberger, and Parpola 1989.

43 Tadmor, Landsberger, and Parpola 1989, 35–37, see also 45.

44 Tadmor, Landsberger, and Parpola 1989, 47. Note that Tadmor considers it more likely to have been composed at the end of Sennacherib's reign or at the beginning of Esarhaddon's (1989, 31). Note that Ann Weaver (2004) and Jennifer Finn (2017, 110) also date the composition to the late reign of Esarhaddon.

45 SAA 3, no. 33: obv. 1': [^{md}30—PAB]—⸢MEŠ—SU *ma*⸣-*al*-⸢*ku*⸣ [*pit-qu-du x x x x x x x x*].

46 See RINAP 3/1, 1, fn. 1 for discussion and literature.

tain the *hīṭu ša Šarru-ukīn*[47] 'sin of Sargon' that caused the gods to turn against him, in order to [*pūtī*] *u pagri itti ili lušēṣi* 'let me save myself with the help of the god'.[48] He gathers the haruspices together, and seems to divide them into groups in order to get multiple independent answers to his query. The question that Sennacherib asks the gods via the haruspices is only partially preserved:

> Saying: 'Was it on account of [...] the gods of [...]. [...] on account of the gods of Babylonia [...] and did not [...] the covenant of the king of the gods, [...] was killed and was not b[uried] in his house?' [...].[49]

Simo Parpola argues that this section should be restored as asking whether Sargon II had '[honoured] the gods o[f Assyria too much]'.[50] Nevertheless, as Eckart Frahm has contended, this would not make historical sense: it therefore seems most likely that the question is whether or not Sargon had wrongly neglected the gods of Assyria in favour of those of Babylonia.[51] The reference to the *adê šar ilāni* 'covenant of the king of the gods', a phrase not found elsewhere, is regrettably only partially preserved. The modern scholarly consensus seems to be that the missing verb is probably *naṣāru* 'to keep, observe', thus Sennacherib is asking if Sargon failed to '[keep] the covenant of the king of the gods'.

The next passage is largely broken, but refers to further extispicy and to the statue of the god Marduk. Sennacherib then invokes the deities Aššur, Mullissu, Sîn and Šamaš, and gives advice in the second person, possibly to Esarhaddon. This advice seems also to refer to the dangers of heeding counsel that is not backed up by extispicy. The terms *puzzuru* 'to conceal' and *šušmû* 'to inform', a verb derived from *šemû* 'to hear' are used. Sennacherib also cites his father, possibly as an example of the consequences of failing to follow this advice. There follows another break of about seven lines. The narrative resumes with Sennacherib still offering advice, recommending that the haruspices be divided into groups.

The final surviving passage of the composition describes Sennacherib's own downfall:

> As for me, after I had made the statue of Aššur my lord, Assyrian scribes wrongfully prevented me from working [on the statue of Marduk(?)] and did not let me make [the statue of Mar-

47 See discussion of Sargon's name in RINAP 2, 19–21.

48 SAA 3, no. 33: obv. 10' and 12'–13'. Note that this expression is difficult to interpret and is only otherwise attested in the *Cuthean Legend of Naram-Sîn* (Tadmor, Landsberger, and Parpola 1989, 42 f; Westenholz 1997, 273 f).

49 SAA 3, no. 33: obv. 17'–20'.

50 Tadmor, Landsberger, and Parpola 1989, 11.

51 Frahm 1997, 228.

> duk, the great lord(?)], and (thus) [shortened(?) my li]fe. [......]. (However), the grand scheme of mine which from times immemorial none of my r[oyal predecessors] had brought into realization, I have (now) communicated to you; [......] Accept what I have explained to you, and reconcile [the gods of Babylonia(?)] with your gods! Aššur, the king of the god[s], has victoriously marched [from sunrise to sunset]; the gods of Heaven and [Earth will prolong] your reign; the shaft of Šamaš and [Adad ......].[52]

The next five lines are highly fragmentary, and about three final lines are broken away entirely. Thus, this composition explores the failings of Sargon and Sennacherib through the lens of their relationships with the gods of Babylonia and Assyria. The importance of the cult images of both Marduk and Aššur are emphasized in this, as is the necessity of specialist scholars who determine the will of the gods. Haruspices are central to the narrative in this regard, and the *ṭupṣarrī Aššurāya* 'Assyrian scribes' mentioned by Sennacherib were apparently instrumental in changing his own relationship with the gods for the worse.

These reflections would have been particularly pertinent to Esarhaddon in the years immediately following the imposition of the succession covenant. It was around this time that Esarhaddon had the Babylonian statue of Marduk refurbished (or perhaps newly made), after its removal from Babylon by Sennacherib.[53] So too, as discussed above, this period saw what was evidently a widespread plot against Esarhaddon, with conspirators in Ashur and Harran, but also in Babylonia, followed by its brutal suppression. Such an incident would surely have added to pre-existing questions concerning Esarhaddon's relationship with the gods.[54] Although the conspiracy may have prompted fears that Esarhaddon could meet a similarly gruesome end to his predecessors, the fact that he had been able to uncover the plot may have led to attempts to ascertain why the gods had spared him and not his father or grandfather. It is also worth noting that, if the surviving evidence is anything to go by, the royal correspondence network, and in particular letters sent by scholars, played a key role in ensuring that Esarhaddon heard of the scheme against him.[55] In addition, there survive many extispicy queries from this time that reveal that this form of divination was used extensively at the time, to determine whether or not a rebellion against the monarch would occur, but also to decide whether particular individuals should be appointed to po-

52 SAA 3, no. 33, rev. 21–26.

53 Tadmor, Landsberger, and Parpola 1989, 50 f.

54 As Karen Radner has argued, Esarhaddon's health problems may have been interpreted as a sign of divine disfavour (2003c, 173).

55 See also Chapter 7.1 and 7.2, as well as Chapter 8.1 and Chapter 8.2.

sitions of trust.[56] It is in this climate of suspicion and upheaval that we can imagine one or more of Esarhaddon's scribes writing *The Sin of Sargon.*

The Sin of Sargon is remarkable in that it seems to reflect openly on divine wrath directed towards the previous two Assyrian monarchs. As the narrative itself makes clear, the aim of this exercise is to identify the mistakes of these previous kings so that Esarhaddon might learn from them. In this way, the author's examination of this issue has a clear supra-individual aim: the preservation of the head of the Assyrian state. Such a measure is probably unsurprising given the violent and untimely deaths of both Sargon II, who died while on campaign, and Sennacherib, who was murdered by one or more of his sons. The composition, then, could be conceived of as an instance of vigilance on its own terms. What makes it particularly interesting when considering the duty of vigilance as recently enacted upon the writer in the form of the succession covenant, however, is the mention of the *adê šar ilāni* 'covenant of the king of the gods'. If Parpola's reconstruction of the missing verb in this sentence is correct, it seems that Sargon's death in an enemy country and lack of burial are attributed in part to his failure *naṣāru* 'to keep, guard' a covenant.[57]

Simo Parpola himself posits that this reference pertains specifically to a bilateral treaty concluded between Assyria and the Babylonian king, Marduk-aplu-iddina.[58] This assumption relies in no small part, however, on the supposition that Sargon is here being criticized for neglecting the god Marduk, when the charge is almost certainly that he neglected Aššur, the 'king of the gods' according to rev. 27'. That Assyrian monarchs themselves were required to keep oaths or covenants sworn by the gods was not a new concept, and, in Sargon's *Letter to the God Aššur,* that monarch describes himself as a *nāṣir samni Enlil Marduk* 'keeper of the oath of Enlil (and) Marduk'.[59] The rare word *samnu* 'oath' here functions to set the oath that Sargon adheres to apart from the oaths and covenants that various client rulers break according to this composition, as well as perhaps serving as a pun: the narrative relays Sargon's eighth campaign, and *samnu* used as an adjective meant 'eighth'.[60] Also interesting is the accusation levied against the king of Urartu in this text: that he 'had not honoured the oath of Aššur, king of the gods'.[61] At the very least, the scribes responsible for composing Esarhaddon's own *Letter to the God*

56 See SAA 4, nos. 149–173 and nos. 139–148.

57 Tadmor, Landsberger, and Parpola 1989, 10 f.: obv. 19'. The reconstruction in SAA 3, 33 is consistent with this.

58 Tadmor, Landsberger, and Parpola 1989, 49.

59 RINAP 2, no. 65: 112.

60 CAD S, 120 s.v. *samnu* adj.

61 RINAP 2, no. 65: 148.

Aššur would have been familiar with this reference, as Esarhaddon's own composition draws on that of Sargon.[62] In Esarhaddon's *Letter to the God Aššur*, meanwhile, he states that he returned the Urartian fugitives that he captured in Šubria to the king of Urartu [*aš*]*šu adê naṣārimma* '[in] order to keep the covenant'.[63]

The suggestion in *The Sin of Sargon* that Sargon himself may have broken a covenant sworn by Aššur inverts the narrative presented in his *Letter to the God Aššur*, and in the royal inscriptions more broadly. Instead of the client kings, hapless and devious by turns, disregarding their oaths and breaking their covenants, it is Sargon himself who is suggested to have done so. In this way, the author of the composition perhaps experimented with the idea that Assyrian monarchs, too, could face divine punishment for the transgression of a covenant. The elevation of the importance of covenant in and around 672 BC may have prompted such reflections.

Importantly, Sargon's *Letter to the God Aššur* also marks an apparent turning point in his religious policy, as it attributes his success not only to Aššur or Marduk, but also to Nabû. The inscription also contains the first extant description of the observation of omens, including the performance of extispicy, determining the outcome of an Assyrian campaign.[64] It seems distinctly possible that the author of *The Sin of Sargon* was inspired by this text in particular. In a similar manner to the way that *Esarhaddon's Apology* and Esarhaddon's *Letter to the God Aššur* alter the spatial dynamics of the covenant narrative in the royal inscriptions, *The Sin of Sargon* shifts the suspicion for breaking a covenant to the king and his advisors. Although none of the preserved sections of the narrative make this explicit, it seems possible that the focus placed on the correct manner in which to use haruspices is designed to imply that Sargon had not employed this technique. If this is so, then the omen scholars who lead Sargon to victory in his *Letter to the God Aššur*, ensuring that he was able to punish transgression, are presented here as the reason that he himself transgressed his covenant, thus dooming himself to divine punishment. Once again, language concerning the client states is applied to the empire's social centre. This time, however, the king himself is presented as being in danger not of direct assassination, but of incurring divine wrath on account of bad advice.

While scholars are not presented in *The Sin of Sargon* as themselves breaking covenants, they are of fundamental importance to the composition as a whole. Both types of scholar mentioned, haruspices and 'Assyrian' scribes, are the sub-

62 As discussed in Chapter 4.2.
63 RINAP 4, no. 33: rev. iii 32'.
64 Robson 2019, 68 f.

jects of suspicion and must presumably be monitored closely in the present day. Although neither group is explicitly linked to the covenant mentioned in the text, the implication seems to be that wrong information from advisors concerning the will of the gods can result in a monarch breaking a covenant. As such, the two notions are loosely associated in this composition.

In the case of the haruspices, Sennacherib's statement that they must be divided into groups seems clearly to imply that merely trusting one group of diviners to accurately relay the wishes of the gods would be dangerous. Thus, the former monarch advocates a scenario in which the king has access to the information as a whole, while the knowledge of the haruspices is limited. Such an approach would have placed the groups of diviners and the monarch in a variant of the vigilant triangle:[65] the haruspices, not knowing how the other groups might answer the king, would fear that any incorrect reply would be discovered. As a result, they would – assuming all went according to plan – have preemptively adapted their own behaviour, giving the correct response on the basis that the other groups may be doing the same.

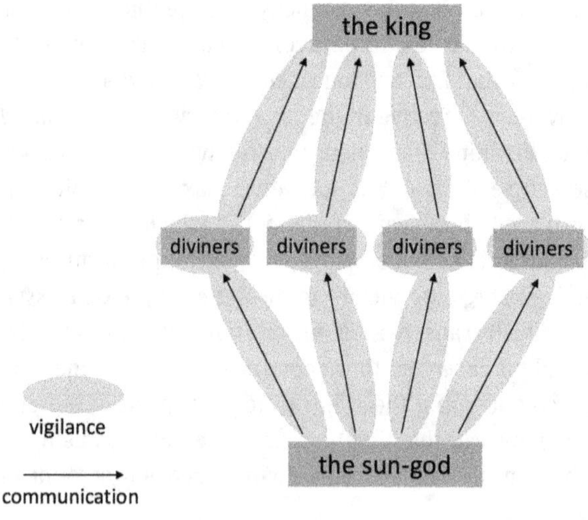

Figure 3: Schema of the vigilant diamond.

Just as in Arndt Brendecke's model of the vigilant triangle, the monarch is unable to perceive what is happening himself. The reason for this is not, as in Brendecke's

65 Brendecke 2016, 111–120. See in particular the schema of the vigilant triangle (Brendecke 2016, 115). See also the discussion of the dynamics of vigilance in Chapter 1.2.1.

schema, the physical distance between the king and the desired object of his knowledge, but rather the fact that in the Assyrian understanding, the will of the god could not be accurately perceived by non-specialists. In further contrast to the vigilant triangle, the diviners' attention was not orientated towards one another. Instead, their collective attention was directed towards the same object, the god, with an attempt to determine his will. The haruspices thus monitored the deity by dissecting and interpreting the entrails of a sheep with the supra-individual goal of ensuring that the king did not act against the will of the gods, a matter of state importance. In the absence of knowledge about the actions of their peers, they would also have directed their attention towards themselves, in anticipation of the king's vigilance. Finally, according to this new schema, the monarch would have been able to survey the answers of these different groups, comparing them and coming to conclusions concerning their veracity. Such a model does not form a vigilant triangle so much as a vigilant diamond (see Figure 3). Particularly worth mentioning here are the relative directions of communication and vigilance in this model. The direction of communication flows from the deity to the specialists, and then from the specialists to the monarch. The substance of the communication here should be the same at both points in the chain and across all of the groups of diviners: a yes or no answer to the question posed to the god by the monarch, by means of the haruspices. Vigilance, meanwhile, flows in the opposite direction: from the diviners to the god, as well as from the monarch to the diviners. Not represented in the diagram, but important to mention, is the fact that deities were presumed to have the ability to observe both the diviners and the king, and to punish them based on what they perceived.

Although it seems possible that this measure was indeed taken under Esarhaddon, there is no further evidence for the division of haruspices into separate groups at the Assyrian court.[66] As such, it seems plausible that this approach may have been suggested and explored around this time as a possible solution to the dubious monopoly on access to the will of the gods held by diviners. Whether or not it was actually put into practice, the recommendation that Esarhaddon do this, as given in *The Sin of Sargon*, is testament to the rife suspicion and distrust at the royal court at the time. That haruspices were indeed operating at the court itself is clear.[67] If one supposes that *The Sin of Sargon* was written during the aftermath of the conspiracy against Esarhaddon, and that its author was not himself a haruspex, it is worth noting that the questions posed to haruspices often concerned the trustworthiness of the king's own officials. Indeed, Ivan

66 Robson 2019, 105; see also Robson 2011.

67 Robson 2019, 105; see also Robson 2011.

Starr has noted that 'by far the largest single group of queries is concerned with matters of internal security, notably the loyalty of various classes of officials and people, as well as individual appointees to office.' He dates this group largely to the years 671 and 670 BC.[68] As such, the author's decision to use the voice of the dead Sennacherib to urge Esarhaddon to be cautious about verdicts given by haruspices may not have been entirely disinterested: while the measure is presented as benefitting the monarch, it would also have benefitted any courtiers afraid that the repercussions against conspirators were getting out of hand.

The author of *The Sin of Sargon* was likely not a haruspex, but he surely would have been an 'Assyrian scribe', the other suspicious group of scholars mentioned in the text. The composition itself does not present scribes as involved in any dynamics of vigilance as such; nevertheless, the very act of writing a narrative in which Sennacherib's scribes are described as preventing the monarch from acting according to divine will (if this is really how we should interpret this section) must be considered to constitute a retroactive act of scrutiny and possibly a disavowal of his professional milieu. The author, in writing the narrative, thus performatively reflects upon the role of his own group in the death of Sennacherib, perhaps explicitly blaming its members. In the absence of any information about the author's biography, it is not clear whether he is implicating himself or, since sons often went into the same profession as their fathers, members of his family in this criticism. Either way, it seems quite possible that the act of creating a literary work scrutinizing and functionally reporting on the misdeeds of members of the scribal class at court, may have been intended to present the author himself as someone who is loyally vigilant on behalf of the crown, alert to crimes both past and present.

6.2.2 The gods watch over the king in prophecy compilations

In addition to actively composing new texts, the king's scholars seem to have been involved in compiling collections of positive prophetic messages sent or told to the king from throughout his reign around 672 BC.[69] Such messages took the form of

68 SAA 4, lxiii.

69 The prophecy compilations are published as SAA 9, nos. 1–4. As Simo Parpola notes, the first compilation tablet has clear thematic connections to *Esarhaddon's Apology*, and was thus quite likely composed, like that narrative, around the nomination of Ashurbanipal and Šamaš-šumu-ukin (SAA 9, lxix; see also Nissinen 2019 for the argument that prophecies featured in SAA 9, nos. 1 and 3 were used in the composition of *Esarhaddon's Apology*). While Simo Parpola considers

direct communications from the gods, oracles, to Esarhaddon. These compilations constitute a total of four tablets and contain statements made by prophets primarily from the ancient cult centres of Arbela and Ashur and were found at Nineveh. They appear to have been drawn up by one person,[70] presumably someone sufficiently close to the king to have access to materials of such a sensitive nature. It is, of course, not certain that all prophecy compilations are based on prophecy reports sent to the monarch. Nevetheless, some seven such reports sent to Esarhaddon or Ashurbanipal have survived,[71] indicating that this practice existed at the time.

Like *The Sin of Sargon*, it seems quite probable that these prophecy compilations were drawn up at close social range to Esarhaddon. He may have specifically requested that they be written, or the scribe who wrote them up may have decided to do so because he thought it would please the king. The compilations, like *The Sin of Sargon*, explore the theme of divine support for the king. It seems clear, therefore, that this was an issue that was much on the minds of those around Esarhaddon. The prophecy compilations serve to repeat the message that the king does indeed enjoy divine support, and – as in *The Sin of Sargon* – the divine support of kings is framed at some points using the language of covenant. Once more, then, it seems that those around Esarhaddon were exploring the link between these concepts, this time by selecting messages on the subject for compilation. That the selection process was in itself an act of narrativization of Esarhaddon's rule is clear from the fact that oracles, while probably relatively common, were not typically set down for the purpose of posterity. Choosing which oracles to include in this collection of messages to Esarhaddon from the entire course of his reign would have been an ideologically charged process.

Two of the twenty-two compiled oracles concern an *adê* 'covenant', and both are inscribed on the same tablet.[72] The first oracle is initially composed in the first person from the perspective of the god Aššur, and seems to pertain to Esarhaddon's succession, as it describes circumstances similar to those set out in *Esarhaddon's Apology:* 'Now then, these traitors provoked you, had you banished, and surrounded you; but you opened your mouth (and cried): "Hear me, O Aššur"'.[73] Aššur goes on to describe the ways in which he helped Esarhaddon, stating that

SAA 9, nos. 2–3 to possibly predate no. 1, this does not seem likely, as no. 3 contains references to events that may have taken place in 670 BC (Pongratz-Leisten n.d., 21).

70 SAA 9, lv.

71 SAA 9, nos. 5–11. Note that SAA 9, no. 7 is addressed to Ashurbanipal while he was still crown prince. See also the letters SAA 16, nos. 59–61, discussed further in Chapter 7.2.1.

72 SAA 9, no. 3.

73 SAA 9, no. 3: ii 10–13. See also Nissinen 2003, 120–21, no. 86.

'I slaughtered your enemies and filled the river with their blood' and ending his monologue with the command 'Let them see it and praise me, for I am Aššur, lord of the gods!'.[74] The perspective then shifts to the third person, and the oracle ends with what are apparently two ritual descriptions:

> This is the oracle of well-being (or: peace) placed before the statue. This covenant tablet of Aššur enters the king's presence on a cushion(?). Fragrant oil is sprinkled, sacrifices are made, incense is burnt and (the tablet) is read out before the king.[75]

Whereas the first ritual description quite clearly relates to the oracle itself, the meaning of the second is more difficult to determine. It seems likely, given that the term *ṭuppi adê* clearly referred to a type of tablet that was distinct from this oracle in many ways, that 'this covenant tablet of Aššur' refers not to the oracle of well-being mentioned in the previous line but to a covenant tablet, such as the succession covenant either of Sennacherib or Esarhaddon. Thus, the scribe compiling these oracles has picked a message to the king from Aššur that links that god's support of Esarhaddon's own succession to a ritual involving the king and a covenant tablet. The relationship in the oracle between covenant, succession and divine favour functions in a similar way to *Esarhaddon's Apology*, which uses a narrative about the king's own succession to highlight the importance of covenants to legitimate succession. In contrast to the covenant composition itself, which presents the king and the gods working in a symbiotic relationship, with the former implementing the covenant and the latter enforcing it, this oracle appears to focus on the fact that a covenant sworn by Aššur itself constitutes a form of worship of the god by the king. Indeed, Esarhaddon's considerable efforts to disseminate holy covenant tablets sealed by Aššur,[76] and thus to communicate that god's greatness, could perhaps be considered a fulfilment of the Aššur oracle's demand quoted above: *lēmurū luna''idūni* 'let them see (it) and praise me'.

The next oracle listed on this tablet contains three references to covenant. This divine message to Esarhaddon comes, like a great many of the oracles in the compilations, from Ištar of Arbela. While this section of the tablet begins 'Word of Ištar of Arbela to Esarhaddon, king of Assyria', in the next line, which appears to be a statement in the first person spoken by Ištar, the goddess addresses her fellow gods: 'Come, gods, my fathers and brothers, [enter] into the coven[ant ...]'.[77] Regrettably, the rest of the line and the one or two that follow are broken. It seems pos-

74 SAA 9, no. 3: ii 22–25.
75 SAA 9, no. 3: ii 27–32.
76 Radner 2017a, 81.
77 SAA 9, no. 3: ii 33–36. See also Nissinen 2003, 121–22, no. 87.

sible, however, that this passage refers to the gathering together of the gods to act as witnesses at a covenant ceremony. If Simo Parpola's suggested restoration of the broken verb of this section as 'enter' is correct, then this divine participation is expressed using the same language that is employed for the human subjects of Esarhaddon. The oracle itself imagines participation in covenant from the point of view of the gods in much more concrete terms than any extant covenant compositions, or indeed any of Esarhaddon's royal inscriptions. When the narrative resumes, it is describing Ištar's actions and statements as part of a covenant ritual:

> On the [terra]ce [...] a slic[e of ...]. She gave them water from a cooler to drink. She filled a pitcher of one seah with water from the cooler, gave it to them and said: 'You say to yourself: 'Ištar is slight!' Then you go into your cities and your districts, eat your own bread and forget this covenant. But every time when you drink this water, you will remember me and keep this covenant which I have made on behalf of Esarhaddon.[78]

This description places Ištar at the centre of the covenant ceremony, with the final clause mirroring the statement in the covenant that Esarhaddon made it 'on behalf of Ashurbanipal'.[79] In this telling, it is the goddess herself who claims to have imposed the covenant, a curious reimagining of the professed dynamics of covenant implementation in Assyria. It is unclear whether she is addressing her fellow gods in this statement, or if she is speaking to the human subjects of Assyria. As I have already mentioned in Chapter 5, Ištar's description of the internalization of memory of the covenant by the means of the water she has given her addressees provides a fascinating insight into one, possibly idiosyncratic, model of the way in which the ritual act of drinking served to ensure iterative awareness of the covenant and its demands in those who concluded it. The inclusion of this description in the oracle collection surely indicates that the writer was himself interested in such ideas.

On the level of *Realpolitik*, meanwhile, it is plausible that the writer of these collections included oracles mentioning covenant in an attempt to reassure Esarhaddon that the covenant he had recently imposed on his people would work. While the composition of this list is not as overt an act of vigilance directed at a particular group as that of *The Sin of Sargon*, it can still be interpreted as a performative act of loyalty. The scribe here presents, rather than information about the treacherous deeds of his fellow men, reports of the concern and love of the gods for the monarch and the assurance that no plot against him will ever succeed. In these tablets, the gods themselves are presented as vigilant on behalf of the king:

78 SAA 9, no. 3: iii 2–15, note that this translation in part follows Nissinen 2003, no. 87.
79 As discussed in Chapter 2.1.

'watching' and 'listening' for signs of his distress, and acting to put down his enemies. In this way, the gods are described as behaving in a manner consistent with the mandated actions of the covenant's stipulations.

6.3 Conclusions

The evidence of responses to the imposition of Esarhaddon's succession covenant on the part of the members of his entourage is scanty, but eloquent. It appears that Esarhaddon's court in the years immediately following 672 BC was largely consumed with two, possibly related, issues: the illness of the monarch and the large-scale plot that was uncovered against him. Responses to the covenant's call to vigilance were refracted through that lens, with the relationship between the monarch and the gods, presented in the covenant itself as so simple and harmonious, being questioned.

In the two letters to Esarhaddon that mention the covenant, the royal scholars Urdu-Nanaya and Adad-šumu-uṣur both attempt to portray it as successful. The letters appear to have been written at similar times, perhaps indicating that there was a period of intense reflection on the efficacy of the covenant in the aftermath of the discovery of the conspiracy against Esarhaddon and during the prolonged phase of the monarch's illness. Urdu-Nanaya does not portray himself as a vigilant subject, and indeed does not comment on dynamics of interpersonal vigilance at all, beyond encouraging his master to be less suspicious of his staff. In Urdu-Nanaya's telling, the divine vigilance ensured by the covenant's imposition renders such distrust on the part of Esarhaddon unnecessary. It is probably relevant that Urdu-Nanaya was himself certainly under suspicion from the monarch at the time, on account of his perceived failure to treat Esarhaddon's illness successfully. In the case of Adad-šumu-uṣur's letter, meanwhile, the advisor does present himself as following the stipulations of the covenant, going so far as to paraphrase them. Interestingly, the third-party report he relays to the king presents a group of people who had failed to report relevant knowledge as dying. If this is an accurate quotation on the part of Adad-šumu-uṣur, then it indicates a more widespread belief in the deadly consequences of not passing pertinent information on to the crown as stipulated in the covenant. Nonetheless, it is Adad-šumu-uṣur himself who makes the connection with the covenant explicit, perhaps showing that he considered it advantageous to represent this incident as a response to the covenant, or felt the need to justify his report as such.

The Sin of Sargon, as well as the prophecy compilation, both indicate that this period saw a wider intellectual exploration of the role of covenant in Assyria, particularly as concerns the relationship between the monarch and the gods. While the covenant portrays the divine approval of the gods for the king as a certainty, these compositions show that, in court circles, the reality was considered far more complicated and contingent. It was considered necessary to reassure Esarhaddon that he did enjoy divine support, and to counsel him on how to keep it. Moreover, *The Sin of Sargon* questions the role of Esarhaddon's own scholars in maintaining this divine approval, an act of vigilance directed presumably towards the composer's own milieu.

Chapter 7: Responses to Esarhaddon's covenant in the provinces

The provincial system of the Assyrian Empire comprised over seventy provinces by the time of the implementation of Esarhaddon's succession covenant in 672 BC. In total, the portion of the Assyrian Empire that was under direct rule in this way equates to roughly the area of modern Spain. As illustrated in the first half of this study, the covenant composition appears to have targeted the inhabitants of the provincial extent of the empire to a greater extent than those of the client states. Nonetheless, it is worth noting that the provinces themselves, despite theoretically having the same status, were a heterogenous group of culturally, linguistically and politically diverse zones.[80] The provinces included the empire's core region, and with it cities such as Nineveh, Ashur and Kalhu, that had been under Assyrian rule since long before the advent of the empire. Approximately two-thirds of the provincial extent of the empire, meanwhile, had been annexed by King Tiglath-pileser III (r. 744–727 BC) between fifty and seventy years before Esarhaddon's reign.[81] Other provinces, such as those created when Esarhaddon annexed what had formerly been the client state of Šubria, for instance, had been integrated still more recently.[82]

The apparent interest of the Assyrian crown in imposing the covenant on the core region in particular is indicative of the difference in status of the various provinces. It stands to reason that the covenant would not only have been implemented in a manner that was highly contingent on the status of the province in question, but also that it would have been interpreted differently depending on the political, cultural and linguistic status of the location. The Assyrian state's programme of population movement, namely the forced deportation and resettlement of conquered populations, would have diluted the cultural character of a particular location somewhat, although it certainly did not do this completely.[83]

The present chapter explores the variety of discourse on covenant that developed in the provinces of Assyria after the imposition of Esarhaddon's succession covenant. Much of the evidence comes from Assyria's religious centre: the city of Ashur. Two relevant letters from the royal correspondence were sent from

80 See further discussion in Chapter 1.2.2.

81 For more on Tiglath-pileser III and his annexations, see among others: Frahm 2017b, 176–189, Dubovský 2004 and Garelli 1991.

82 The annexation of Šubria in 673 BC is discussed at length in Chapter 4.2.

83 On multilingualism in the Assyrian Empire, see recently Radner 2021. For recent discussion of Assyrian deportation practices, see Radner 2018b, Sano 2020 and Valk 2020. See also Oded 1979.

 https://doi.org/10.1515/9783111323435-010

Ashur, one of which explicitly references covenant, while the other seems implicitly to frame itself as a response to Esarhaddon's succession covenant. In addition to this, the private archives of scholarly families at Ashur provide insights into the scholarly reception of covenant beyond the king's entourage. Relevant curses of later covenants were extracted – possibly as scribal exercises – and covenant found its way into the experimental literary work, *The Underworld Vision of an Assyrian Prince.*

Beyond the Assyrian heartland, several letters, including anonymous denunciations, mention the succession covenant in the context of events taking place in Harran and Guzana, two western provinces in modern Turkey and Syria respectively. These communications cite the covenant in a range of different ways, often helping to explain the sender's decision to report on those around him.

Finally, a clause in private legal documents mentioning the *adê ša šarri* 'covenant of the king' as an agent of punishment of those who had broken a contract appears to have begun to circulate during the latter portion of Esarhaddon's reign. This clause is widely attested in the provincial extent of the empire (see Figure 4).

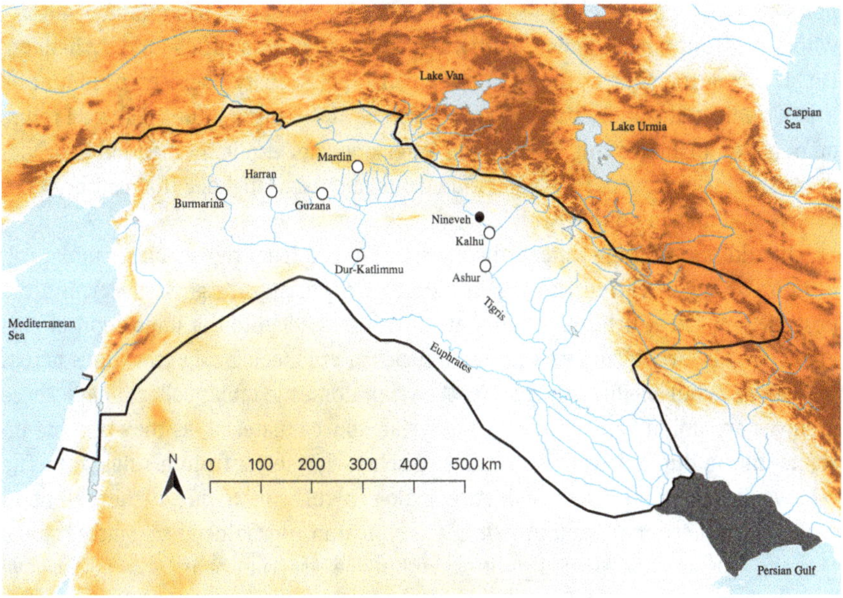

Figure 4: Map of locations of responses to Esarhaddon's succession covenant from the provincial extent of the empire: Ashur, Burmarina, Dur-Katlimmu, Guzana, Harran, Kalhu, Mardin region (Girnavaz) and Nineveh. The unknown site of Mallanate is not shown. The capital, Nineveh, is marked in black. The shaded area indicates the ancient coastline.

7.1 Responses from Ashur

Beyond the letters sent by members of the king's entourage, the only known missives sent to the Assyrian crown during the reign of Esarhaddon that pertain to the core region of the empire and include reference to an *adê* that could be the succession covenant of Esarhaddon come from the city of Ashur. As was established in the first half of this study, the succession covenant had a particular bearing on Ashur in various ways. Firstly, the covenant tablets bore the seals of the god Aššur, the head of the Assyrian pantheon, whose importance is also stressed in the text. As such, the covenant manuscripts had both an ideological and physical connection to the empire's religious centre. Beyond this, however, the *Letter to the God Aššur*, which would have been read publicly in the city, sent a clear message to the local population that it was not possible to escape punishment for disloyalty to the crown.

The first missive discussed here is sent by an individual whose name has broken off the tablet, but can probably be identified as a priest at the Aššur temple.[84] He uses a reference to the covenant to assure the monarch of his loyalty, and reports on a priest of the god Ea-šarru. The second letter comes from Nabû-ušallim,[85] whose profession is unknown, but who may have been an official of Esarhaddon. He accuses a member of the local administration, the city overseer, of mounting a plot against the king, that, in the words of the letter's editor, Eckart Frahm, amounts to 'high treason'.[86] While the letter does not explicitly mention Esarhaddon's succession covenant, it refers to events and uses language that can be interpreted as doing so obliquely.

In addition to these letters, the source material from the scholarly tablet collections found in various homes in Ashur also provide insights concerning the possible ways in which the inhabitants of Ashur reacted to the imposition of covenant. In the early twentieth century, a series of archaeological excavations across the city of Ashur yielded several Neo-Assyrian domestic tablet collections.[87] These archives are unique among the Neo-Assyrian source material, as they provide insights into multiple private collections of scholarly tablets from a single city. The relative lack of other such evidence makes it difficult to judge the extent to which the collections found at Ashur are typical of a provincial scholarly milieu. It seems certain that Ashur's position within the Assyrian heartland would have had an impact on the intellectual pursuits of its inhabitants; similarly, Ashur's sta-

84 Published as SAA 13, no. 45.
85 YBC 11382, published in Frahm 2010.
86 Frahm 2010, passim.
87 Catalogued in Pedersén 1986.

tus as an erstwhile capital city and the seat of the god Aššur likely afforded it a particularly strong – or at least unusual – cultural tradition. Despite this, the private libraries found at Ashur are a rare opportunity to investigate responses to the royal covenant in a scholarly community outside of Nineveh, the empire's capital.

Six domestic collections of scholarly tablets are known from the city of Ashur, and two of them contain tablets that can be viewed as relevant to the question of the manner in which people responded to the imposition of covenant in 672 BC. These family archives are here referred to by the designations given to them by Olof Pedersén: N4 and N6. Both tablet collections belonged to professional, literate families active in the seventh century until the fall of Ashur in 614 BC. The two tablets differ from each other significantly in terms of their genre, however, and the archives also contrast with each other substantially. The first tablet is a short excerpt of an apparent covenant of King Sîn-šarru-iškun (r. ca. 627–612 BC) found at the N4 archive, the largest of the private libraries found at Ashur with some 800 scholarly tablets. The second tablet is the only known copy of *The Underworld Vision of an Assyrian Prince*, a literary composition found at the N6 archive, an otherwise small and fairly unremarkable archive with some fifteen scholarly tablets.

Table 3: References to the succession covenant in the royal correspondence from Ashur.

Publication	Relevant extract	Sender	Date	Provenance
1. SAA 13, no. 45	The king's word is now fixed in my mouth, and I am a keeper of the king's covenant.	[...]	Post-672 BC. Likely reign of Esarhaddon	Ashur
2. YBC 11382[88]		Nabû-ušallim	ca. 671 BC	Ashur

7.1.1 A priest of the Aššur temple

The beginning of the letter published as SAA 13, no. 45 is not preserved, and thus the identity of the letter's sender is not known. The contents of the remainder of the letter, however, primarily concern the affairs of the Aššur temple at Ashur, which probably indicates that the writer was a priest of that deity. As his precise identity is not known, it is not possible to be sure where in the temple hierarchy he was placed, although his role as a priest of Assyria's imperial deity would surely

88 Frahm 2010.

have conveyed him considerable status. Events at the Aššur temple would, of course, have been considered of substantial relevance to the crown, and letters to the Assyrian monarch are known from multiple of its priests.[89] It is not possible to know whether or not the current correspondent was a priest who wrote to the monarch frequently.

The precise historical context of this letter is obscure. The lack of information concerning the sender is compounded by the problem that, although he mentions two other people by name, Binunî and Nergal-belu-uṣur, they are not obviously attested in any other sources.[90] As such, they cannot be used to date the document, which may stem from either the reign of Esarhaddon or to that of Ashurbanipal. As such, it is not possible to assert unequivocally that the covenant that the priest references in the letter is Esarhaddon's succession covenant, even if this does seem quite likely. Nonetheless, the missive does supply information about the priest's understanding of covenant, and the duties that it brought with it.

In his letter, the Aššur priest appeals to the covenant by claiming that *bēl adê ša šarri anāku* 'I am a keeper of the king's covenant'.[91] The statement is immediately preceded by the claim that:

> I confirmed the king's word, and gave (what was due) to the king. Now then, Nergal-belu-uṣur, the chief cook, can report on me. The king's word is now fixed in my mouth.[92]

Here, the priest assures the king that he has carried out the orders given to him in the form of the *abat šarri* 'king's word', namely a direct command from the monarch.[93] The demand seems to have involved giving something to the king. It seems unlikely that this is an allusion to the covenant stipulations themselves, but rather it is likely to refer to some other, more specific order. Indeed, it appears to be quoted in the first and second preserved lines of the letter, which are not entirely complete, but probably state that something must be taken from the Aššur temple.[94] The Aššur priest goes on to claim that the chief cook, Nergal-belu-uṣur, can confirm his claims, as he is *bēl ṭēmēya* 'the keeper of my report'. The priest therefore seems

---

89 The Aššur priest for whom the most letters are known is Akkullanu (PNA 1/1, 95–96 s.v. Akkullānu).

90 PNA 1/2, 345 s.v. Binūnî; PNA 2/1, 943 s.v. Nergal-bēlu-uṣur.

91 SAA 13, no. 45: obv. 7'–8'.

92 SAA 13, no. 45: obv. 2'–7'.

93 On the institution of the 'king's word', see among others Postgate 1974a, Postgate 1980, Garelli 1989 and Radner 2003a, 887.

94 SAA 13, no. 45: obv. 1'–2'.

to invoke his status as one bound by covenant in connection with his continued obedience to what are presumably more recent orders sent by the king. Here, then, the priest frames the bond of the covenant as a reason that he remains generally loyal and obedient to the king, and perhaps an indication that the statements that he is making are truthful.[95] The priest even recommends that the monarch confirm this by questioning a third party. In this way, he could be interpreted as fulfilling the covenant's demands for wholehearted loyalty and total transparency,[96] as well as generally supporting its demands for vigilance. As far as it is possible to glean from the letter, however, he does not seem to be doing this in the context of a rebellion or a plot against the monarch, but in a more quotidian situation. In the priest's conception, therefore, obedience to the monarch appears to be synonymous with the act of keeping the king's covenant.

It is probably significant that the priest's reference to covenant is directly followed by a report on the behaviour of someone else, specifically a priest of the god Ea-šarru named Binunî. The sender quotes an apparently threatening statement made by Binunî regarding the chariots of the Aššur temple, and urges the king to investigate the matter.[97] The letter ends with the statement that – in contrast to Binunî, who puts his trust in gold – *annuku ina muhhi šarri bēlīya takkulāka* 'I trust in the king, my lord'.[98] As such, while the sender does not explicitly frame his report on Binunî as a direct consequence of his duty to the covenant, his description of himself as someone who is bound by the covenant and who acknowledges the necessity that others report on him to confirm his loyalty, may well be intended to support his report on Binunî.

7.1.2 Nabû-ušallim reports from Ashur

The letter YBC 11382,[99] sent by a certain Nabû-ušallim to Esarhaddon, probably in 671 BC, does not explicitly reference Esarhaddon's covenant of 672 BC, but the author seems to draw on the wording and concepts of the covenant composition so strongly that it renders this letter relevant for discussion in this section. Nabû-ušallim appears to have been in the habit of reporting to the king, and thus he may

95 This interpretation accords with that of Baker and Gross 2015, 78, who cite this letter as an example of 'dissemination of the concept of the *adê*' evident in the royal correspondence 'where adherence to the *adê* is cited as a means of professing loyalty.'

96 Discussed in Chapter 3.

97 SAA 13, no. 45: obv. 9'–rev. 3.

98 SAA 13, no. 45: rev. 9–10.

99 Published in Frahm 2010.

have been a royal official of some kind, but his status is unknown.[100] He writes his letter from Ashur, reporting on an apparent plot in the upper echelons of that city. In his letter, Nabû-ušallim names the main intriguers as Abdâ, the city overseer of Ashur, as well as the conspirator Sasî.[101] Particularly interestingly for the present purposes, Nabû-ušallim informs Esarhaddon that Abdâ has sworn 120 elite soldiers from Ashur to a covenant of their own.[102] This letter's find location is unknown. It is probable that is comes from Nineveh, but it may have been found at Ashur. The latter contingency would presumably mean that the letter was never sent.[103]

The activities upon which Nabû-ušallim is reporting very clearly go against the terms of Esarhaddon's succession covenant, which explicitly forbids the implementation of covenants by individuals or groups beyond the Assyrian crown. It is perhaps interesting both that the plotters themselves have convened a covenant ceremony as a means of advancing their activities against the monarch, and that Nabû-ušallim stresses this fact in his report. It is worth considering that the former action could itself be construed as a reaction to the succession covenant, namely as a repudiation of its stipulations. Nabû-ušallim's focus on this element of the plot, meanwhile, may seek to highlight the gravity of their actions by implicitly accusing them of breaking the covenant. The brief period that had elapsed between these events and the implementation of the succession covenant, as well as the high status and central location of those involved, renders it very likely that both Nabû-ušallim and those whom he accuses would have been aware of its stipulations.

Nabû-ušallim implicitly references his faithful adherence to the covenant as a reason that in Ashur *nišē ma'dūte izērrūni ina muhhi duākīya idabbubu* 'many people hate me and talk about killing me'.[104] He does this by stating that the desire to kill him comes from *ina muhhi mīni ša ammarūni ašammûni ana šarri bēlīya aqabbûni* 'that which I see, hear, and say to the king, my lord'.[105] The framing of this statement mirrors that of the covenant, as well as other letters discussed below that explicitly reference the covenant, and thus make it seem probable that Nabû-ušallim intended the formulation to portray him as a loyal subject who adheres to the covenant and who is in danger because of it. If this formulation is intended as a reference to the covenant, however, it is worth noting that in addition to referencing *šemû* 'hearing' and *qabû* 'speaking', two verbs used frequently in the

100 See the discussion in Frahm 2010, 111–114.
101 YBC 11382: obv. 13–18.
102 YBC 11382: rev. 7.
103 Frahm 2010, 90.
104 YBC 11382: obv. 7–8.
105 YBC 11382: obv. 5–6.

covenant composition, Nabû-ušallim also refers to *amāru* 'seeing', a term not used in that text in the context of the duty to report.

The letter was published by Eckart Frahm, who also notes that the possible statement in the lines 24–25 seems to bear some resemblance to the general idea conveyed in the curses of the covenant: 'Šini will hear you and will pour (liquid) lead(?) into your mouth'.[106] Frahm draws a parallel in particular to the covenant curse 'may tar and pitch be your food'.[107] Nevertheless, it is notable that the term used for 'lead' in the letter, *abāru*, does not appear in the curses of any known covenant composition. As such, one must either conclude that this is a reference to something else, perhaps more specific to the situation, or that, if is it intended to evoke the covenant, the sender has misremembered the curse, and perhaps that he is not sufficiently acquainted with the curses of the covenant to quote them from memory. This may also imply that he does not have easy access to a covenant tablet.

7.1.3 A covenant excerpt in the house of the exorcists (N4)

The scholarly tablets that constitute the N4 archive were found in a single room in a large courtyard house, located approximately 350m south of the Aššur temple. The tablet collection contains a wide array of scholarly compositions, mostly incantations, rituals and medical texts.[108] This is unsurprising considering the family profession: the owners of the N4 archive were *āšipu*s, the same profession as that held by Adad-šumu-uṣur. The contents of the library make clear that the family was professionally linked to the Aššur temple, with both administrative and religious duties there,[109] while they also likely served private clients, either at their home or that of their patient.[110] So too, the library was used to store the works of apprentices,[111] and it is probable that the family was engaged both in training their

106 YBC 11382: obv. 24–b.e. 25. Note that Frahm's interpretation of this passage is tentative (Frahm 2010, 102 f).

107 SAA 2, no. 6, § 56: 490; Lauinger 2012, vi 73.

108 Robson 2019, 130–132.

109 Maul 2010, 201.

110 That the *āšipu* went to the home of private patients in order to treat them is clear from textual evidence (Scurlock 1999, 79, esp. fn. 63). The *ašipus* of N4 probably also performed rituals to treat patients at their homes: a room in the house was found with red plastered walls and deposits of apotropaic figurines and other objects buried at ritually significant points (Pedersén 1986, 41–76, N4; Robson 2019, 130).

111 As demonstrated by Maul (2010, 200).

own progeny and in giving a general scribal education to the sons of some other local families.

In addition to ritual and medical texts, the library also housed some hymns and prayers. In her recent analysis of the archive, Eleanor Robson connects these documents, as well as royal rituals to be performed in the temple, to the family's position at the Aššur temple. Nonetheless, not all of the compositions have immediately obvious practical applications in the temple, and can thus perhaps be taken as indicative of a wider scholarly interest in the concerns of the Assyrian crown.[112]

In this grey zone, one finds the only tablet bearing a portion of a covenant that can be attributed with certainty to the archive, or indeed to any Neo-Assyrian private archive. It is, of course, perfectly possible that some of the various other covenant fragments found at Ashur may have come from these archives, but as the findspots of these tablets are unknown it is impossible to make conclusions on the subject.[113] The tablet found at the N4 archive contains a short excerpt of a covenant of King Sîn-šarru-iškun:[114]

> If you (pl.) should sin against this covenant of Sîn-šarru-iškun, king of Assyria, your lord, his sons, (or) his grandsons, may Nergal, the perfect lord, (pour out) your (pl.) blood into ditches and ravines.[115]

The protasis references the generalized hypothetical of breaking the covenant in any way, using the verb *haṭû* 'to do wrong, sin' in the second person plural. Supposing that this is indeed an excerpt from a covenant – now lost – that Sîn-šarru-iškun imposed on his subjects, these may indeed be consecutive lines of that composition.[116] It is also possible, however, that the scribe lifted the protasis and apodosis from different sections of the original composition, or even that he may have composed this short protasis himself, paraphrasing the many clauses of the covenant into one that he felt captured the broad sentiment of the whole. Indeed, as A. Kirk Grayson notes, protases such as these are not generally directly followed

---

112 Note that the *Exorcist's Manual*, a list of works that an *āšipu* theoretically ought to master as part of his training, contains mainly rituals and incantations, pharmacological texts and omen compendia (Schwemer 2011, 421 f.).

113 Frahm 2009a, nos. 66–71.

114 SAA 2, no. 12. See also Grayson 1987, 154 and Grayson 1980, 124 f.

115 SAA 2, no. 12: obv. 1–7.

116 Compare SAA 2, no. 6, § 71: 555–559; Lauinger 2012, vii 48–53. One exemplar, ND 4349U, includes a brief protasis even though the curse that follows is not at the beginning of the curse section.

by curses in other covenant compositions,[117] and the syntax of the clause is also rather clumsy,[118] rendering it perhaps unlikely that these lines were to be found in exactly this form and relative position in an original covenant composition.

Whether the protasis was copied from an original, paraphrased or maybe even invented by the scribe, its lack of specificity suggests that the apodosis, a curse of the god Nergal, lord of the underworld,[119] is the primary focus of the section. Perhaps surprisingly, the curse excerpted here is not attested in any other extant covenant.[120] If this curse is indeed taken from such a composition, its unusual imagery may have increased the degree of scholarly interest in the clause, rendering it worth writing down. It is not impossible, of course, that this curse formulation was an original composition of the scholar drafting the tablet.

Whatever the relationship between this tablet and any original covenant upon which it was based, it is clear that the scribe who drew it up was concerned in particular with the concrete consequences of breaking the covenant, as opposed to the specific circumstances under which the covenant might be broken. That just one curse is excerpted – the reverse of the tablet is blank, illustrating that the scribe chose to include only one – and that the god inflicting the curse is Nergal is significant. An *āšipu* often performed incantations and rituals intended to dispel the evil forces that made a person sick. Illness, in the Mesopotamian conception, had various possible causes, including deities, demons, ghosts and witchcraft.[121] Nergal and other underworld gods often play a key role in incantations designed to neutralize these forces, as demons and ghosts needed to be sent back to the underworld, while these gods also had the power to imprison witches and thus break an evil spell.[122] One may conjecture, therefore, that the decision to write down this curse stemmed, at least in part, from a professional interest in Nergal. If this is the case, then the excerpt tablet points to a personalized form of engagement with a royal covenant, with the scribe concentrating on a deity that was particularly relevant to him. This form of engagement with a covenant perhaps suggests a degree of vigilance directed towards the self.

---

117 Compare also the wording of the only other known treaty of Sîn-šarru-iškun (SAA 2, no. 11), where the clause that comes directly before the curses differs substantially from that found here.

118 Grayson 1987, 155.

119 On Nergal, see RlA 9, 215–223 s.v. Nergal A. Philologisch.

120 The Nergal curses found in other covenants do not resemble this one, although they do resemble one another, referring to Nergal as the 'strongest among the gods' and referencing plague and petulance (SAA 2, no. 4: rev. 26'; SAA 2, no. 6, § 49: 455 and SAA 2, no. 9: rev. 18'). In SAA 2, nos. 5 and 6, a curse of the goddess Gula references blood, but this is where the similarity ends.

121 For an overview, see Schwemer 2015.

122 Schwemer 2011, 430–433.

The circumstances under which this tablet was written are the subject of debate. Grayson describes it as 'a schoolboy's exercise in writing a treaty clause', stating that is 'does not say much, and is inept in style and grammar'.[123] Simo Parpola and Kazuko Watanabe, meanwhile, argue against this characterization on the grounds that scribal exercises are 'typically clumsily shaped and round or ovoid' in format. They therefore consider the tablet to be in keeping with 'the standard format of excerpt tablets'.[124] Robson has since pointed out that excerpt tablets were the typical format of first-millennium scribal exercises and thus supports Grayson's assessment.[125] Indeed, the tablet's horizontal orientation, 2:1 ratio and small size are all typical of the Neo-Assyrian scribal exercises found at Ashur. If it is an exercise tablet, however, it is the only one known from Ashur that bears a single excerpt, or a covenant clause. Nevertheless, the corpus of scribal exercises from Ashur is very small,[126] and thus it seems possible that covenant clauses were indeed excerpted or composed more generally as part of a scribal education. It is regrettable that the other covenant fragments found at Ashur do not have a recorded find context, as it would certainly be illuminating to know if they, too, belonged to private collections.[127]

That an extract from a covenant of a later monarch was kept in the library of a family of high-ranking *āšipu*-healers can, of course, not be interpreted as a response to the Esarhaddon's covenant of 672 BC. Nonetheless, it is interesting that the owners of the N4 tablet collection considered it relevant to study covenant compositions, and apparently did so with a particular interest in Nergal, a god present in their daily lives. Furthermore, they may have considered it appropriate for those receiving their scribal education to excerpt portions of a royal covenant. This provides insights into the manner in which a scholarly family living in Ashur may have engaged with the imposition of a covenant that demanded loyalty to the Assyrian crown. It seems possible that responses to Esarhaddon's succession covenant may have been similar, although this remains unknown. The tablet also illustrates that later Assyrian monarchs continued to impose covenants on the people of Ashur, and that the owners of the N4 collection engaged with them.

---

123 Grayson 1987, 155.

124 SAA 2, 1.

125 Robson 2019, 146, fn. 169. Robson cites Gesche (2001, 174–184) as evidence of this.

126 The Neo-Assyrian scribal exercises from Ashur are published in Maul and Manasterska 2023.

127 In particular the 'short version' of the succession covenant of Sennacherib, see Frahm 2009a, no. 69 as well as SAA 2, no. 3.

7.1.4 *The Underworld Vision of an Assyrian Prince* in a house of scribes (N6)

The tablet collection of N6, found in a domestic dwelling in the southern part of the main city, is generally fairly unremarkable, containing a small group of lexical lists and another of incantations.[128] These incantation tablets have led some to regard the inhabitants of N6 as *āšipu*-healers, but the extant colophons do not support this. Only two colophons are preserved from the N6 library, written by two different individuals described as *šamallû ṣehru* 'junior apprentice', one the son of a *ṭupšarru* 'scribe' and the other the son of a *ṭupšarru aššurû* 'Assyrian scribe'. As such, it seems likely that the owners of the N6 archive were likely scribes,[129] perhaps working in the local administration. The tablet collection is certainly less extensive than the N4 archive, and it seems likely that its owners were somewhat lower in status, despite still belonging to the educated scholarly milieu of the city of Ashur. The tablet bearing the literary composition termed by modern scholars *The Underworld Vision of an Assyrian Prince* is, like *The Sin of Sargon*, only known from one copy. Its precise use and status within the N6 tablet collection is unclear, as are the circumstances of its original composition. Nonetheless, like the excerpt tablet attributed to the N4 archive, it indicates that the people who owned it were interested in covenant, and in particular its implications for them specifically.

The tablet of *The Underworld Vision of an Assyrian Prince* is badly broken. It is an unusual narrative and thus difficult to reconstruct.[130] The obverse of the tablet is particularly poorly preserved. The beginning of the text refers to actions that perhaps make most sense if they are carried out by a king or prince. He interacts with diviners, as well as governors and magnates, and is described as hoarding wealth and generally acting badly. Portions of the narrative appear to take place in Nineveh and Ashur. Probably while in Ashur, the protagonist appears to have dealings with *mār ṭupšarri* 'the son of a scribe' (i.e. a scribe).[131] About halfway through the obverse of the tablet, something happens that distresses the protagonist. There is mention of the underworld and a royal banquet. At this point, the name Kummaya is preserved on the tablet for the first time.[132] It is probable that he is the protagonist of the earlier lines, although they are so broken as to ren-

128 Pedersén 1986, 81f.; Fadhil 2012, 7f.

129 Fadhil 2012, 13–16.

130 My reconstruction largely follows that of Alasdair Livingstone in his edition of the tablet as SAA 3, no. 32. See also Benjamin Foster's more recent translation (2005, 832–839). I also thank Eckart Frahm for sharing various readings of the tablet with me.

131 SAA 3, no. 32: obv. 17.

132 SAA 3, no. 32: obv. 27.

der this uncertain. Kummaya is later described as a prince *mār rubê*, literally 'son of the ruler'.[133] Towards the end of the obverse of the tablet, Kummaya invokes Nergal and his consort, Ereškigal, and later has a vision of the underworld. The description of this forms the majority of the reverse of the tablet, which is much better preserved than the obverse. The narrative describes an array of terrifying composite beings, before describing *qarrādu Nergal ina kussî šarrūti* 'the hero Nergal seated on a regal throne'.[134] Somewhat confusingly, the narrative here switches to first person. It seems probable that the narrator is Kummaya, although this is not certain.[135] In either case, Nergal is angry and almost kills the narrator, until the god Išum intercedes on his behalf. Convinced by Išum, Nergal satisfies himself with issuing a warning, which particularly stresses the fact that those whom he addresses have not obeyed a monarch who is now dead: 'This [corpse] which (lies) buried in the underworld, is that of the proud shepherd who fulfilled the wishes of my father [Aššur], the king of the gods'.[136] As several scholars have already noted, the following description of the king is consistent with the deeds of Sennacherib.[137] Nergal accuses those whom he addresses of 'closing his ear to his (i. e. the king's) speech', and claims that 'the luminous splendour of his terrifying majesty will throw you down instantly'.[138] The narrator awakens and the narrative shifts back to the third person. Kummaya laments and 'in his pain he praised before the peoples of Assyria the mighty deeds of Nergal and Ereškigal, who had come to the aid of the prince.'[139]

Perhaps surprisingly, the narrative ends not with an explicit reference to Kummaya or his fate, but rather with three lines that mention the scribe and also introduce a covenant into the narrative:

> But also that scribe, who previously had accepted bribes, who occupied the post of his father, with the wise understanding which Ea had given him, he took the words of praise to heart and spoke to himself: 'So that the (curses of) the covenant may not come near me to do (me) evil, and may not threaten me, let me always carry out my actions as [Nergal] has ordered!' He went and repeated it to the palace, saying: 'Let this be my expiation!'[140]

---

133 SAA 3, no. 32: rev. 13.
134 SAA 3, no. 32: rev. 11.
135 It is also possible that it is the scribe mentioned at the end of the narrative.
136 SAA 3, no. 32: rev. 22.
137 This connection has been made by almost all scholars who have dealt with the text. See most recently Finkel 2021, 207.
138 SAA 3, no. 32: rev. 27.
139 SAA 3, no. 32: rev. 32.
140 SAA 3, no. 32: rev. 33–35.

These three lines reframe the entire composition as the *namburbû* 'expiation' of a previously corrupt scribe.[141] Specifically, the narrative is not told in order to prevent Nergal or the dead king from punishing the scribe, but rather in order to protect him from an *adê*-covenant. As a scribe is mentioned as being involved in dubious activities in the body of the narrative, it is tempting to equate the scribe of the last lines with this character. The scribe of the final lines is also, however, clearly framed as the composer of the text, and indeed later scribes who may have copied it would also have been placed in this role.

Modern scholarship on the *Underworld Vision* has often been interested in associating the characters of the narrative with real historical figures. It ought to be noted that some stories that dealt with events that took place under Sennacherib and Esarhaddon incorporated fictional, or semi-fictional characters. This is very likely the case of the *Story of Ahiqar*, for instance, which is discussed in Chapter 8.4, where the fictional Ahiqar may well be an amalgamation of various historical figures. As such, I do not consider it necessarily the case that Kummaya, or indeed the scribe, are historical figures. Nonetheless, it is worth discussing the various suggestions that have been made. Wolfram von Soden suggested that Kummaya be identified as Ashurbanipal, and several scholars have followed him in this.[142] There is not much in the narrative that accords with such an identification, however, and the reference to covenant in this context would be particularly confusing. More recently, other suggestions have been put forward. Irving Finkel has identified the prince as Šamaš-metu-uballiṭ, another son of Esarhaddon, whom Finkel argues was probably sickly (the prince's name means 'Šamaš has revived the dead') and may have taken part in a ritual involving the underworld.[143] Finkel also takes up von Soden's suggestion that the scribe of the narrative may himself be identifiable as Urdu-Gula, the son of the *ašipu*-healer Adad-šumu-uṣur, who is known to have fallen out of royal favour and sent various pleading letters to the king.[144] Finkel's interpretation relies on the assumption that Kummaya is a son of Esarhaddon who is not Ashurbanipal. It also perhaps overstates the impact of the expulsion of one

141 CAD N/1, 224–25 s.v. *namburbû*; AHw, 726 s.v. *namburbû*. The CAD translates the term 'ritual for warding off a portended evil, apotropaion, apotropaic ritual', while AHw translates it as 'Löseritus'.
142 von Soden 1936. Other scholars who have suggested that Kummaya may be Ashurbanipal more recently include Helge S. Kvanvig (1988, 434), Alasdair Livingstone (SAA 13, xxviii), Benjamin Foster (2005, 833; 2007, 97), Seth L. Sanders (2009, 163–165) and Jennifer Finn (2017, 104–106).
143 Finkel 2021, 208 f.
144 von Soden 1936, 11 f.; Finkel 2021, 203–206. Benjamin Foster states that the scribe 'calls to mind' Urdu-Gula, although he notes that the connection is speculative (2005, 833). Note that Wolfram von Soden's suggestion that the scribe be identified as Urdu-Gula is based in part on parallels with the *Story of Ahiqar*: see discussion in Chapter 8.4.

particular scholar from the royal entourage, conceiving of it as sufficiently important and unusual to have warranted inclusion in this literary composition. In actuality, however, it seems highly likely that this was actually a common occurrence and indeed the system of royal patronage of scholars in the later Assyrian Empire was inherently precarious.[145] As such, this identification strikes me as improbable. Finally, Eckart Frahm has recently put forward the compelling theory that Kummaya may in fact reference Urdu-Mullissu.[146] This final suggestion resolves various difficulties with the interpretation of the composition as referring to a son of Esarhaddon, such as the clear parallels between the dead king in the underworld and King Sennacherib.

The inclusion of the scribe in the composition, as well as the inclusion of the covenant itself, seems particularly compatible with the narrative around the death of Sennacherib found in the latter years of Esarhaddon's reign. *Esarhaddon's Apology*, for instance, stresses the importance of Sennacherib's succession covenant in ensuring the demise of Urdu-Mullissu's plot. So too, one is reminded by the mention of a scribe of the apparently guilty 'Assyrian scribes' cited in connection with Sennacherib in *The Sin of Sargon.*[147] As such, I consider Frahm's suggestion of a link between Urdu-Mullissu and Kummaya fundamentally convincing, while also considering it important to stress that those who composed and read the narrative need not necessarily have considered it to be wholly based on any one historical event.

The composition date of the narrative is unknown, though it is clearly a seventh–century Assyrian composition. Several scholars have suggested that it was composed under Ashurbanipal, although this is in part based on the widespread assumption that he was to be equated with Kummaya. Nonetheless, the inclusion of what have been argued to be Egyptianizing elements in the text may suggest a date after Esarhaddon's successful invasion of Egypt in 671 BC, as Egyptians were deported to the core region during this period.[148] Several scholars, including recently Seth Sanders and Jennifer Finn, meanwhile, have argued that the narrative has various parallels with royally-commissioned compositions from Ashurbani-

145 See the discussion in Chapter 6.1.
146 Eckart Frahm put this theory forward in his presentation at the 2018 *Rencontre Assyriologique Internationale* in Innsbruck. I thank him for discussing his interpretation and sharing his unpublished materials on the subject with me.
147 See the discussion in Chapter 6.2.1.
148 On Egyptianising elements, see Ataç 2004, 69, 71; Radner 2009, 226 and Loktionov 2016. Note, however, that some Egyptians were already resident in the core region much before this date, as is evident from the document StAT 2, no. 164 discussed below (see also StAT 2, xvi).

pal's reign.[149] In addition to this, following, Eckart Frahm's theory, the thematic overlap with *Esarhaddon's Apology* and *The Sin of Sargon* may perhaps suggest a composition date not before the final years of Esarhaddon's reign.[150] Nonetheless, if the narrative does indeed address the murder of Sennacherib, it is possible that it may have been written earlier than previously supposed.

In either case, the *Underworld Vision* can certainly be viewed as ruminating on the meaning of the duty to report to the monarch inherent in a succession covenant, as well as the dangerous consequences of not doing so. If it was composed around or after the imposition of Esarhaddon's succession covenant, then it can be seen as a response to this, even though it may well be framed as referring to an *adê* imposed by Sennacherib. Such a situation would be consistent with some of the evidence from the royal correspondence concerning the provinces, in which individuals reporting from the western provinces, discussed below, refer to Sennacherib's succession covenant. This may perhaps be taken to indicate that the portrayal of Sennacherib's succession covenant as having played a key role in thwarting Urdu-Mullissu, as found in the *Apology*, played a substantial role in the understanding of the function of such covenants in some parts of the provincial system. If the *Underworld Vision* was composed earlier, it may reveal that the imposition of Esarhaddon's covenant mapped onto and augmented a pre-existing discourse among certain groups.

Who were these groups and where were they located? Mehmet-Ali Ataç has posited that the *Underworld Vision* was the work of a 'Ninevite intellectual milieu',[151] but he does not explain his assumption. Although it is certainly possible that the *Underworld Vision* was the work of a member of the Assyrian court under Esarhaddon or Ashurbanipal, it seems equally possible that it was composed by scholarly elites in another city, most likely Ashur. That the only known copy of the *Underworld Vision* was found in Ashur is, of course, not necessarily evidence that it was composed there. Nonetheless, while there is reference to Nineveh in the composition, Ashur is also mentioned and several references in the text would have been as pertinent to the literate elite of Ashur as to those of Nineveh: The New Year (*akītu*) house and the dead body of the king,[152] for instance, as well as Sennacherib's succession covenant.

---

149 Sanders 2009, 164 f.; Finn 2017, 108 f. On intertextuality in the text more generally, see also Kvanvig 1981 and Bach 2018.

150 Note that Cristina Barcina has argued that the *Underworld Vision* was written under Esarhaddon (2017, 111).

151 Ataç 2004.

152 Sennacherib reintroduced the New Year (*akītu*) festival at Ashur (Pongratz-Leisten 2015, 416; on Sennacherib's religious reforms, see Frahm 1997, 20 and 282–288 and Vera Chamaza 2002, 111–

Assuming that the *Underworld Vision* was written by a member of the intellectual elite of either the Assyrian court or Ashur, the composition parallels *The Sin of Sargon* in its role as an act of vigilance directed towards oneself and one's own group. Like *The Sin of Sargon*, it appears to portray a scribe as acting poorly, this time in a way that explicitly contravenes a covenant. In contrast to *The Sin of Sargon*, however, the *Underworld Vision* may be interpreted as absolving the scribe: by obeying the stipulations of the covenant and reporting what he has heard, he exonerates himself. If the composition should indeed be viewed in the context of a wider claim that the scribes of Sennacherib were responsible for his downfall, then the *Underworld Vision* accepts this view, while also offering the prospect of forgiveness, something that contrasts sharply with the narrative of total revenge found in *Esarhaddon's Apology*.[153] So too, the *Underworld Vision* may present a more complex vision of the ability to act wrongly and then observe the stipulations of the covenant than the succession covenants of either Esarhaddon or Sennacherib, in which any treacherous activity will be punished directly.[154]

Even if the owners of the N6 tablet collection were not themselves the original composers of the *Underworld Vision*, although this is not excluded, it is telling that they possessed such a composition. As the *Underworld Vision* is known in only one copy, like *The Sin of Sargon*, it seems reasonable to suppose that this was not a widely circulated work of literature. Instead, it may well have appealed to a small social set, for whom its content seemed relevant, and indeed its apotropaic qualities may have felt necessary. As in the case of the excerpt of Sîn-šarru-iškun's covenant, the god Nergal is central to the narrative, in a way that he certainly was not in the succession covenant compositions of Esarhaddon or indeed Sennacherib.[155] The family who owned the N6 archive were probably not professional *ašipu*-healers, but they did own various works of *ašipūtu* 'lore of the *ašipu*-healer' and thus clearly had some expertise on the subject.[156] While it could be coincidence, the fact that both the owners of the N4 archive and N6 possessed texts that reference covenant and portray Nergal as the agent of punishment for contravention of its terms may indicate that these scholars viewed the wrath of Nergal as particularly relevant to them. The owners of the N6 archive are associated in a colophon with the term *ṭupšarru aššurû* 'Assyrian scribe', which is perhaps reminiscent of the term *ṭupšarrī Aššurāya* 'Assyrian scribes'

167) and Assyrian kings were buried in Ashur (Mofidi-Nasrabadi 1999, Lundström 2009 and Hauser 2012).

153 See the discussion in Chapter 4.1.

154 By the gods, as discussed in Chapter 2.3., and by Assyria's subjects, as discussed in Chapter 3.

155 Compare the discussion of the deities in Esarhaddon's succession covenant in Chapter 2.3.

156 Fadhil 2012, 15 f.

found in *The Sin of Sargon*. Their identity as scribes also certainly overlaps with that of the scribe in the *Underworld Vision*. That such people kept the *Underworld Vision* in their archive can certainly be considered an act of vigilance directed towards the self and one's milieu. The structure of the composition, which places the scribe who copies it in the role of the scribe in the narrative, thus reiterating the scribe's repetition of that which he has heard and witnessed, reinforces the importance of adherence to a covenant's stipulations to report not only in its contents but also in its form and structure.

7.2 Responses concerning the western provinces

Four letters pertaining to covenant sent to King Esarhaddon, and one that was probably sent to the crown prince, Ashurbanipal, refer to events taking place in two western settlements located in the provincial extent of the empire: Harran and Guzana, with one letter also mentioning events in Que and the land of the Cimmerians (Gimir). Harran and Guzana were both provincial centres located between the Khabur Triangle and the Euphrates, up to some 300 km from the Assyrian capital. Both of these provinces had long been under Assyrian rule by the reign of Esarhaddon. Guzana, modern Tell Halaf, had been integrated into the empire under Ashurnasirpal II (r. 883–859 BC).[157] Harran, meanwhile, had become its own province during a restructuring of the provinces in the eighth century, but the area itself had also been brought under Assyrian control by the end of Ashurnasirpal II's reign.[158] As is noted in Chapter 2, Harran was the only location outside the core region and Babylonia to be mentioned in the list of divine witnesses in the covenant composition, presumably indicating its importance both generally and for the aims of the covenant in particular.

The letters discussed in this section were all edited by Mikko Luukko and Greta van Buylaere, who characterize all of these letters as 'denunciations', which Luukko elsewhere typifies as containing 'criticism and accusation'.[159] Several of these letters appear to have been written in the context of the 671/670 conspiracy against Esarhaddon.[160] Three of the letters are sent by a named author,

157 RlA 11, 51 s.v. Provinz. C. Assyrien.

158 RlA 11, 54–55 s.v. Provinz. C. Assyrien.

159 SAA 16, xxix; Luukko 2018, 166.

160 Note that Mikko Luukko and Greta van Buylaere connect the high number of petitions and denunciations sent to Esarhaddon with the succession covenant: 'The great number of denunciations extant from the reign of Esarhaddon is rather extraordinary when compared with those sent

Nabû-rehtu-uṣur, and the remaining two are sent anonymously. Despite the formally anonymous character of the letters, however, it is worth noting that the contents indicate that they were not functionally anonymous: the senders expected the recipient, namely the king or crown prince, to know their identity.[161] As such, it seems most likely that these individuals were specifically tasked with reporting on events in these locations. In contrast, the contents of Nabû-rehtu-uṣur's letters give the impression that his reports may have been unsolicited.

Table 4: References to the succession covenant concerning the western provinces in the royal correspondence.

Publication	Relevant extract	Sender	Date	Provenance
1. SAA 16, no. 59	Nikkal [has revealed(?)] those who sinned against [your] father's goodness, and your [father's] and your own covenant.	Nabû-rehtu-uṣur	ca. 671 BC	Harran
2. SAA 16, no. 60	Those who sin against [your father's goodness, yo]ur fat[her's and] your own covenant, and who [plot against yo]ur [life...] [I am a keeper of the covenant of the king my lord]; I cannot c[onceal the words of ...].	Nabû-rehtu-uṣur	ca. 671 BC	Harran
3. SAA 16, no. 61	[Those who sin against] your father's [goodness, your father's and your own] covenant, and who [plot] against your life... [I am] a keeper of the covenant of the ki[ng, my lord; I cannot conce]al the word[s of...]	Nabû-rehtu-uṣur	ca. 671 BC	Harran
4. SAA 16, no. 71	[...... who] brought [...], says: 'I am a servant of the king; his [fat]her [made] me enter the covenant. You will h[ear w]hatever [I he]ar in Gi[mir(?)].	[Anon.]	ca. 671 BC	Harran (Que and Gimir mentioned)

to his predecessors. It probably results directly from several provisions in his succession treaty (SAA 2, no. 6), concluded in 672' (SAA 16, xxix).

161 See the discussion below.

Table 4: References to the succession covenant concerning the western provinces in the royal correspondence. *(Continued)*

Publication	Relevant extract	Sender	Date	Provenance
5. SAA 16, no. 63	They answered of one [accord]: 'We have eaten the slice(?) of our sons and daughters, and [that of] Aššur-zeru-ibni too, but we are keepers of the c[ovenant of the king], we are devoted to Esarhaddon.'	Anon.	ca. 672–669 BC	Guzana

7.2.1 Nabû-rehtu-uṣur reports on Harran

The three extant letters sent to the king by Nabû-rehtu-uṣur may stem from the city of Harran, although it is also possible that they were written when Nabû-rehtu-uṣur was elsewhere.[162] In either case, they certainly focus on activities in Harran itself, and it is therefore clear that Nabû-rehtu-uṣur had access to information on events taking place in that location.[163] Nabû-rehtu-uṣur mentions covenants in all three of his letters, which renders him the correspondent of the Assyrian crown for whom the most such communications survive. The reports all centre around the conspiracy of 671/670 BC, and Martti Nissinen suggests that the letters date between the months Ṭebetu (X) 671 BC and Nisannu (I) 670 BC.[164] Unfortunately, not much is known about the identity of Nabû-rehtu-uṣur beyond what can be gleaned from his correspondence. His letters betray a familiarity not only with Esarhaddon's succession covenant, but also with the language of prophecy, perhaps indicating that he can be linked with such practices. He also seems to have access to knowledge about high-ranking individuals in Harran but also in the monarch's entourage. Beyond these letters, a man named Nabû-rehtu-uṣur is attested in an administrative document as a member of the staff of the queen mother, either during the reign of Esarhaddon or Ashurbanipal. The name also appears in

162 Published as SAA 16, nos. 59–61.
163 The sender's references to the deities Nikkal and Nusku, both associated with the mood-god, Sîn, whose temple at Harran was an important cult centre, further corroborate this link. On Harran, see Chapter 2.3.2., as well as Gross 2014 and Novotny 2020. Esarhaddon himself visited the Sîn temple at Harran in 671 BC, as discussed in Radner 2003c, 171.
164 Nissinen 1998, 128.

some legal documents from Nineveh.[165] It is possible that one or more of these attestations refers to the author of the three letters discussed here, though this is far from certain.

Nabû-rehtu-uṣur refers to covenant in a similar way in all three of his surviving letters to Esarhaddon. All of the reports concern one of the exact situations outlined in the covenant, namely a planned insurrection. The scenario described in the letter SAA 16, no. 59 also exhibits a close resemblance to the third scenario (§ 10) of Esarhaddon's succession covenant. This scenario explicitly highlights the potential role of *raggimu* 'prophets' and *mahhû* 'ecstatics' in a conspiracy:

> If you hear any evil, improper, ugly word which is not seemly nor good to Ashurbanipal, the great crown prince designate, son of Esarhaddon, king of Assyria, your lord ... from the mouth of a prophet, an ecstatic, an inquirer of oracles, or from the mouth of any human being at all, you shall not conceal it but come and report it to Ashurbanipal, the great crown prince designate, son of Esarhaddon, king of Assyria.[166]

According to Nabû-rehtu-uṣur, meanwhile, a slave woman is *sarhat* 'enraptured(?)'[167] and has given the following prophecy: 'It is the word of Nusku: The kingship is for Sasî. I will destroy the name and seed of Sennacherib!'[168] It seems likely that Nabû-rehtu-uṣur himself would have been aware of this parallel. It is perhaps interesting, however, that Nabû-rehtu-uṣur does not use the same vocabulary as the covenant to describe the prophetess of his letter. As such, he is evidently not quoting from the covenant directly. It is probably also relevant to note that, in contrast to the instructions given in the covenant composition, Nabû-rehtu-uṣur does not address his letters to Ashurbanipal.

Nabû-rehtu-uṣur appears in this letter to react to the prophecy of the slave woman by referring to an oracle of the goddess of Nikkal, the consort of the moon-god Sîn:

---

165 SAA 7, no. 9: obv. ii 11'. See also Nissinen 1998, 108, fn. 423 and PNA 2/2, 861–862 s.v. Nabû-rēhtu-uṣur.

166 SAA 2, no. 6, § 10: 108–122.

167 The meaning of *sarāhu* in the G stem is uncertain according to CAD S, 171 s.v. *sarāhu.* CDA, however, defines it as 'to go around, turn; search; tarry', and states that it can be translated as 'seek out, i.e. bewitch', in the case of magic (CDA 311, s.v. *sarāhu*). Compare also SAA 19, no. 229, obv. 13.

168 SAA 16, no. 59: rev. 4'–5'. Note that Karen Radner has argued that the reference to the 'seed of Sennacherib' may suggest that Sasî was a descendant of Sargon II (Radner 2003c, 173). If this is correct, then the situation accords with yet another portion of Esarhaddon's stipulations, namely the mention of 'any descendant of former royalty' as a potential threat, see Chapter 3.1.

> Those who have sinned against [your] father's goodness, your [father's covenant] and your covenant, Nikkal [has revealed(?) (them)]. Destroy their [peopl]e, name and seed from your palace! [May] she cast [......]! [May] the accomplices of Sasî [die quickly]![169]

Nikkal herself is not mentioned in Esarhaddon's succession covenant, but is included in SAA 2, no. 3, Sennacherib's succession treaty, and probably in SAA 2, no. 4, Esarhaddon's accession treaty.[170] It seems probable that Nikkal is included in the letters as the goddess who is sending a message to the king largely due to Nabû-rehtu-uṣur's likely location, in Harran, a cult centre of the moon god. As the covenant composition states that subjects must swear by their local gods, Nikkal is implicitly included in Esarhaddon's succession covenant.[171] It is also probably significant that she is present in covenants designed to secure the safety of Esarhaddon and his position as king. That Nabû-rehtu-uṣur is aware of other covenants, and specifically of SAA 2, no. 3, which is mentioned in *Esarhaddon's Apology*, is clear from the fact that he mentions – in all three letters – not only *adêka* 'your covenant' (i.e. that of Esarhaddon) but also, prior to this, *ṭābti ša abīka* 'your father's goodness' and *adê ša abīka* 'the covenant of your father' (i.e. that of Sennacherib). As such, Nabû-rehtu-uṣur stresses that these people are disregarding the wishes of Sennacherib, and are breaking the terms of a covenant sworn to him, something that he frames as at least as important as the fact that they are breaking Esarhaddon's own covenant.

In the lines of the letter quoted above, the verb following the goddess's name is broken away. Nevertheless, it is clear that Nabû-rehtu-uṣur frames her as intimately involved in either bringing to light the devious sins of the conspirators, or perhaps in punishing them directly. In either case, she plays an active and central role, it seems, in ensuring the success of the covenant. If the translation of the broken verb as 'has revealed' posited by Luukko and van Buylaere is correct, then the dynamics of vigilance expressed in this letter are rather interesting. Here, the goddess appears to have been vigilant, monitoring those who have entered the covenant and then communicating this both to Nabû-rehtu-uṣur himself and also to Esarhaddon through Nabû-rehtu-uṣur. This image of a deity policing the actions of those who have entered the covenants of Esarhaddon and Sennacherib is not particularly surprising given the role of the gods in the covenant itself. However, the notion of a deity reporting this to the king is not stated explicitly in the covenant composition. Thus, Nabû-rehtu-uṣur frames himself as reporting on the con-

169 SAA 16, no. 59: obv. 4–7.
170 Note that Nikkal may perhaps be interpreted here as a manifestation of Ištar, as Mullissu often is in Neo-Assyrian prophecies (Nissinen 1998, esp. 119).
171 See Chapter 2.3.2.

spirators as a result of his need to report a prophecy of Nikkal, and it is in this context that he mentions the covenant.

The imperative statement, 'destroy their [peopl]e, name and seed from your palace', is another explicit allusion to the covenant composition, which repeats 'destroy their/his/your name and seed from the land' four times in the stipulations.[172] The concept of erasing or destroying the *šumu* 'name' and *zēru* 'seed' of a person who breaks the covenant is also found in four of the curses.[173] Thus Nabû-rehtu-uṣur quotes the covenant composition here, but his quotation is not exact. The combination of 'people, name and seed' is not found in any of the covenant compositions, nor does it appear to be a reference to a different type of royally-commissioned work, such as the royal inscriptions. Nabû-rehtu-uṣur's use of the phrase *issu libbi ēkallīka* 'from your palace' may perhaps draw again on stipulations in the covenant. In particular, the command that:

> If anyone in the Palace makes an insurrection, whether by day or by night, whether on a campaign or within the land against Esarhaddon, king of Assyria, you must not listen to (i. e. obey) him.[174]

Here it appears possible that Nabû-rehtu-uṣur is using his knowledge of the covenant composition to reference multiple stipulations that he considers relevant to the circumstances, as well as elements of various curses.

The references to the covenant in the letters SAA 16, nos. 60 – 61 differ from those in SAA 16, no. 59 in various ways. The letters are near duplicates, which is itself intriguing:

> Those who sin against your father's [goodness, yo]ur fat[her's and] your own covenant, and who [plot] against your life, shall be placed in [yo]ur hands, and you shall delete their name (var. 'seed')[175] [from As]syria and from your [pa]llace. This is the word of Mullissu; the king, my lord, should not be ne[glectful] about it.[176]

The formulation 'shall be placed in your hands, and you shall delete their name/ seed' conveys a similar message to the imperative formulation in SAA 16, no. 59 – in both cases, the king is required to act upon the message conveyed to him by a goddess through Nabû-rehtu-uṣur. The use of *šumu* 'name' in SAA 16, no. 60

---

172 SAA 2, no. 6, §§ 12, 13, 22 and 26. See also Lauinger 2012, iv 1 – 2. Reference to the name and seed of those who swear to the covenant is also made in § 57, the oath section of the composition.

173 SAA 2, no. 6, §§ 45, 62, 66 and 105. See also Lauinger 2012.

174 SAA 2, no. 6, § 18: 198 – 201.

175 SAA 16, no. 61: obv. 7.

176 SAA 16, no. 60: obv. 5 – 9; SAA 16, no. 61: obv. 4 – 9.

and *zēru* 'seed' in SAA 16, no. 61 in otherwise nearly identical formulations, once again seems to reference Esarhaddon's succession covenant composition. Once more, he makes the statement *issu libbi ēkallīka* 'from your palace', and adds the phrase *issu Māt–Aššur* 'from Assyria', a statement that, while it does not quote the covenant composition exactly, does align with its use of *ina māti* 'from the land'.

In these letters, Nabû-rehtu-uṣur attributes the divine message to ^{d}NIN.LÍL, Mullissu, rather than ^{d}NIN.GAL, Nikkal. Mullissu, the consort of the god Aššur is a goddess frequently attested in the few known prophecies of the Neo-Assyrian period as a manifestation of Ištar.[177] As such, she is perhaps a more typical deity to mention in a prophecy than Nikkal. Unlike Nikkal, Mullissu is also attested several times in the covenant composition.[178]

After stating that he has had a vision, Nabû-rehtu-uṣur justifies his decision to report by citing the covenant:

> [I am] a keeper of the covenant of the ki[ng, my lord]; I cannot c[once]al the word[s of ...].[179]

Assuming that the *dibbī* 'words' to which he refers here are those of Mullissu, Nabû-rehtu-uṣur evidently considers it his duty, as one of those who is bound by the covenant (a *bēl adê ša šarri* 'keeper of the king's covenant'), to report divine statements to the king. This is not explicitly mandated in the covenant, but the statement in scenario three that 'evil, improper and ugly word' concerning Ashurbanipal, and in particular any 'from the mouth of a prophet, an ecstatic, an enquirer of oracles', is to be reported, may have served to reinforce the notion that visions and omens were to be conveyed to the king or crown prince. One may also wonder, however, whether the narrative that corrupt scholars withholding divine communications were culpable for the deaths of Sargon and Sennacherib that was circulated during Esarhaddon's reign also informed Nabû-rehtu-uṣur's framing of his report.[180]

---

177 Nissinen 1998, 119.

178 See Chapter 2.3.2.

179 SAA 16, no. 60: obv. 11–12 and no. 61: obv. 11–12.

180 See the discussion of *The Sin of Sargon* in Chapter 6.2.1.

7.2.2 An anonymous informer of the crown prince

The letter published as SAA 16, no. 71 is sent by a writer who, based on his orthography and epistolary conventions, can be identified as the sender of two other anonymous letters addressed to the crown prince, Ashurbanipal, namely SAA 16 nos. 69–70.[181] The initial lines of this letter are broken away, but it seems likely that it, too, was an anonymous missive addressed to the crown prince. The three letters all deal with similar topics, including a certain Sasî, presumably the same man who played a leading role in the conspiracy of 671/670 BC. The sender deliberately does not include his own name, nor does he explicitly state his location in the portions of the letter that survive. Nonetheless, the wording of a statement made by somebody else that he reports to the prince includes the phrase *kî ša šarru ana Harrāni išpurannāšīni* 'just as the king wrote to us in Harran',[182] which implies that he either is or had been located there. If the use of the first-person plural in this statement includes the writer, then this segment of the report also suggests that the crown was itself in active contact with the writer of these letters. Esarhaddon and Ashurbanipal were thus probably aware of his identity, signifying that his letters were not functionally anonymous, even though they did not include the sender's name.[183]

Mikko Luukko and Greta van Buylaere consider it most likely that the anonymous informer was a scholar of some kind.[184] They also note that he lays out the letters in a 'report' format. Indeed, the purpose of these letters seems to have been to gather and relay the statements of others to the crown prince. As such, it seems probable that the writer was tasked with acting as an informant on Ashurbanipal's behalf. In this way, his letters – unlike the others discussed in this chapter[185] – are consistent with the command of the covenant composition to report information to Ashurbanipal specifically.

The reference to covenant in the letter is not made by Ashurbanipal's anonymous informer himself, but rather by one of the people whose statement he is reporting. The name of this person is not preserved: he is described as someone

---

181 SAA 16, xxxv.

182 SAA 16, no. 71: obv. 6'–7'.

183 As noted in Luukko 2018, 170, esp. fn. 56, the decision to write denunciations anonymously may sometimes have been linked on a practical level with the dangerous possibility of a letter being intercepted.

184 SAA 16, xxxv.

185 See the letters written to the crown prince of Babylon, Šamaš-šumu-ukin, discussed in Chapter 8.1.1.

'[who] brought [...]'.[186] He states that 'I am a servant of the king; his [fat]her [made] me enter the covenant' and claims that he is planning to report to the anonymous informant: 'You will h[ear w]hatever [I he]ar in Gi[mir(?)]'.[187] Thus he professes loyalty to Esarhaddon, mentions what is probably Sennacherib's succession covenant, and states that he will report from Gimir, the region in which the Cimmerians live.[188] In this instance, in contrast to the letters discussed in the previous sections, then, the appeal to covenant as a sign of loyalty is not addressed to the crown itself, but to a fellow royal subject, the writer of the letter. The anonymous informant presents this as a statement that he is reporting verbatim. If this is accurate, then the statement is evidence that Assyrian subjects did not merely appeal to covenant in order to profess loyalty to the crown, but that they also addressed such statements to one another. Of course, both men were likely socially close to the Assyrian crown, in as far as this can be gleaned from the fact that they were both active as informants, and it should not necessarily be assumed that the subjects of Assyria at large would express themselves in such a way. Nonetheless, such declarations made between peers can be interpreted to some degree as evidence that the duty of vigilance was shaping the language of interactions between some individuals at least.

It is, of course, interesting that the speaker seems to reference Sennacherib's succession covenant as opposed to that of Esarhaddon. This parallels to some degree the language of Nabû-rehtu-uṣur, who refers to both covenants, stressing that the plotters are breaking both. The claim may imply that the speaker and Nabû-rehtu-uṣur both differentiated between loyalty to the king, as expressed through reference to Sennacherib's succession covenant, and loyalty to the new crown princes of Assyria and Babylonia, which was to be expressed through reference to Esarhaddon's covenant. This would also imply an understanding that the stipulations of Esarhaddon's succession covenant were designed primarily to protect the crown prince of Assyria. It may also be an indication that the stress placed on the importance of Sennacherib's succession covenant in the period approaching 672 BC was successful in raising its profile. In this letter, however, the speaker states explicitly that he himself entered Sennacherib's covenant. As it seems likely that fewer people entered that covenant than that of Esarhaddon,[189] it is also possible that this statement would in part have conveyed the speaker's status as a member of a relatively select group. In addition, of course, it highlights his longstanding loyalty and proximity to the crown for at least a decade.

186 SAA 16, no. 71: rev. 2.

187 SAA 16, no. 71: rev. 5–6.

188 On the Cimmerians during the reign of Esarhaddon, see Adalı 2017. See also Fuchs 2023.

189 See the discussion in Chapter 4.1.

7.2.3 The 'enigmatic' anonymous informer

The final letter-writer to send a missive pertaining to the western provinces also wrote anonymously, this time addressing the monarch himself.[190] On the basis of his distinctive orthography and tablet size, Mikko Luukko and Greta van Buylaere have been able to ascribe a total of seven letters to him.[191] They have also attempted to identify this scribe by comparing the orthography of his anonymous reports to that of other letters of Esarhaddon's reign. These efforts were unsuccessful, however, suggesting that the anonymous author was not someone for whom other letters have survived until the modern day, if he did in fact ever write letters to the king in his own name.[192] As in the case of the anonymous informant to the crown prince discussed above, Luukko and van Buylaere suggest that he was a scholar.[193] So too, it is clear that the identity of the enigmatic informer was known to Esarhaddon, and that the monarch wrote to him directly. The 'enigmatic' informer frequently gives Esarhaddon advice, implying that he was socially close to the king.

The letter that directly references covenant, published as SAA 16, no. 63, refers to events taking place in the western province of Guzana. Nonetheless, the informer's other letters do not give the impression that he was primarily based in Guzana, and indeed Luukko and van Buylaere suggest that he may have been a member of Esarhaddon's entourage. It seems more likely that he either wrote letter SAA 16, no. 63 while he was visiting Guzana, or perhaps on the basis of information given to him by informants of his own.

The letter is a lengthy report on several people based in Guzana and relates to *hīṭānišunu* 'their sins/crimes'.[194] The exact nature of the events in Guzana is not entirely clear from the contents of the letter, but Luukko and van Buylaere note the possibility that it pertains to the 671/670 BC conspiracy in Harran.[195] As in SAA 16, no. 71, the enigmatic informer does not himself refer to the covenant. Instead, he presents it as being referenced by people whom he quotes directly in the

---

190 The designation of the writer as 'the enigmatic anonymous informer' follows SAA 16, xxx.

191 SAA 16, xxx–xxxv.

192 Luukko and van Buylaere note that Mar-Issar is the most likely candidate that they have looked at, although the substantial differences between the orthography and style of his letters and those of the anonymous sender render it very unlikely that they are the same person (SAA 16, xxxiv–xxxv).

193 SAA 16, xxxiii–xxxiv.

194 Note that the noun used is *hīṭu*, the same that is used in *The Sin of Sargon* (see Chapter 6.2.1.).

195 SAA 16, xxxi: Luukko and van Buylaere also discuss the difficulties of interpretation for this letter here.

letter. In contrast to SAA 16, no. 71, this statement is not one that was delivered directly to the writer of the letter. Instead, he is reporting something that has been reported to him: *iqabbûninni mā* 'I am told that'.[196]

The portion of the letter that includes the reference to a covenant concerns the elders of Guzana, who are reported to have been questioned by the governor and to have responded to the question 'To whom are you [devoted]?' with the answer: 'We have eaten the slice(?) of our sons and daughters, and [that of] Aššur-zeru-ibni too, but we are keepers of the cov[enant of the king], we are devoted to Esarhaddon.'[197] Here, then, the elders of Guzana invoke the covenant when speaking to a member of the provincial administration, specifically the *pīhātu* 'governor', namely the first person mentioned in the provincial version of the succession covenant. This clearly implies that appeals to covenant as an assurance of loyalty were not only the purview of those writing to the king or crown prince, but rather had become a more common means of asserting to others one's own rectitude and loyalty to the king. The fact that this statement was addressed to the governor, who may have been responsible for implementing the covenant in Guzana itself,[198] is likely significant.

Perhaps interestingly, the elders of Guzana appear in their statement to admit that they have done something wrong, if that is indeed what is meant by 'we have eaten the *hirṣu* 'slice(?)' of our sons and daughters'.[199] Nevertheless, they insist that they each remain a *bēl adê ša šarri* 'keeper of the king's covenant',[200] a term is also used to indicate general loyalty in the letter from a priest of Aššur, SAA 13, no. 45, and in the letters of Nabû-rehtu-uṣur (SAA 16, nos. 60–61). While the exact meaning of this statement on the part of the elders of Guzana is rather unclear, this may perhaps be an instance of using the covenant to differentiate between actions which one may censure, but are not serious crimes, and much more serious transgression of the covenant stipulations. While the elders of Guzana have done the former, they are not guilty of the latter. Again, the use of a reference to covenant in an exchange that the monarch was not expected to witness implies that its imposition had an impact on interactions between high-status Assyrian subjects in the western provinces. This may imply that a dynamic akin to that of Arndt Brendecke's 'vigilant triangle' existed within this group to some degree.[201]

196 SAA 16, no. 63: b.e. 33.

197 SAA 16, no. 63: b.e. 33–34; SAA 16, no. 63: rev. 3–5.

198 See the discussion in Chapter 5.2.2.

199 CAD H, 199 s.v. *hirṣu*; AHw, 341 s.v. *he/irṣu*.

200 See also the discussion of this term in Scurlock 2012, 181.

201 Brendecke 2016, 115. See also Chapter 1.2.1.

7.3 The covenant of the king as the punisher of private legal parties

Neo-Assyrian legal documents have been found at locations across the Assyrian provincial system, particularly in the heartland and at provincial capitals in the west of the empire. These documents are some of the extant sources of the period that are closest to the daily lives of those inhabitants of the Assyrian provincial system whose lives existed at a remove from the Assyrian crown. While some forms of legal transactions, such as sales of people or land, were frequently documented using Assyrian cuneiform on clay tablets, the Assyrian legal system also relied heavily on documents written up in the Aramaic language and alphabetic script.[202]

Although they are formulaic and frequently taciturn about the particular circumstances under which they were written, these documents reveal various aspects not only of the legal and economic lives of people living under Assyrian rule, but also serve to illuminate multiple features of their lives and beliefs. Despite the access that these sources can provide to those living at a greater social remove from the monarch, it is worth noting that legal archives often belonged to wealthy individuals or members of the imperial administration. As such, these people are over-represented in these sources, as is the case for other written evidence from Assyria.[203]

There is fairly little evidence concerning how exactly the formulae and formats of these legal documents were promulgated, but it is difficult to explain the impressive uniformity of the documents throughout the provinces without some recourse to a top-down approach. Neo-Assyrian legal documentation was fairly simplistic, with three basic templates: loan, conveyance and receipt, which could be modified to accommodate a wide range of purposes. Karen Radner has pointed out that the two forms of property, real estate and humans, that were included in conveyance documents would likely have been of interest to the imperial administration for the purposes of taxation.[204] Similarly, members of the administration tasked with settling legal disputes sometimes accepted such documents as evidence.[205] Beyond this, it cannot have escaped the notice of the top rung of the

---

202 Fales 2017a, 415; Radner 2011b and Radner 2021.

203 A good example of this is the archive of Šulmu-šarri at Dur-Katlimmu, modern Tell Sheikh Hamad. He was an extremely wealthy man and would have known King Ashurbanipal personally (for the archive see Radner 2002 and Röllig 2014, for a discussion of the world of Šulmu-šarri, see Radner 2015a, 38–44 and Radner 2017b, 220 f.).

204 Radner 2011b, 397.

205 Faist 2020, 154–158.

Neo-Assyrian administration that these legal documents were an effective way to convey information to the population and, in particular, to various demographics that might otherwise have been hard to reach.

Two new legal clauses in the Assyrian language, both concerning covenant, appear in the legal documentation of the Assyrian Empire in the seventh century: *adê ša šarri ina qātēšu luba''û* 'may the covenant of the king call him to account (lit. will look into his hands)' and *adê ša šarri lū bēl dēnīšu* 'may the covenant of the king be his legal opponent'.[206] The clauses are both penalties, and were not included in all legal documents, indicating that either the interested parties or the scribes themselves could choose to include them or not. This is significant because, although one or both of them may have been innovations of the Assyrian imperial administration, the decision to add it to a particular legal document would have been the prerogative of those immediately involved in the transaction.

The earliest attested clause that involves the *adê ša šarri* 'covenant of the king' may pre-date the implementation of Esarhaddon's succession covenant. The clause, *adê ša šarri ina qātēšu luba''û* 'may the covenant of the king call him to account (lit. will look into his hands)',[207] is attested in a marriage contract found at Ashur, published as StAT 2, no. 164. The document in question needs to be collated. It is attributed to the N31 archive, found in a private house belonging to a family of Egyptians.[208] The marriage contract is the earliest dated document of that archive and is one of two such contracts attributed to the archive.[209] StAT 2, no. 164 docu-

---

206 These clauses are discussed, with a list of attestations, in Radner 2019, 322–324. Note that Radner does not include legal documents published in StAT 2, as they have not been collated. These tablets are: StAT 2, nos. 33 (year broken), 145 (year broken), 146 (year broken), 164 (675 BC), 169 (641* BC), 242 (612? BC), 266 (648* BC) and 272 (647* BC), all from Ashur. Other attestations not included in the list are: Faist 2020, no. 12 (VAT 20691; 651 BC) from Ashur; CTN 2, no. 221 (year broken) and CTN 6, no. 7 (year broken) from Kalhu; ACP, no. 28 (649 BC) from Mallanate; and the Hasankeyf tablet (Toptaş and Akyüz 2021), which documents a sale of people and land in Ilhina, in the province of Guzana (Bagg 2017, 244 s.v. Ilḫini), although the scribe who drew it up came from Ashur. The document likely dates after the fall of that city.

If all these attestations are added to those identified by Karen Radner, the total number of attestations from each location is as follows: Ashur, 29; Kalhu, 10; Nineveh, 4; Dur-Katlimmu, 2; Mallanate, 1; Mardin region (Girnavaz), 1; Guzana(?), 1; Unknown, 1. It is worth noting, however, that this clause did not only exist in the Assyrian language, but was also translated into Aramaic, as in the case of a tablet from Burmarina (Fales, Bachelot, and Attardo 1996; the document is not dated).

207 Note that this formulation was fairly common and the subject of this verbal phrase was typically a deity, although it is sometimes attested in relation to kings (CAD B, 360–365 s.v. *bu'û*; AHw, 145 s.v. *bu''û*).

208 Pedersén 1986, 125–129.

209 The other marriage contract is StAT 2, no. 184.

ments the transfer of the woman, Mullissu-hammat, together with her dowry, by her father, Pabba'u, *mukīl sīsê* 'horse keeper' of the goddess Ištar of Arbela,[210] to her husband, Auwa, son of Tapnahti. Mullissu-hammat is herself a *šēlûtu ša Issar ša Arbail* 'votaress of Ištar of Arbela', a status that seems to provide her with certain privileges in her relationship with her husband.[211] The clause appears twice, stated as a potential consequence of first Pabba'u, his relatives or his prefect, breaking the contract, and then of Auwa or those associated with him doing the same. The father and daughter, then, were associated with Ištar of Arbela, and several of the witnesses to the transaction were also associated with the temple of a god whose name is broken.[212] While the bride's name is Assyrian, both her father and her father-in-law have Egyptian names, as do some of the witnesses.[213] In spite of the Egyptian heritage of the legal parties, therefore, either they or the scribe who drew up the document evidently considered it desirable to include this penalty clause. It is interesting that this transaction took place in Ashur, and that several of the participants were part of the temple administration of the Assyrian heartland.

It is worth noting that StAT 2, no. 164 is the only tablet in the N31 archive that dates to a year outside the period 650–612 BC.[214] If the dating of this document to late 675 BC is correct, then the initial composition of this clause was not a response to the imposition of the succession covenant. Instead, the clause was introduced a minimum of two and a half years before the covenant's imposition (as there may be earlier instances of the clause that do not survive). If the clause was composed and disseminated centrally, with Esarhaddon and his advisors playing a role in the process, then this may be evidence of an earlier attempt during that monarch's reign to stress the importance of covenant to Assyrian rule. Nonetheless, all of the other attestations of the clause that can be dated with certainty were drawn up in 651 BC or later, suggesting that the adoption of the clause took place in earnest only during the reign of Ashurbanipal.[215] Karen Radner has argued that this may be linked to the revolt of Šamaš-šumu-ukin of 652–648 BC.[216]

It has already been observed that the depiction of the manner in which the covenant of the king functions in the legal clauses differs from that found in the

---

210 PNA 4/1, 65 s.v. *mukīl sīsê*.

211 Radner 1997, 209: 'Weder ihr Ehemann noch seine Verwandten haben Autorität über sie, da sie der Issār geweiht ist'.

212 StAT 2, no. 164: rev. 15–17.

213 PNA 3/1, 977 s.v. Pabbā'u; PNA 3/1, 1311 s.v. Tap-nahte. In PNA (1/2, 433 s.v. Awa), it is suggested that the name Auwa may be West Semitic.

214 StAT 2, xvi.

215 The tablet drawn up in 651 BC is Faist 2020, no. 12.

216 Radner 2019, 325.

covenant compositions themselves. Firstly, rather than the gods of the covenant punishing the transgressor, it is the covenant itself that has agency.[217] This mirrors the depiction of the covenant in the *Underworld Vision*, where it is the covenant that may 'come near me to do (me) evil'. This perhaps indicates a broader interpretation of the covenant on the part of those employing the clause, who perhaps saw the covenant as a deity or took it to stand for the divine wrath of a host of deities.[218] Such a view may or may not have been encouraged by the Assyrian crown. In this way, the covenant appears to have been viewed in this context as itself a vigilant actor, or as the catalyst that made the gods vigilant on the monarch's behalf.

In addition, and perhaps more surprisingly, the legal clause portrays the covenant not as punishing those who contravene the stipulations of the covenant itself, but instead those who break the terms of a private contract.[219] This suggests that the composers and users of these clauses viewed the covenant as an integrated part of the Assyrian legal system, and specifically as something that upheld it. In this way, the covenant is vigilant not directly on behalf of the crown, although the maintenance of legal order can be viewed as benefiting it. Instead, the covenant is observant and acts in order to protect the legal parties involved in that particular private transaction. Neo-Assyrian legal documents generally cast doubt on the legal party who stands to gain from falsely claiming that the legal transaction was not carried out lawfully, generally the seller, debtor or losing party in a judicial dispute. However, they frequently also explicitly state that other individuals and groups, namely the family and associates of that legal party, will not succeed in making a false claim. The formulation generally includes the male members of the person's family as other potential litigants, but it also sometimes refers to other individuals who may have a conceivable claim over the property of that person, such as his military or administrative superiors. StAT 2, no. 164, for instance, refers to *šakanšu* 'his prefect' in this list.[220] Thus, the vigilance of the covenant potentially applied to those directly involved in a legal transaction, as well as to their family and, perhaps most significantly, members of the local administration.

When thinking about the significance of this clause, it is necessary to consider the possible context of its composition and proliferation. It seems logistically sensible to suppose that these legal clauses were composed centrally, but it is worth

217 Karen Radner characterizes the covenant in these clauses as 'an avenging angel, a fury' and states that it is conceptually close to the Greek Erinyes (Radner 2019, 325, fn. 84).
218 Note the variant clause from 627* BC in SAA 14, no. 155: rev. 3–5. 'May Aššur, Sîn, Šamaš, Bel, Nabû, the gods of the king, call him to account'.
219 Radner 2019, 322.
220 StAT 2, no. 164: rev. 9.

noting that local variation between the documentation of different Assyrian settlements during the Neo-Assyrian period implies that legal scribes learnt to write contracts to some degree according to local traditions.[221] As such, the inclusion of particular clauses – in addition to presumably being a matter of personal choice – was sometimes also a question of local preference. Assyrian legal documentation, then, did not function purely according to a simple top-down model. It is therefore possible that the clause was composed by a legal scribe independently of the Assyrian crown, in which case it would reflect personal and local understanding of the 'covenant of the king'. Whether or not this is the case, however, the continued use of the clause, as well as the small variations in its phrasing, would likely have taken place at the discretion of local legal scribes and their clients.

Considering this, it is perhaps significant that the first known attestation of a clause involving the covenant of the king comes from Ashur, the place in which Sennacherib's succession covenant was probably enacted.[222] The majority of subsequent attestations of the clauses (59 %) also originate from that city, indicating that it was integrated into the local tradition of legal scribal practice there. In one case, a scribe from Ashur even included the clause in a sale document of land in the province of Guzana, presumably drawn up after the fall of Ashur in 614 BC. The sales document includes mention of Cyaxares the Mede as *šarru* 'king',[223] and yet still includes the clause stating that the king's covenant will call to account those who break the contract. Even though the scribe likely wrote the document up at considerable geographical remove from his home, and in a political situation very different from that of the time in which the clause was first used and proliferated, he had apparently been taught the clause and still considered its inclusion meaningful.

The covenant clauses were certainly not unique to Ashur and its scribes, however, and it is attested elsewhere in the core region, namely Kalhu (20 %) and Nineveh (8 %). Beyond the heartland and its scribes, the clauses are also found on tablets from Dur-Katlimmu (4 %), Mallanate (2 %) and modern Girnavaz (2 %), as well as translated into Aramaic in Burmarina. The Aramaic attestation is particularly significant, as it shows that not only Assyrian speakers would have known or understood this clause. Rather, at least some of the Aramaic speakers of the provinces, of which there were many in the western region particularly, would also have been acquainted with it. It is clear that the clause comes from the Assyrian because the phrase 'call him to account (lit. look into his hands)' is translated

221 As argued in Tushingham 2019.

222 See the discussion in Chapter 5.2.1.

223 Roaf 2021; Toptaş and Akyüz 2021.

literally into Aramaic.[224] As such, the Aramaic version of the clause conserves the claim that the covenant is acting in a vigilant fashion, reminiscent of a deity, on behalf of the legal party.

Regrettably, the tablet is not dated, although its first editor, Frederick Mario Fales, states that it dates to the late Assyrian phrase, from or after the reign of Esarhaddon.[225] The tablet documents the pledge of a man by three members *zy kṣr mlk*' 'of the king's cohort' who were likely stationed at Til Barsip, even though they are stated originally to come from Bit-Zamani some 300 km away.[226] The creditor is a local merchant named Še'-'ušnî, and it is he whom the clause was presumably intended to protect. The witnesses bear Aramaic and mixed Assyrian-Aramaic names, and Fales states that 'it is of interest to note that both linguistic affiliations bear references to the gods of the Moon-cult based in nearby Harran'.[227] Whether this is merely an indication of the broad influence of that temple, or if it means that some of the witnesses may have been involved in the cult more directly, is unknown. In either case, one may wonder whether this connection is relevant to the choice to include the clause, given the explicit mention of Harran in the covenant composition.

It is also worth dwelling briefly on the two documents from Dur-Katlimmu that reference covenant. Like the scribe from Ashur discussed above, the scribes who drew these documents up were likely doing so after the fall of Ashur in 614 BC.[228] The clauses do not reference the covenant of the king, but rather *adê ša mār–šarri* 'covenant of the crown prince', which Radner has argued refers to Aššur-uballiṭ II of Assyria, the successor of Sîn-šarru-iškun (r. ca. 627–612 BC). As Aššur-uballiṭ II came to power after the fall of Ashur, he was never coronated in that city and so, as Radner argues, was probably seen in Assyria as the crown prince rather than the monarch.[229] Again, despite the deterioration of Assyrian hegemony by this time, those documenting and participating in these legal transactions appear to still have considered royal covenant a meaningful agent of vigilance and retribution, able to act on their behalf. While it is quite probable that these people were loyalists of the Assyrian royal family, this may also indicate that they believed in the power of the covenant to affect their lives far beyond the constraints of the true power of the Assyrian crown at the time.

224 Fales, Bachelot, and Attardo 1996, 100 f.
225 Fales, Bachelot, and Attardo 1996, 108.
226 Fales, Bachelot, and Attardo 1996, 109.
227 Fales, Bachelot, and Attardo 1996, 109.
228 The tablets in question are Radner 2002, nos. 128 and 199.
229 Radner 2018a.

7.4 Conclusions

The available evidence regarding conceptions of and reactions to covenant in sources from and concerning the provincial extent of the Assyrian Empire is piecemeal but revealing. Several letters from the royal correspondence mention covenants as a means to affirm loyalty to the king, as well as in order to refer specifically to the plot of 671/670 BC. It is clear that the conspirators were considered to have broken not only Esarhaddon's succession covenant but also Sennacherib's succession covenant, something that would cause them to be divinely punished. The references to covenant in Esarhaddon's correspondence regarding the provinces parallel the broader evidence from the king's entourage, that the imposition of Esarhaddon's succession covenant along with the composition of texts such as *Esarhaddon's Apology* also increased the conceptual importance of Sennacherib's succession covenant. Nabû-rehtu-uṣur's reference to 'your father's goodness, and your father's covenant and your covenant' suggests that, while these covenants were considered distinct, some thought of them as serving complementary purposes. The letters also indicate that both of the covenants were invoked in similar ways in conversations beyond those with members of the Assyrian royal family, implying that it was not unusual to reference them, at least among provincial elites around Harran and Guzana.

The tablets from the N4 and N6 archives at Ashur also imply that covenant was considered and discussed beyond the Assyrian crown's direct social sphere. The *Underworld Vision* from the N6 archive suggests that members of the scholarly milieu of Ashur were reflecting on the significance of the duty to report to the Assyrian crown, as well as the danger of the covenant's own vigilance. Although this may have been prompted by the imposition of Esarhaddon's succession covenant, or even the accession of Ashurbanipal, the murder of Sennacherib provides historical context in which these themes are narrativized. The fact that the Assyrian scribes of N6 kept a composition in which a scribe acted wrongly but ultimately adhered to the covenant imposed upon him implies that the text was in part intended to have a didactic function. So too, the later covenant excerpt from the N4 archive indicates that the family of *ašipu*-healers who owned the tablet collection considered it appropriate to write out and perhaps even teach clauses of royal covenants.

It is possibly significant that both tablets refer to Nergal as the deity who will punish contravention of the covenant. The deity does not feature prominently in the covenant composition, but was an important figure for *ašipu*-healers. This was the profession of the owners of N4, and even though those of the N6 archive were scribes, they had some training in the lore of the *ašipu*-healer. This may reflect that they were interested in the possible ramifications of covenant as inflicted by a deity relevant to them, perhaps implying a vigilance directed towards the self. The N4 tablet, which contains an extract of a covenant of Šîn-šarru-iškun, can certainly not be considered a response to the succession covenant of Esarhaddon. Rather, it indicates that Assyrian monarchs continued to impose covenants on the population of Ashur after the reigns of Sennacherib and Esarhaddon, and that the scholarly milieu of that city engaged with them to some degree.

The covenant clauses found in legal documents from the reign of Esarhaddon can also be regarded as an instance of the augmentation of the importance of covenants imposed upon the provincial population under that monarch. While the earlier clause pre-dates the covenant's implementation, implying that it may refer to an earlier covenant, it is first attested not long before 672 BC. The covenant clauses are far better attested from midway through Ashurbanipal's reign, specifically during the period of Šamaš-šumu-ukin's rebellion in Babylonia (652–648 BC). It is possible that these clauses refer to Esarhaddon's succession covenant here, but Ashurbanipal also imposed covenants of his own during this period, so it may refer to them. Later, after the Assyrian crown had lost control of the core region, it seems that scribes in Dur-Katlimmu adapted the clause to reflect the situation, implying that they conceived of the covenant as belonging to the current Assyrian ruler. Thus, even though the initial use of the covenant in legal documents may be tied in with the planned imposition of the succession covenant under Esarhaddon, these clauses more generally should not be viewed as a direct response to Esarhaddon's succession covenant specifically. Despite this, they provide insights into the manner in which some inhabitants of the provinces would have encountered the concept of covenant and the way that they may have conceived of it. While these clauses do not refer to the duty of vigilance, they portray the covenant as vigilant of behalf of the Assyrian legal system, as well as the individual parties involved in a particular transaction. Perhaps interestingly, the legal protection – or threat – of the covenant is not limited only to directly involved legal parties. Instead, it pertains to their family and associates, sometimes including their superiors, perhaps promoting both lateral and bottom–up vigilance.

The majority of the source material comes from Ashur, with a significant proportion also stemming from or pertaining to Harran and nearby settlements, such as Guzana. The covenant clauses are also fairly well represented in the legal documentation from Kalhu. This geographical distribution, especially in the case of Ashur, can certainly be largely explained in terms of the availability of source material from particular locations more generally. As such, the lack of evidence from other regions should not necessarily be taken as evidence of a lack of response by the people there. Nonetheless, it is probably significant that three of the locations that the analysis of Part One of this study has argued were particularly important to Esarhaddon and his advisors, namely Ashur, Kalhu and Harran, do seem to have yielded both direct responses and broader awareness and discourse.

In terms of the social groups that generated these responses and discourse, it is notable that several of the people who wrote or owned the documents discussed in this chapter were affiliated with temples, especially those in Ashur and Harran. As the covenant tablets were set up in temples, it seems possible that these groups would have been particularly aware of them. In the case of the scholarly milieu of Ashur, as well as the legal parties and scribes who used covenant clauses, the people involved were likely not in personal contact with the crown. Nonetheless, they were frequently affiliated with the state administration in some way. As such, it is difficult to ascertain whether the covenant and its duty of vigilance permeated far beyond such groups.

Chapter 8: Responses to Esarhaddon's covenant across the client states

As established in the first half of this study, King Esarhaddon and those who aided him appear to have concerned themselves more with imposing the succession covenant on the provincial system than the client states. Despite this, it is clear from the covenant manuscripts found at Kalhu that Esarhaddon and his subordinates made the substantial effort of commissioning an individual divinely-sealed document for each of the client rulers of the Assyrian Empire. So too, several sources written in various languages – Aramaic, Assyrian, Babylonian, and Hebrew – attest that the imposition of the succession covenant did indeed provoke direct responses from the client states, in the form of letters, as well as prompting discourse around the concept of covenant and the duty to report to the Assyrian monarch. As such, the source material speaks eloquently of multifaceted and at times subversive reactions to the demands of Assyrian covenant in client states (see Figure 5).

The first texts discussed in this chapter are six letters pertaining to events in Babylonia, an administrative anomaly in the Assyrian Empire. Assyria's southern neighbour had been subjected to multiple different strategies to bring it under Assyrian rule during the Neo-Assyrian period. During Esarhaddon's reign, the region was directly under that monarch's control, and can therefore not be considered an ordinary client state. When Esarhaddon imposed his succession covenant, however, he made clear that he intended Assyria and Babylonia to be ruled in future by separate monarchs, albeit both members of the Assyrian royal family. As such, from 672 BC onwards, the Babylonian crown, made up of Esarhaddon and his successor Šamaš-šumu-ukin, differed from that of Assyria. As such, I include my analysis of the correspondence sent from Babylonia in this chapter. One of the letters included here was sent to Esarhaddon from Šamaš-šumu-ukin, and shows that some subjects in Babylonia considered it their duty to report on seditious activity to their new crown prince. The remaining five letters appear to be addressed to Esarhaddon, and three of them stem from Uruk, perhaps indicating that certain Babylonians were responsibilized to report to the monarch in a similar fashion to their Assyrian counterparts.

In addition to the letters sent to the king and the crown prince of Babylon by some Babylonians, a response to the covenant was sent from another inhabitant of a client state: a royal delegate stationed in Phoenicia, specifically the island of Arwad situated off the coast of modern Syria. This official, writing to the monarch, accuses the client ruler and local merchants of acting wrongly, implying that they are not respecting their obligations to Assyria. Perhaps interestingly, this letter can be read less as a report on petty corruption perpetrated by the foreign sub-

 https://doi.org/10.1515/9783111323435-011

jects of the Assyrian monarch, and more as an accusation directed at Esarhaddon's court officials. It is these people whom the writer, Itti-Šamaš-balaṭu, seems to perceive as the greater threat.

This chapter next examines the view that portions of the Hebrew Bible can be construed as containing evidence of responses in the client state of Judah to the imposition of Esarhaddon's succession covenant. The nature and extent of the influence of Esarhaddon's succession covenant on, in particular, Deuteronomy chapters 13 and 28, have been the subject of much debate. The focus of the present discussion will be on the possible rejection and inversion of the succession covenant's duty of vigilance in Deuteronomy 13. This chapter of Deuteronomy seems to indicate that a group at a substantial geographical remove from the Assyrian monarch nonetheless had a highly sophisticated reaction to the demands of the Assyrian crown. The chapter is particularly interesting as it can be interpreted as containing implicit critiques of the succession covenant and its duty of vigilance.

Finally, the chapter ends with an examination of another narrative tradition that long outlived the Assyrian Empire, and that appears to have been geographically widespread: the tales of Ahiqar. Through an examination of the Aramaic *Story of Ahiqar*, attested on 5th century BC papyri found at Elephantine in modern Egypt, I argue that this tradition appears to have offered some points of criticism of the climate of distrust at the Assyrian court at this time. In particular, the Aramaic *Story of Ahiqar* highlights the potential for abuse inherent in a regime that reifies vigilance and denunciation, something that may be viewed as a response to the covenant's imposition.

8.1 Responses from Babylonia

Given the special status of Babylonia, it is perhaps unsurprising that it is the client state concerning which the most letters mentioning covenant are sent. As in the provincial system, the years immediately following the imposition of Esarhaddon's succession covenant appear to have been turbulent in Babylonia, with some figures there involved in a plot against the monarch.[230] The different status of Babylonia to the Assyrian provincial system render the manner in which correspondents evoke covenant particularly interesting, as the succession covenant has different implications for the subjects of Babylonia than it does for all other inhabitants of the Assyrian provinces or client states.

230 See Chapter 1.

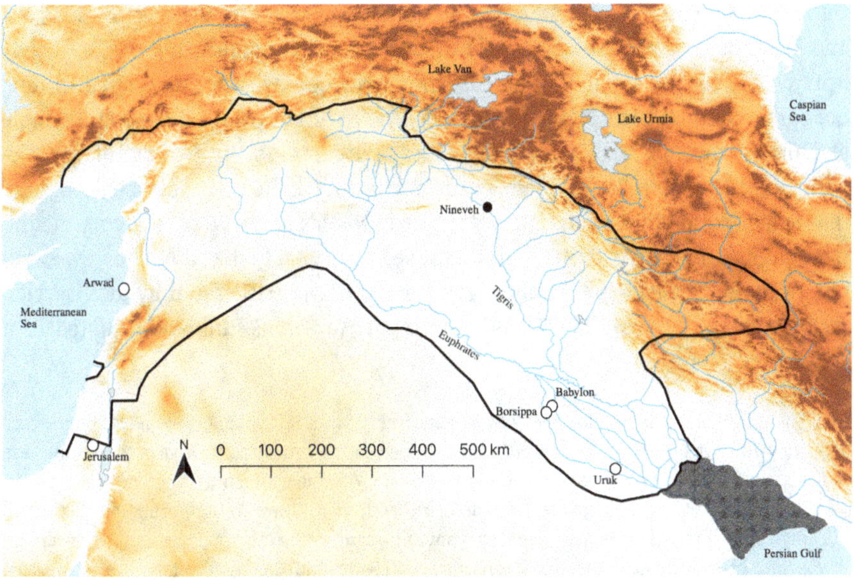

Figure 5: Map of locations of responses to Esarhaddon's succession covenant from client states of the empire: Arwad, Babylon, Borsippa, Jerusalem, Uruk. The capital, Nineveh, is marked in black for reference. The shaded area indicates the ancient coastline.

8.1.1 Šamaš-šumu-ukin, crown prince of Babylon

Some four letters from Šamaš-šumu-ukin to his father, Esarhaddon, are currently known to modern scholars.[231] These letters indicate that the crown prince of Babylon, like Ashurbanipal, had a significant role in the administration of the empire.[232] Indeed, Karen Radner has argued that Ashurbanipal and Šamaš-šumu-ukin may have been nominated to their positions in part so that Esarhaddon would have additional support in ruling his empire, something that may have been necessary due to the monarch's ill health at that time.[233] In the letter that mentions a covenant, SAA 16, no. 21, Šamaš-šumu-ukin reports to his father on the contents of two letters. The first letter was sent to him by three men, two *mār Bābili* 'citizens of Babylon', Šaridu and Nabû-ahhe-ereš, and one *mār Barsip*

231 SAA 16, nos. 21–24; PNA 3/2, 1215 s.v. Šamaš-šumu-ukīn.

232 On Ashurbanipal's involvement, see SAA 16, xxvii–xxviii.

233 As argued in Radner 2003c; see further discussion in Chapter 1.1 and Chapter 6.1.

'citizen of Borsippa', Bel-iddina.[234] This missive contains references to covenant and denounces the astrologers Bel-eṭir and Šamaš-zeru-iqiša and the haruspex Aplaya. The second letter that Šamaš-šumu-ukin quotes is not addressed to him, but rather sent by Urdu-Nabû, presumably the Assyrian priest of the Nabû temple at Kalhu known by that name from other sources,[235] to the two astrologers denounced in the first letter. Simo Parpola, the first editor of the tablet, suggests that Šamaš-šumu-ukin likely confiscated Urdu-Nabû's letter after receiving the initial denunciation.[236] As Parpola argues, it seems probable that the letter was written in 670 BC,[237] placing it firmly in the context of the crisis faced by the Assyrian crown during the years 671/670 BC. The report to Šamaš-šumu-ukin quoted in it states that:

> The king concluded a covenant with us concerning you: 'Tell your lord whatever you hear!' Now, Bel-eṭir (and) Šamaš-zeru-iqiša have neglected the order the king gave them (and) are acting on their own. Aplaya, whom the king sent (with the command): 'Go (and) set up sanctuaries in Babylon!', has made common cause with them. They are observing the stars (and) dissecting lambs, but he does not report anything concerning the king, our lord, or the crown prince of Babylon. Aplaya alone is a haruspex; Bel-eṭir (and) Šamaš-zeru-iqiša are astrologers, they watch the sky day and night. Moreover, he has assembled the people who captured Aššur-nadin-šumi (and) delivered him to Elam, and has concluded a covenant with them, adjuring them by Jupiter (and) Sirius. We have now heard (about it) and informed the crown prince of Babylon.[238]

Table 5: References to the succession covenant concerning Babylonia in the royal correspondence.

Publication	Relevant extract	Sender	Date	Provenance
1. SAA 16, no. 21	(Quoting another letter): The king concluded a covenant with us concerning you: 'Tell your lord whatever you hear!'	Šamaš-šumu-ukin	670 BC	Babylonia
2. SAA 18, no. 80	It is written in the covenant: 'Write to me (about) whatever you see or hear.'	Itti-Marduk-balaṭu	672–669 BC	Uruk
3. SAA 18, no. 81	I[t is written in] the covenant: 'Write to me (about) whate[ver] you se[e] and h[ear]!'	[...]	672–669 BC	Uruk

234 SAA 16, no. 21: obv. 6–7; PNA 2/2, 794 s.v. Nabû-ahhē-ēreš, PNA 1/2, 312 s.v. Bēl-iddina. Šaridu is not included in PNA.
235 As argued in Parpola 1972, 30; PNA 3/2, 1408 f. s.v. Urdu-Nabû.
236 Parpola 1972, 33.
237 Parpola 1972, 27; Luukko and van Buylaere also suggest a date of 670 BC (SAA 16, xviii).
238 SAA 16, no. 21: obv. 9–rev. 8.

Table 5: References to the succession covenant concerning Babylonia in the royal correspondence. *(Continued)*

Publication	Relevant extract	Sender	Date	Provenance
4. SAA 18, no. 83	We entered [into] a covenant with the king, your father, [and] we have entered [in]to a covenant with the king, our lord. Furthermore, the [k]ing has written to us, saying: 'Write to me (about) whatever you see or hear.'	Anonymous	672–669 BC	Uruk
5. SAA 18, no. 102	I am a keeper of the watch and guard of the covenant of the king, my lord.	[...]	672–669 BC	Babylonia
6. SAA 8, no. 536	[...] ... [......] ... of the king my lord [......] before me, [......] when I entered the covenant [......] is good(?) so I did not write to the king my lord.	[...]a, son of Bel-ušallim	Reign of Esarhaddon, possibly Ashurbanipal	Babylonia

The message of these informers, at least as Šamaš-šumu-ukin tells it, begins with a direct reference to what is presumably Esarhaddon's succession covenant. The informers state that *adê šarru ina muhhīka issēni issakan* 'the king concluded a covenant with us concerning you'.[239] This phrasing mirrors that of the covenant composition, as the colophon refers to it as *ina muhhi* 'concerning' both Ashurbanipal and Šamaš-šumu-ukin.[240] Nonetheless, the main body of the text frames the covenant as concerning Ashurbanipal, rather than Šamaš-šumu-ukin.[241] So too, the versions of the covenant composition that are currently known do not stipulate that they report simply to *bēlikunu* 'your lord', as stated in the letter, but rather to Ashurbanipal. It is, of course, possible that the informers are referring to a different version of the succession covenant that featured Šamaš-šumu-ukin more prominently and was imposed only on the inhabitants of Babylonia. However, it also seems possible that the three letter-writers have interpreted the known version of the main body of the covenant composition as decreeing that they report to their own crown prince, that of Babylon. This mirrors the various letters that invoke what appears to be Esarhaddon's succession covenant and yet are addressed to the monarch himself, rather than to the crown prince of Assyria. If this inter-

239 SAA 16, no. 21: obv. 9–10.

240 SAA 2, no. 6, § 107: 664–670; Lauinger 2012, viii 63–71. See also the discussion in Chapter 2.1.

241 The only section of the main body of the covenant in which Šamaš-šumu-ukin is mentioned is SAA 2, no. 6, § 7: 83–91. See also the discussion in Chapter 2.1.

pretation is correct, then it is perhaps an indication of variation in the interpretation of the demands of the covenant within Babylonia.

As in the letter to Esarhaddon from Adad-šumu-uṣur,[242] the informants claim to quote the covenant, but in fact paraphrase it. Once more, this may indicate that the informants, while aware of the broad message of the covenant, were not in fact familiar with the details of its stipulations. Alternatively, they may have known the stipulations, but considered this an accurate rendition of their fundamental import. As in the case of the letter from Adad-šumu-uṣur, they use the verbs *šemû* 'hear' and *qabû* 'speak', which are also found in the covenant stipulations on reporting. They omit the injunction against concealment, as well as the statement that they are to 'come' and report what they have heard. As in Adad-šumu-uṣur's letter, the person to whom the report must be made in their telling differs from the stipulations of the succession covenant.

The people that these informants are denouncing, and the actions that they are relaying to Šamaš-šumu-ukin are noteworthy. The diviners are 'observing the stars (and) dissecting lambs', but are failing to report the implications of these observations for the monarch or Šamaš-šumu-ukin. Unlike the scenario set out in the letters of Nabû-rehtu-uṣur, which mention a form of divination that is cited in the covenant composition, namely prophecy, the eventuality described here is not explicitly set out in the covenant.[243] Nonetheless, it appears to parallel the suspicion of diviners evident in the *The Sin of Sargon*, as well possibly as *The Underworld Vision of an Assyrian Prince.*[244] The accusation here indicates that this was not merely a narrative projected onto the reign of previous monarchs, but that some subjects of the Assyrian crown were mistrustful of such scholars in the present.

As in the report from Nabû-ušallim in Ashur,[245] the informants also report that the haruspex, Aplaya, has concluded a rival covenant, something that is explicitly forbidden in Esarhaddon's covenant.[246] It is perhaps significant that both of these details mirror aspects either of the text of the covenant itself or of the wider discourse around the aftermath of the covenant's imposition. The wording of the report indicates that the informants and the denounced parties are located in Bab-

---

242 SAA 10, no. 199; see Chapter 6.1.2.

243 Note that the covenant composition refers to prophets and another type of diviner, the *mār šā'ili amat ili* 'dream interpreter', see Chapter 3.1. On the letters from Nabû-rehtu-uṣur (SAA 16, nos. 59–61) and this section of the covenant, see Chapter 7.2.1.

244 See the discussion of these works in Chapters 6.2.1. and 7.1.4.

245 YBC 11382, published in Frahm 2010. See Chapter 7.1.2.

246 SAA 2, no. 6, § 13: 153–156. See also Chapter 3.3.

ylonia. However, both Urdu-Nabû and Bel-eṭir are associated with the core region of Assyria: the former in his capacity as a priest of Nabû in Kalhu, while Bel-eṭir appears to have been active at Esarhaddon's court.[247] So too, Aplaya is described as having been sent to Babylon by Esarhaddon, perhaps from Assyria. As such, despite their location in a client state, the plotters themselves are possibly courtiers. This geographical dynamic: informants in a client state reporting on corruption by people with social but also physical access to the monarch, is also evident in the report of Itti-Šamaš-balaṭu, discussed below.

8.1.2 Itti-Marduk-balaṭu, 'chief administrator' of the Eanna temple and anonymous information from Uruk

There are three letters sent from Uruk and mentioning covenant that were found at Nineveh and probably date to the reign of Esarhaddon.[248] The first, SAA 18, no. 80, is sent by Itti-Marduk-balaṭu, the *šatammu* 'chief temple administrator' of the Ištar temple in Uruk, Eanna, who took up this position in ca. 674 BC.[249] He is referenced in other letters of Esarhaddon's reign and appears under the name of Balaṭu in several legal documents dating between 674 and 669 BC. He is not attested in any dated documents beyond 669 BC.[250] Grant Frame has suggested, however, that SAA 18, no. 80 dates to ca. 666 BC, early in the reigns of Ashurbanipal and Šamaš-šumu-ukin.[251] He bases this suggestion on the reference at the end of the letter to horses: '(As to) the horses which the king sent [to] Uruk, I have harnessed [...] and [...]'.[252] He links this statement to a letter from a governor of Uruk under Ashurbanipal, Nabû-ušabši, who states that a *teppir*-official of the Elamite king has sent horses to Uruk,[253] positing that the letters are referring to the same incident. Nonetheless, as Frances Reynolds has established, there is no space in the broken portion of the tablet after the word *šarru* 'king' to restore 'of Elam', and therefore

247 Bel-eṭir seems to have been active at Esarhaddon's court at some point: PNA 1/2, 299 s.v. Bēl-ēṭir, see also the discussion in Frame 1992, 117 f. See also PNA 1/1, 116 s.v. Aplāia or Apil-Aia and PNA 3/2, 1221 s.v. Šamaš-zēru-iqīša,

248 These letters are published as SAA 18, nos. 80–81 and 83.

249 PNA 2/2, 589 s.v. Itti-Marduk-balāṭu. The translation of *šatammu* follows PNA 4/1, 185 s.v. *šatammu.*

250 See references in PNA 2/1, 589 s.v. Itti-Marduk-balāṭu.

251 Frame 1992, 127, fn. 138.

252 SAA 18, no. 80: rev. 12'–r.e. 14.

253 ABL 268; note that Matthew Waters argues that ABL 268 was more likely written shortly before the Teumman campaign in 653 BC (Waters 2022, 258, fn. 28; see also Waters 1999a and Waters 1999b, 476).

there are no grounds on which to link these letters.[254] As such, as Reynolds states, it seems most likely that the letter was written during the reign of Esarhaddon,[255] the period when Itti-Marduk-balaṭu is otherwise attested.[256]

Itti-Marduk-balaṭu was himself from Babylonia.[257] In his role as chief administrator of the Eanna temple, Itti-Marduk-balaṭu would have been at the very top of the social hierarchy in the city of Uruk. The Ištar temple at Uruk was ancient by the Neo-Assyrian period, something that the inhabitants of the city well knew. In addition to the high status that his position would have conferred on him, Itti-Marduk-balaṭu's role as chief administrator of the vast Eanna temple complex would have also given him access to significant wealth and manpower. Indeed, Itti-Marduk-balaṭu's predecessor in the role under Esarhaddon, Nabû-naṣir, was the father of a governor of Uruk, and the grandfather of King Nabopolassar (r. 625–605 BC), founder of the Neo-Babylonian Empire.[258] Even though Nabopolassar's rise to the throne was almost fifty years away when Itti-Marduk-balaṭu wrote letter SAA 18, no. 80, Nabû-naṣir's legacy highlights the position's political and social potential, both within Uruk and beyond.

The letter itself is brief and, regrettably, only partially preserved. Itti-Marduk-balaṭu addresses the monarch and blesses him with the statement 'may the gods of all the lands bless the king, my lord!'.[259] Itti-Marduk-balaṭu continues the letter with a report on a festival at the temple, in which the *ilāni ša šarri* 'gods of the king' have set out in procession. There follows a break, and the next preserved line mentions covenant. As in Adad-šumu-uṣur's letter and the missive quoted by Šamaš-šumu-ukin, Itti-Marduk-balaṭu professes to cite a covenant stipulation: *adê iššaṭir umma mala tammarā u tašemma' šuprāni* 'It is written in the covenant: 'Write to me (about) whatever you see or hear.''[260] If this statement does indeed

254 SAA 18, 63, no. 80: rev. 12'.

255 SAA 18, xxiv.

256 Note that Itti-Marduk-balaṭu is frequently stated to be attested between 674 and 666 BC (SAA 18, xxiv, PNA 2/1, 589 s.v. Itti-Marduk-balāṭu). As far as I can tell, however, the only evidence for the 666 BC date is Grant Frame's suggestion that SAA 18, no. 80 was written when Nabû-ušabši had already taken up his post as governor of Uruk (see discussion above). Nabû-ušabši is himself first attested in this role in 661 BC, but his predecessor is last attested in 666 BC, and so this is the earliest that Nabû-ušabši could have taken the post (Frame 1992, 127; see also PNA 2/2, 901f. s.v. Nabû-ušabši). As Frances Reynolds has provided a convincing rebuttal of this dating, however, the documents referencing Itti-Marduk-balaṭu actually date between 674 and 669 BC (see PNA 2/1, 589 s.v. Itti-Marduk-balāṭu).

257 He wrote in Neo-Babylonian script and language, and was active there (see above).

258 Robson 2019, 168.

259 SAA 18, no. 80: obv. 3–5.

260 SAA 18, no. 80: rev. 2'–5'.

refer to Esarhaddon's succession covenant, then this characterization of the reporting stipulations is particularly interesting. Itti-Marduk-balaṭu includes not only the verb *šemû* 'to hear', which does appear in the covenant composition, but also the verb *amāru* 'to see'. So too, Itti-Marduk-balaṭu uses the verb *šapāru* 'to write' instead of *qabû* 'to speak'. Thus, while the general meaning of the injunction to report to the Assyrian crown is preserved in Itti-Marduk-balaṭu's letter, the form of perception that is demanded in his telling – seeing and hearing – differs significantly from the wording of the succession covenant. The mandated response to the things that Itti-Marduk-balaṭu perceives, writing to the king, also contrasts to the text of the covenant.

One could view these changes as reflecting the manner in which Itti-Marduk-balaṭu has interpreted the covenant composition, but this is complicated by two other references to covenant found in letters sent to Esarhaddon from Uruk. One of these letters, first published by Frances Reynolds as SAA 18, no. 81, is preserved only very fragmentarily. The name of the sender is not preserved, although it is possible that it was written by Itti-Marduk-balaṭu, as the wording is very similar to that of SAA 18, no. 80. In addition, the denunciation appears to refer to temple personnel, which implies that the author is a priest at Eanna.[261] The preserved portions of the letter's reference to covenant matches that of SAA 18, no. 80 almost verbatim: [*ina libbi*] *adê iš*[*šaṭir*] *umma mim*[*ma mala*] *tamma*[*rā*] *u ta*[*šemma'*] *šuprāni* 'I[t is written in] the covenant: 'Write to me (about) whate[ver] you se[e] or h[ear].".[262] The words *ina libbi* are not preserved, and are reconstructed by Reynolds, meaning that the only certain difference between the two statements is the addition in SAA 18, no. 81 of the word *mimma* 'anything'. This additional word does not alter the meaning of the phrase. As such, it seems that either Itti-Marduk-balaṭu wrote at least two letters to Esarhaddon in which he quoted covenant in this way, or Itti-Marduk-balaṭu and one of his colleagues at the Eanna temple did so. The former seems to me more likely, but the latter is not to be ruled out. In the case of SAA 18, no. 81, the writer appears to be reporting on activities at the temple itself.[263] In SAA 18, no. 80, meanwhile, Itti-Marduk-balaṭu reports on the movements of the Aramean *nasīkāti* 'sheikhs' who had been dwelling in Harmašu, which was probably located near the Elamite-Babylonian

---

261 The report seems to concern a certain 'Nabû-kina-[...], foreman of the ob[lates ...]' (SAA 18, no. 81: obv. 8'–9'). On the Babylonian term *rab širki* 'master of temple oblates', see PNA 4/1, 115 s.v. *rab širki*.

262 SAA 18, no. 81: obv. 4'–8'.

263 SAA 18, no. 81: obv. 8'–rev. 1: '[Now then] Nabû-kina-[...], foreman of the ob[lates ...], [has ...ed] to [...]'.

border.[264] He informs the monarch that they have come to Uruk and that some have gone to Elam.[265] That he views this report as satisfying the demand of the covenant is strongly implied by the statement with which he concludes this portion of the letter: 'I h[ave] (now) wri[tten] to the king, my lord.'[266] In contrast to a report such as that which was sent to Šamaš-šumu-ukin, Itti-Marduk-balaṭu does not attempt to explain why, or even whether, these activities are potentially problematic. It is possible that he considers this obvious, or perhaps he considers the contents of what he is reporting less important than the act of reporting in itself.

The final letter sent from Uruk mentioning covenant is SAA 18, no. 83. It contains a similar formulation to SAA 18, nos. 80–81. In contrast to these letters, however, it is written anonymously.[267] Like the other two letters, the tablet is written in Neo-Babylonian language and script. It is set out in landscape format, rather than the portrait format used in most letters. The initial half of the document is written in the first person plural, and the second is in the first person singular, suggesting that it is written by one person perhaps on behalf of a group. The document mentions Uruk, as well as the son of Ina-teši-eṭir, a man from Uruk.[268] Perhaps interestingly, the contents of the letter claim that 'we' have brought various criminals from Uruk into the presence of the king to be questioned. It continues to request that a royal messenger be sent to question the author of the letter concerning other matters that are 'relevant to the king which I have heard'.[269] The letter closes with the request that the monarch question someone else, a certain Zera-ukin, concerning something he has heard.[270] In this way, the letter does not contain denunciations so much as request that the monarch question various parties, who themselves will report on concerning matters. In this way, this letter differs significantly from those previously discussed in this study, in which the issue at hand is reported directly.

The text does not contain the usual introductory formulae of a letter, beginning instead with a reference to covenant:

264 Note that Reynolds locates this settlement in Assyria (SAA 18, 204), for a discussion see, however, Bagg 2020, 262 s.v. Harmasa. On the term 'sheikh' see PNA 4/1, 74 s.v. *nāsiku*.

265 SAA 18, no. 80: rev. 7'–10'.

266 SAA 18, no. 80: rev. 10'–11'.

267 Mikko Luukko includes the letter in his list of anonymous denunciations of the Neo-Assyrian period (2018, 166). See also the discussion in SAA 18, xxiv–xxv.

268 SAA 18, xxv; PNA 2/1, 541 f. s.v. Ina-tēšî-ēṭir.

269 SAA 18, no. 83: rev. 2–3.

270 PNA 3/2, 1444 s.v. Zēru-kēn.

> We entered [into] a covenant with the king, your father, [and] we have entered [in]to a covenant with the king, our lord. Furthermore, the [k]ing has written to us, saying: 'Write to me (about) whatever you see or hear.'[271]

This portion of the letter is illuminating. The quotation is the same as that of SAA 18, no. 81 and almost the same as that of SAA 18, no. 80. In contrast to those letters, however, this document attributes the statement not to the covenant itself but apparently to a separate document sent by the monarch. It is possible that all three letters were written by Itti-Marduk-balaṭu, which would explain the consistency of the quotation across the documents. As in the case of the anonymous denunciations from the western provinces,[272] the writer of the letter clearly believes that the king knows who he is: he references a letter sent to 'us' and also mentions that he and the other members of the group have brought 'criminals and witnesses into the king's presence'.[273] In the second half of the letter, he states that 'a royal messenger should come out to question me.'[274] It is clear, then, that the individual writing the letter is in contact with the monarch, and was known to him, a description that fits Itti-Marduk-balaṭu.

Whether or not the sender of the letter is Itti-Marduk-balaṭu, the author is evidently someone who swore to a covenant under both Sennacherib and Esarhaddon. By mentioning both the king's covenant and the covenant of *abīka* 'your father', the author of this letter frames his duty to the Assyrian crown in a manner similar to the informants on Harran, Nabû-rehtu-uṣur and the anonymous author of SAA 16, no. 71. The author of this letter appears to link these covenants with the duty to report. Instead of the covenants, however, he ascribes the demand to report to an additional written statement sent to 'us' by the monarch. This implies that, in addition to imposing the covenant, Esarhaddon separately commanded that some people report their observations to him. If this is the case, then the author of this letter appears to view these two acts as closely linked. Whether Esarhaddon himself made an explicit link between them is not stated.

It seems possible, then, that the statement made in SAA 18, no. 83 is accurate: that the formulation 'write to me (about) whatever you see or hear' is indeed a verbatim quotation from an order given by Esarhaddon. One may perhaps imagine, given the context, that the authors of all three of the letters, assuming they are not one person, would have received this order. Thus, the author (or authors)

---

271 SAA 18, no. 83: obv. 1–5.

272 See the discussion of the anonymous denunciations SAA 16, nos. 63 and 71 in Chapter 7.2.2 and 7.2.3.

273 SAA 18, no. 83: obv. 9–rev. 1.

274 SAA 18, no. 83: rev. 3–4.

of SAA 18, nos. 80–81 may perhaps have misattributed this statement to Esarhaddon's covenant. If this reconstruction is correct, then this confusion would indicate the awareness of these parties of the general demands of the covenant stipulations that they report to the crown, surely an indication of their successful mobilization on the part of Esarhaddon. However, it also implies that Itti-Marduk-balaṭu and the author of SAA 18, no. 81 were not sufficiently aware of the covenant stipulations to quote them accurately. If this is indeed an instance of the fusion in the minds of some Uruk residents of an order sent by Esarhaddon with the covenant stipulations, then it bears comparison with the letter from a priest of Aššur, SAA 13, no. 45, in which the writer equates obeying the *abat šarri* 'king's word' with keeping the covenant.[275] This suggests that this view existed among temple personnel in both Ashur and Uruk, and also gives some hints as to why that may have been.

8.1.3 Miscellaneous references to covenant from Babylonia

Two further letters written in Neo-Babylonian script and possibly dating to the reign of Esarhaddon contain references to covenant. They are published as SAA 18, no. 102 and SAA 8, no. 536. Letter SAA 18, no. 102 clearly pertains to treasonous activities in Babylonia. The denunciation implicates Aplaya son of Nadinu, who according to the writer is held as 'father of the Chaldeans', but is also 'an insolent cad and a traitor'.[276] The author of the letter is Babylonian, as his letter is written in Neo-Babylonian language and script, and its contents pertain to Babylonia. Nonetheless, he states that 'I am [a ... of ...], my lady'.[277] Frances Reynolds suggests that the reference to 'my lady' refers to the influential queen mother, Naqi'a,[278] in which case the writer may have been in her employ and thus had ties to the royal house.

In her recent discussion of the letter, Karen Radner describes the author as 'very much on message' as pertains to the covenant. He states that:

> [I put my trust] in the king, [my] lor[d ...]. I am [a ... of the queen mother], my lady; [I have co]me up a hundred, (nay, a) thousand [times], and I have not abandoned the feet of Assyria that I have grasped. I am a keeper of the watch and a guard of the king my lord's covenant.

275 See the discussion in Chapter 7.1.1.

276 SAA 18, no. 102: obv. 9'–10'. See also PNA 1/1, 118 s.v. Aplāia or Apil-Aia, note that it is here suggested that the letter was sent from Borsippa.

277 SAA 18, no. 102: obv. 4'.

278 SAA 18, 83, no. 102: obv. 4'.

> The king, my lord, said: 'Remove the evil one and instigator of rebellion from the country!' Aplaya son of Nadinu is an insolent cad and a traitor.[279]

The phrase *bēl maṣṣarti u nāṣir adê ša šarri belīya anāku* 'I am a keeper of the watch and a guard of the king my lord's covenant' mirrors to some degree the statements found in SAA 13, no. 45 and SAA 16, nos. 60–61 and 63, in which the writers use the term *bēl adê* 'keeper of the covenant'. Here, the phrase is *nāṣir adê* 'guard of the covenant', and the writer instead refers to himself as a keeper of the *maṣṣartu* 'watch'.[280]

The term *maṣṣartu* means 'watch, guard', and is frequently attested in Neo-Assyrian letters. Frederick Mario Fales has argued that the term has two distinct meanings: one refers to an astrological-astronomical watch, or vigil. The other meaning, according to Fales, is 'vigilance' more generally and includes the responsibility of subjects 'to keep their eyes and ears open' and report to the king.[281] This suggestion is compelling, and the term *maṣṣartu* was certainly associated with the concept of alertness on behalf of the Assyrian crown. Nonetheless, instances where the word *maṣṣartu* unequivocally refers to monitoring the actions of the king's subjects for individual acts of potentially seditious activity are lacking. As such, I consider more probable the argument put forth by Heather Baker and Melanie Groß that its general meaning refers to 'assigned duties or service', which had 'connotations of vigilance in their performance'.[282] Thus, although the term *maṣṣartu* is certainly related to ideas of attention and protection, its meaning was probably more specific than generalized watchfulness. In either case, it seems unlikely that it had precisely the same implications as the covenant and its specific duty of vigilance. This notwithstanding, the author of this letter evidently connects the concepts of *maṣṣartu* and *adê*, and considers his claim to keep and protect these institutions to be evidence of his loyalty to the crown. This connection presumably reflects a belief, at least on the part of some, that these two demands on the part of the crown were related to some degree.

The final communication to Esarhaddon that mentions covenant is published as SAA 8, no. 536.[283] Perhaps interestingly given the discussion above, it is an astrological-astronomical report written by a Babylonian subject. The writer's name

---

279 SAA 18, no. 102: obv. 3'–9'.

280 On this term, see SAA 10, xxi–xxiv, Fales 2011 and Baker and Gross 2015, 80 f.

281 Fales 2011, 369.

282 Baker and Gross 2015, 80 f.

283 Note that the vast majority of datable astrological-astronomical reports were written between 679 and 665 BC (SAA 8, xx).

is mostly broken away: it is [...]a, son of Bel-ušallim.[284] As this communication is a report on celestial omina, he was presumably a trained astrologer, a *ṭupšar Enūma Anu Ellil*, literally 'scribe of (the omen series) 'When Anu, Ellil (and Ea established in council the plans of heaven and earth)''.[285] This designation generally referred to a secondary specialism, and therefore the writer of this report was likely also active in his local community in Babylonia in other capacities, perhaps of a scholarly or priestly nature.[286] The report opens with the writer's celestial observations. There follows a break, and the note referring to a covenant is preserved on the last lines of the reverse before the astrologer signed his name: '[......] when I entered the covenant [......] is good(?) so I did not write to the king my lord'.[287] This attestation is only relatively poorly preserved, but the astrologer here references covenant in a familiar way, namely by stating that he *erēbu* 'entered' into it, language that is also included in SAA 18, no. 83. What is different in this case, however, is that the writer seems to be citing this fact in order to justify his failure to write to the king. In this report, it is possible that the reference to the covenant refers to the actual instance of the astrologer entering it, which is why he did not report on his astronomical-astrological observations. Nonetheless, it is also possible that, since he has entered the covenant and knows only to report under certain circumstances, he was aware that he need not write to the king, as nothing concerning had occurred. The reference to something being *ṭābu* 'good', mirrors the term used in the covenant, as subjects are to report anything they hear that is not good.

8.2 Letters from Itti-Šamaš-balaṭu, royal delegate in the Phoenician client state of Arwad

There is only one surviving letter of the Assyrian royal correspondence that explicitly mentions the succession covenant of Esarhaddon and stems from a client state that is not Babylonia. This is not particularly surprising, as, with the exception of Babylonia, only a handful of letters from client states survive from the reign of Esarhaddon.[288] The document is sent by a certain Itti-Šamaš-balaṭu, about whom no evidence survives beyond three, or possibly four, letters that he addressed to

284 PNA 1/2, 337 s.v. Bēl-ušallim.
285 On this term, see Robson 2019, 4f.
286 Robson 2019, 139.
287 SAA 8, no. 536: rev. 4'–5'.
288 See the discussion on foreign affairs in the correspondence in SAA 16, xxi–xxvi.

Esarhaddon.[289] From these documents, however, it is possible to glean that he was an Assyrian official stationed in northern Phoenicia during the final years of Esarhaddon's reign. He never explicitly mentions his specific role in these missives, but it seems likely that he was a *qēpu*-official: this term literally means 'trusted one' and denotes an Assyrian royal delegate at a foreign court.[290] All client states would likely have had a *qēpu*-official in residence, and it seems reasonable to suppose that they would have been expected to send reports to Esarhaddon, particularly if something was wrong.

Table 6: References to the succession covenant from western client states in the royal correspondence.

Publication	Relevant extract	Sender	Date	Provenance
1. SAA 16, no. 126	As [it is said] in the covenant: '[May iron swords consume him] who go[es] to the south [and may iron swords consume him] who g[oes] to the north. May your waterskins b[reak] in a place of [severe] t[hirst]' – [by] the gods of the king, [I have don]e just as [it is said] in [to covenant].	Itti-Šamaš-balaṭu	670 BC	Arwad

In his correspondence, Itti-Šamaš-balaṭu mentions Ikkilû (i. e. Yakin-Lû), king of the client state of Arwad,[291] on several occasions, and as such it seems probable that he was stationed there for at least some of his posting as a *qēpu*-official (see Figure 5). Arwad, and Phoenicia at large, were of significant interest to Esarhaddon, as they were wealthy trading centres, as well as having access to important raw materials, such as cedar wood.[292] It is therefore possible that Esarhaddon would have selected a particularly experienced and trusted official to act on his behalf in that location, although, in the absence of more evidence concerning Itti-Šamaš-balaṭu's background, this supposition must remain purely speculative. That Itti-Šamaš-balaṭu's reports to the king were taken seriously, however, may

289 PNA 2/1, 589 s.v. Itti-Šamaš-balāṭu. The letters are published as SAA 16, nos. 126–129. The name of the sender of the final letter, SAA 16, no. 129, is not preserved and the editors attribute it to Itti-Šamaš-balāṭu based on its 'idiolect and writing conventions' (SAA 16, 115, no. 129).

290 On these officials, see Dubovský 2012, for a list of persons associated with this title, see also PNA 4/1, 86 f. s.v. *qēpu*.

291 PNA 2/1, 488 f. s.v. Iakīn-Lû.

292 On Phoenicia under Esarhaddon, see Fales 2017b, 238–244.

be illustrated by the existence of an extispicy query asking whether or not Ashurbanipal should send a message to Ikkilû via a certain Nabû-šarru-uṣur, the *rab–muggi* 'military governor'.[293] This text, drawn up during the first month of 670 BC, may well document a direct reaction to information that Itti-Šamaš-balaṭu had provided.[294] It is possibly significant that Ashurbanipal, the crown prince of Assyria, was here responsible for reacting to the information sent to the monarch by Itti-Šamaš-balaṭu in early 670 BC. The editors of Itti-Šamaš-balaṭu's correspondence, Mikko Luukko and Greta van Buylaere, suggest that Esarhaddon may have been on campaign in Egypt during that time, implying that they consider the query to date to 671 BC. As Frederick Mario Fales has already argued, however, it seems likely that the enquiry was made in the first month of 670 BC.[295] If this is correct, then it is possible that Esarhaddon was simply too ill to perform this task at the time, and had thus entrusted it to his son.[296] Whatever the case may be, it is worth noting that Itti-Šamaš-balaṭu himself addressed his letters to the king, and therefore was presumably working on the assumption that his letters were being read by Esarhaddon himself.

Despite this evidence that the crown was reacting to the information that he relayed, the tone of Itti-Šamaš-balaṭu's letters appears to convey significant concern on his part and possibly a lack of confidence that he enjoys the trust of the crown. He begins his surviving letters with copious compliments and blessings for Esarhaddon. The letter in which he mentions a covenant begins as follows:

> To the king, my lord, the righteous, sincere, and beloved of his gods: your servant Itti-Šamaš-balaṭu. Good health to the king, my lord! May Aššur, Šamaš, Bel, Nabû, Nergal, Ištar to Nineveh, Ištar of the Kidmuri temple, (and) Ištar of Arbela very generously give to the king, my lord, long days, everlasting years, happiness, physical well-being (and) joy.
>
> Just as the king, my lord, is truthful to god and man, and the command of the king, my lord, is good to god and man (and) 'the black-headed people,' in the same manner the powerful gods of the king, my lord, who raised the king, my lord, from childhood till maturity, will fully carry out (these blessings) and render them to the king, my lord. And [they will brin]g all the enemies [of] the king, my lord, [to submission] before the feet of the king, [my] lord.[297]

Itti-Šamaš-balaṭu here emphasizes Esarhaddon's divine approval and associates his own wishes for the monarch's good health with this assurance. Such a claim

293 On the identity of Nabû-šarru-uṣur, see Fales 1988. On the term *rab–muggi*, with reference to this text, see Radner 2002, 12f.

294 SAA 16, xxv; the document is published as SAA 4, no. 89.

295 Fales 1988, 107, fn. 16, as well as 117.

296 See further discussion in Chapter 6.

297 SAA 16, no. 126: obv. 1–18.

may suggest that Itti-Šamaš-balaṭu, like Esarhaddon's court scholars, was aware that there was concern regarding the degree of divine support that the monarch enjoyed at this time. Such an insight would suggest that Esarhaddon's woes were known among Assyrian officials beyond the royal court. While socially close enough to the monarch to correspond with him, Itti-Šamaš-balaṭu resided at a considerable physical distance from Assyria's royal centre: Arwad is 680 km from Nineveh as the crow flies. The statement that the gods will bring to submission 'all the enemies [of] the king' certainly seems to imply that Itti-Šamaš-balaṭu knew that the monarch was facing significant challenges, although what exactly he believed them to be is not specified.

Whether or not Itti-Šamaš-balaṭu had been told of the divine aspect of the crisis that had gripped the Assyrian court, however, he appears to have wished to use his assurances of divine support for the monarch to ingratiate himself with Esarhaddon. Although these statements are not preserved in SAA 16, no. 126, the letter that mentions covenant, the two other letters that can be securely attributed to Itti-Šamaš-balaṭu include pleas to be allowed to return to Esarhaddon's palace:

> May the king, my lord, not leave me in their hands. I am (but) a dead dog. May I wield the brooms in the palace of the king, my lord! [May I decorate the inte]rior of the palace of the king, my lord! May the gods of heaven and earth bless the king, my lord![298]

In both cases, this seems to be due to Itti-Šamaš-balaṭu's fears for his own safety in his current posting. In one of the letters, SAA 16, no. 127, he claims that the ruler of Arwad, Ikkilû, has not been letting boats into the port of the Assyrian king, but instead has 'turned the whole trade for himself'.[299] Ikkilû has been claiming that this behaviour has been authorized by the Assyrian crown. Itti-Šamaš-balaṭu also draws Esarhaddon's attention to a man 'from Ṣimirra' who 'goes back and forth to Assyria, finds out in detail whatever matter (and) news there is, and goes and tells it to him (i.e. Ikkilû).'[300] Itti-Šamaš-balaṭu states that he is unable to arrest this man, as the man is in Ikkilû's presence. Particularly interestingly, Itti-Šamaš-balaṭu ends his description of this state of affairs by claiming that:

> The king, my lord, should know that there are many in the entourage of the king, my lord, who have invested silver in this house – they and the merchants are systematically scaring me. I (however) put my trust in the king, my lord. I don't give one shekel (or even) half a shekel to anybody but the king, my lord. The king, my lord, should know (this).[301]

298 SAA 16, no. 128: rev. 13–e. 3.
299 SAA 16, no. 127: obv. 17.
300 SAA 16, no. 127: b.e. 25–rev. 2.
301 SAA 16, no. 127: rev. 6–14.

Thus, while Itti-Šamaš-balaṭu himself is stationed in a foreign state and is reporting in this letter primarily on the activities of the local client king, his letter casts aspersions on members of Esarhaddon's own entourage, the *manzāz–pāni ša šarri,*[302] who are interfering in foreign affairs in order to engage in corrupt activities. Indeed, it is these people, rather than Ikkilû, whom Itti-Šamaš-balaṭu claims to fear.

The second letter in which Itti-Šamaš-balaṭu asks to return to the palace of the king, SAA 16, no. 128, appears to describe the same situation. The section is less well preserved than that of SAA 16, no. 127, but provides some additional details nonetheless:

> [... There are many] in the entourage of the king, my lord, who have invested silver together with the merchants in this house and [there is] a strong alliance over the distance [...]. They are systematically scaring me. There is no fr[ien]d of mine here. I put my trust in the king, my lord. I do not give one shekel (or even) half a shekel to anybody but the king, my lord.[303]

This letter contains much of the same contents as SAA 16, no. 127, along with the statement that those in the king's entourage and the local merchants have *salāmu dannu ana muhhi rūqi* 'a strong alliance over the distance'. Itti-Šamaš-balaṭu's accusations do not include any references to the Sasî conspiracy, nonetheless the revelation that some of the members of Esarhaddon's immediate circle were engaged in schemes to enrich themselves, if it was believed, would surely have come as a blow to the monarch and deepened his distrust in his cohort. In this way, Itti-Šamaš-balaṭu's correspondence adds to the chorus of claims of seditious activity from locations across the Assyrian Empire during the final years of Esarhaddon's reign. That the royal delegate describes those closest to Esarhaddon as carrying out their activities abroad, and as doing it for economic reasons, adds an aspect of disloyalty not found in other letters on the plots of this time, in which the dissent seems to be largely ideological.[304] If these letters are to be seen as in any way connected to the plot to overthrow Esarhaddon, then they may provide insights into the potential financial advantages of doing so.

The letter in which Itti-Šamaš-balaṭu mentions covenant does not contain the same accusations, nor does it seem to include the pleas to return to court. It is possible that the letter dates to a period in which Itti-Šamaš-balaṭu's situation was less

---

302 On the designation *mazzāz pāni* and its translation as 'courtier', see Gross and Pirngruber 2014. See also Gross 2020, 561 f.

303 SAA 16, no. 128: rev. 1'–13'.

304 Compare Radner 2016, 53: 'The motivation of the 670 insurgents was apparently ideological. As far as we can see, economic reasons play no role at all.'

acute. Despite this, his mention of *mari nagarūtīni* [*ša*] *šarri* 'all the enemies [of] the king' seems to imply that Esarhaddon is facing a period of trouble, and so it is perhaps reasonable to suppose that the enemies to whom Itti-Šamaš-balaṭu alludes are the same people that he complains of in his other missives. The letter is relatively poorly preserved, in particular the reverse of the tablet. As such, the section concerning the covenant, which follows the introductory matter and well-wishes quoted above, is not followed by much elucidating context. The mention of the covenant comes immediately after the wish for the gods to bring Esarhaddon's enemies under his control:

> As [it is said] in the covenant: '[May iron swords consume him] who go[es] to the south [and may iron swords consume him] who g[oes] to the north. May your waterskins b[reak] in a place of [severe] t[hirst]' – [by] the gods of the king, [I have don]e just as [it is said] in [the covenant].[305]

Particularly interesting about this reference to covenant is the direct quotation from two of the curses found in that composition.[306] As the four lines across which the quotation is written are not completely preserved, the editors of Itti-Šamaš-balaṭu's correspondence, Mikko Luukko and Greta van Buylaere, have reconstructed them in part from reference to the succession covenant, which may give the unintended impression that these lines are more faithful to the original than is the case. Nonetheless, the letter has the clear distinction of, rather than merely paraphrasing the injunction to report, as various other letter-writers did in this period, including a direct quotation from the covenant curses. The two curses are indeed both found in Esarhaddon's succession covenant, although they are not consecutive in the original text. The surviving section of the quotation seems to indicate that the two statements were quoted verbatim. Itti-Šamaš-balaṭu has in each case excerpted a small portion of the curse exactly, discarding the rest. The sections of the succession covenant from which he quotes are as follows:

> If you should forsake Esarhaddon, king of Assyria, Ashurbanipal, the great crown prince designate, (his brothers, [sons by the same mother] as Ashurbanipal, the great crown prince designate, and the other sons, the offspring of [Esa]rhaddon, king of Assyria), going to the south or to the north, may iron swords consume him who goes to the south and may iron swords likewise consume him who goes to the north.[307]

305 SAA 16, no. 126: obv. 19–26.

306 As discussed in Steymans 2006, 337 f.

307 SAA 2, no. 6, § 96: 632–36; compare Lauinger 2012, viii 26–32, in which the description of the brothers differs slightly.

> Just as (this) waterskin is split and its water runs out, so may your waterskin break in a place of severe thirst; die of thirst![308]

It seems likely that this quotation on the part of the royal delegate alludes to the enemies of Esarhaddon, a group that may be identical to those about whom Itti-Šamaš-balaṭu complains in his other letters. If this is the case, then his choice of these two curses found in the succession covenant is surely deliberate. The former passage refers to movement of the cursed person *ana imitti* 'to the right' and *ana šumēli* 'to the left', translated by Simo Parpola and Kazuko Watanabe as 'south' and 'north', respectively. This could perhaps be interpreted, in Itti-Šamaš-balaṭu's usage, as a reference to the members of the monarch's entourage and the long-distance activities that they and the merchants in their employ are carrying out. By following this quotation with the statement that the cursed person's waterskin should break *ina kaqqar ṣumāmīt laplaptu* 'in a place of severe thirst', Itti-Šamaš-balaṭu also draws on a section of the covenant that appears to refer to misfortune in a remote place. The curses of the covenant composition that come between these two sections do not have similar associations. In this way, it is possible that Itti-Šamaš-balaṭu is here adapting the composition to fit the situation at hand. Itti-Šamaš-balaṭu would surely himself have been sworn to the covenant's oath. That he was able to quote it accurately suggests either that he had memorized this very long text, or that he had access to a copy of it. According to the covenant's stipulation, Ikkilû's succession covenant tablet should of course have been on display in Arwad.

It is worth noting that the succession covenant was probably not the only *adê* that was in effect in Arwad: Esarhaddon had subjected the Phoenician city-state of Tyre to a bilateral treaty in the year 676 BC,[309] and it seems probable that a similar one was also concluded with the ruler of Arwad. This is relevant in particular because the stipulations of Esarhaddon's treaty with King Ba'al of Tyre contain passages that govern maritime trade.[310] As such, Itti-Šamaš-balaṭu's claim that Ikkilû is not paying dues to Assyria would mean, if such an *adê* existed with Arwad, that he was violating its stipulations. The succession covenant, in contrast, does not explicitly include financial crimes in its provisions: the *pirru* 'tax-collection point' is mentioned,[311] but as a potential place in which a rebellion against Ashurbanipal may take place, rather than the site of theft from the crown. That Ikkilû may be act-

308 SAA 2, no. 6, § 102: 652–55; Lauinger 2012, viii 52–54.
309 SAA 2, no. 5. On Ba'al of Tyre, see PNA 1/2, 242–243 s.v. Ba'alu.
310 SAA 2, no. 5: rev. iii 1'–30'.
311 On *pirru*, see Postgate 1974b, 163–166 and Jursa and Radner 1996, 95.

ing against such a treaty is possibly relevant to Itti-Šamaš-balaṭu's reference to covenant. Nonetheless, it seems clear that, if he does have another *adê* in mind, it can only be in addition to Esarhaddon's succession covenant. Firstly, the covenant with Ba'al does not contain the curses from which Itti-Šamaš-balaṭu quotes. Indeed, that document contains only 'traditional' curses, while the ones that Itti-Šamaš-balaṭu references are 'ceremonial' in type.[312] It is, however, possible that the curses for a treaty concluded with Ikkilû would have been different to those used for the treaty that Esarhaddon imposed on Ba'al. Beyond this, it is clear that the *adê* to which Itti-Šamaš-balaṭu is referring is one that does not only apply to the people of Arwad. If the reconstruction offered by Luukko and van Buylaere of the fragmentary passage that comes after the covenant quotation is correct, then Itti-Šamaš-balaṭu himself professes to have obeyed the covenant. A bilateral treaty with Arwad would not apply to Itti-Šamaš-balaṭu personally, in contrast to the succession covenant, to which he had recently sworn. In addition, if this letter refers to an at all similar situation to the one that Itti-Šamaš-balaṭu describes in his other communications, then it may be the disloyal actions of members of Esarhaddon's cohort that he is discussing here. These individuals would also be bound by the succession covenant and not by a treaty with one specific client state.

Unfortunately, the way in which Itti-Šamaš-balaṭu has acted is not preserved, and therefore no further insights into the actions that he thought constituted keeping the covenant can be gained. Nonetheless, the manner in which Itti-Šamaš-balaṭu uses the covenant in his letter is telling. His precise quotations seem to imply an intimate knowledge of the composition, and possibly easy direct access to a covenant manuscript. While the information that Itti-Šamaš-balaṭu is reporting to Esarhaddon may indeed pertain to those who were resident in northern Phoenicia, it seems distinctly possible that Itti-Šamaš-balaṭu also considered it to be related to members of the royal entourage. If that is the case, then this lone example of an Assyrian in a client state referencing the covenant is actually to be viewed as a report on a group of people based at Esarhaddon's court.

8.3 Reframing the duty of vigilance in Judah: Deuteronomy 13

Comparing Esarhaddon's succession covenant with the stipulations of Deuteronomy 13 and curses of Deuteronomy 28, one is struck by the strong similarities between these chapters of Deuteronomy and the succession covenant that Esar-

312 SAA 2, no. 5: rev. iv 1'–r.e. 19.

haddon would have imposed on his client, Manasseh of Judah (r. 687–642 BC), in 672 BC. The stipulations found in Deuteronomy 13 are particularly relevant to the study of responses to the covenant's duty of vigilance, as they mirror some of the very stipulations in Esarhaddon's succession covenant most clearly designed to impose this duty. Nonetheless, the terms of Deuteronomy 13 differ in many significant conceptual particulars from those of Esarhaddon's succession covenant. Most importantly, it is not Ashurbanipal, but rather the god Yhwh on whose account those bound by the covenant must act. As in the succession covenant, where those swearing to it must acknowledge that Ashurbanipal is the only legitimate claimant to the Assyrian throne, the danger in Deuteronomy 13 is the acknowledgement of a rival deity. Nonetheless, these demands are expressed in a manner that closely resembles the contents of Esarhaddon's succession covenant.

Why do these similarities exist? The reason for the parallels between these texts remains the subject of much debate. Many scholars have argued for a connection between Esarhaddon's succession covenant and Deuteronomy 13 and 28.[313] Nonetheless, this is far from a consensus, and various scholars have argued against literary dependency.[314] The 2019 special issue of the journal *Hebrew Bible and Ancient Israel* (HeBAI 8/2), which was dedicated to 'Perspectives of the Treaty Framework of Deuteronomy', offers an instructive encapsulation of the current debate on the subject, including articles both for and against a direct relationship.[315] In his contribution, Hans Ulrich Steymans uses Deuteronomy 13 as a case study to argue that the similarity between the texts 'is so unique that there must be a direct connection'.[316] He posits that a scribe writing a draft of an early version of Deuteronomy during the reign of Manasseh of Judah would have been inspired by his exposure to Esarhaddon's succession covenant to 'start dreaming... of a better loyalty – to Yhwh instead of Esarhaddon'.[317] In Steymans's telling, therefore, Deuteronomy 13 is a direct response to the imposition of the succession covenant on Judah: an instance of emulation of but also resistance to the Assyrian crown.

313 Eckart Otto has been one of the strongest advocates of this position, arguing that portions of Deuteronomy amount to translations of the succession covenant (E. Otto 1999, 68). See also E. Otto 1998, as well as recently E. Otto 2016, and E. Otto 2017. Many other scholars have also argued for a relationship between the texts, see for instance Frankena 1965; Levinson 2010; Radner 2006; Rüterswörden 2006; Steymans 1995; Steymans 2003; Steymans 2013; Steymans 2019.

314 Skeptical voices include Koch 2008, Pakkala 2009, Crouch 2014 and Morrow 2019. See recently also Arnold and Shockey 2022, who argue for 'conceptual borrowing', but against direct literary borrowing.

315 See Edenburg and Müller 2019, with bibliography.

316 Steymans 2019, 112.

317 Steymans 2019, 131.

In contrast, William Morrow writes in the same issue that the 'thesis that Deuteronomy's scribes intended to subvert the claims of Neo-Assyrian hegemony… can no longer be sustained on the basis of Deuteronomy 13 and 28.'[318] He bases this assertion on parallels between these sections and other such Ancient Near Eastern treaty compositions, although he acknowledges that there are 'some convincing connections' between Esarhaddon's succession covenant and these chapters of Deuteronomy.[319] As such, his major contention is that the Deuteronomistic author who penned these sections, whom he considers would have been writing no earlier than the reign of King Josiah of Judah (r. 640–609 BC), and possibly significantly later, would have been unaware of Esarhaddon's succession covenant.[320] As such, he proposes a model of mediated dependency, with the composers of Deuteronomy 13 and 28 drawing on a Judean loyalty oath written by earlier scribes who did draw on Esarhaddon's succession covenant.[321]

Juha Pakkala, meanwhile, accepts that Judean scribes drew on the Ancient Near Eastern treaty tradition, and probably specifically on client treaties between Judah and either Assyria or Babylonia, when drawing up these sections of Deuteronomy. He argues, however, that the scribes would have been working in the post-monarchic period, shortly after the fall of the Kingdom of Judah in 587 BC.[322] He posits that the Judean scribes who did this would have been steeped in this broader treaty tradition, and used it to develop their theological beliefs.[323] As such, while he does not exclude the possibility that Esarhaddon's succession covenant was one of the treaties that informed the scribes, he does not appear to consider it necessarily to have had a special status within this tradition.

Finally, Jacob Lauinger points out that Judeans may well have been aware – and indeed resentful – of Esarhaddon's succession covenant long after the end of Esarhaddon's reign, as it was likely intended to be used for many years after that monarch's death.[324]

When reflecting on the scribe(s) who would have composed Deuteronomy 13, then, these various arguments present four possible scenarios: 1) the scribe(s) were aware of Esarhaddon's succession covenant, and wrote Deuteronomy 13 in the years immediately after 672 BC; 2) the scribe(s) were aware of Esarhaddon's

---

318 Morrow 2019, 135.
319 Morrow 2019, 154.
320 Morrow 2019, 154.
321 Morrow 2019, 156 f.
322 Pakkala 2019, 164 f.
323 Pakkala 2019, 180 f.
324 Lauinger 2019; note that Jacob Lauinger frames his article in part as a response to the arguments advanced by Carly Crouch (2014).

succession covenant, but were writing during a later period (at least fifty years after its imposition and possibly much longer); 3) the scribe(s) were not aware of Esarhaddon's succession covenant, but knew of a different such document that was dependent on it, and were writing in a later period; 4) the scribe(s) were not aware of Esarhaddon's succession covenant, but were steeped in a broader tradition of Ancient Near Eastern treaties, and were writing in a later period.

It is worth noting that all of these theories allow for potential influence of Esarhaddon's succession covenant on Deuteronomy 13. Indeed, each one allows for an interpretation of these portions of Deuteronomy to be viewed as evidence of a response to Esarhaddon's succession covenant to some degree. Nonetheless, the nature of this response varies significantly depending on which hypothesis one accepts, and as such it is worth reflecting on their relative likelihood and respective implications. Each of the four suggestions is possible, but the fourth suggestion – that the similarities between the two texts are evidence of little more than a shared tradition – strikes me as least probable. As has been established in this study and elsewhere, Esarhaddon's succession covenant was highly specific and innovative within the confines of its genre.[325] As such, it should not be conflated with the diffuse concept of treaty tradition, as much of its content is unique among known documents of this type.[326]

The relative likelihood of the remaining three possibilities depends in no small part on the much-debated question of the composition date of Deuteronomy specifically, and of the Deuteronomistic History (i.e. the books of Deuteronomy, Joshua, Judges, Samuel, and Kings) in general. In his influential 1943 study, Martin Noth posits that one writer living during the exilic period (586–538 BC), known in modern scholarship as the Deuteronomist, wrote the Deuteronomistic History by incorporating various diverse sources into a single work.[327] Noth identified what he termed the *Urdeuteronomium* (proto-Deuteronomy; Dtn. 4:44–30:20), the original core of the Book of Deuteronomy, which he argued the Deuteronomist reframed as the word of Moses.[328] In the intervening decades since the publication of Noth's work, however, many scholars have critiqued and honed his thesis. Although many diverse opinions exist regarding the date and composition context of the Deuteronomistic History, scholars frequently argue for a more complex and protracted redaction history for these books, both as a group and individually,

325 As discussed in Chapters 2 and 3. See also Watanabe 2015.

326 As is argued by Hans Ulrich Steymans (2019).

327 Noth 1957; see also the English translation of the first part of this book, Noth 1991.

328 See also Eckart Otto's discussion of the portrayal of Moses and its relationship to Neo-Assyrian royal ideology (E. Otto 2009).

than was suggested by Noth.[329] Unsurprisingly, therefore, the dating of Deuteronomy remains the subject of controversy, and a clear consensus it lacking. As such, while the first argument listed above, that at least some parts of Deuteronomy 13 were composed shortly after the imposition of Esarhaddon's succession covenant in 672 BC, has the advantage of parsimony,[330] it also seems to me perfectly possible that the second or third scenarios may be correct.

The relationship between the scribe writing Deuteronomy 13 and Esarhaddon's succession covenant differs in each of these three scenarios. The first case is a clear instance of emulation and subversion: the scribe was apparently inspired by the framing of the succession covenant to imitate it, but did so in such a way that rejected the demand of loyalty to the Assyrian royal house in favour of loyalty to Yhwh.

In the second instance, the nature of the response is the same, using the language of a foreign oppressor to reject their demands,[331] but its political import is somewhat different. William Morrow argues that rejection of Assyria would not have been relevant after 622 BC, and those writing Deuteronomy 13 and 28 at this time or later would not have had any reason to subvert Esarhaddon's succession covenant.[332] Jacob Lauinger's arguments, as well as the results of this study, challenge this assumption, however, as they indicate that the covenant may have been politically relevant after Esarhaddon's reign.[333] It is also worth stressing that, although it was Esarhaddon who imposed his succession covenant, it pertains to Ashurbanipal (r. 668 – ca. 631 BC), whose long reign brings us much closer in time to Morrow's *terminus post quem*. As such, I consider it possible that the scribe(s) writing Deuteronomy 13 may still have wished to subvert the terms of Esarhaddon's succession covenant even during this later period. Indeed, Morrow's comment that Assyria was losing influence during the reign of Josiah could be construed as supporting this reconstruction: the diminishment of Assyrian power could have rendered both the Judean monarch and elite scribes more able to defy their erstwhile overlords. Indeed, 2 Kings claims that Josiah did just this, re-

329 As is clear from discussion of subsequent scholarship on the Deuteronomistic History, see for instance Knoppers 2010.

330 Steymans 2019, 112.

331 William Morrow has previously characterised this form of subversion as a 'paradox' (Morrow 2009).

332 Morrow 2019, 154.

333 See the discussion in Chapter 5.3., as well as Lauinger 2019.

belling against the Assyrians,[334] while also implementing major religious reforms and imposing a covenant on his people.[335] Whether or not this is entirely accurate, it seems quite possible that Josiah shrugged off the yoke of Assyrian imperial rule, and he was certainly remembered as a reformer. If Deuteronomy 13 was a part of this, then perhaps the scribe(s) could be construed as taking part in a wider scale rejection of Assyrian rule. If Deuteronomy 13 was written much later than this, and the scribe(s) chose to allude to Esarhaddon's succession covenant over fifty years, and possibly much longer, after 672 BC, then such a decision can probably be taken as evidence that its demands had a significant impact in the client state of Judah.

The third possibility, that Deuteronomy 13 was written after the reign of Esarhaddon and was influenced by his succession covenant only indirectly, is the most complex in terms of reconstructing the possible intentions of the scribe(s) who composed it. William Morrow posits that the scribe(s) who wrote Deuteronomy 13 would have based it on a different Judean 'loyalty oath', that was itself inspired by Esarhaddon's succession covenant.[336] In this scenario, then, Judean scribes at some point utilized Esarhaddon's succession covenant in order to compose a Judean covenant. This document was then repurposed later by the Deuteronomistic scribe(s) who composed Deuteronomy 13. The nature of this intermediary loyalty oath, if it existed, is, of course, unknown. It is possible that the composition incorporated elements of the stipulations and curses of Esarhaddon's succession covenant not in relation to Yhwh, as in Deuteronomy, but rather in relation to a Judean monarch.[337] It could be argued that this scenario would be more a case of simple emulation than subversion, as it probably represents a smaller change to the original contents of Esarhaddon's succession covenant than a covenant concerning Yhwh.[338] Despite this, the act of repurposing portions of Esarhaddon's succession covenant to pertain to the royal house of Judah instead of the Assyrian crown may well have been intended to subvert the demand of Esarhaddon's succession covenant that subjects of the client states be loyal to the Assyrian crown and become vigilant on its behalf. Such a composition, if it ever existed, would presumably have been composed by elite Judean scribes at the behest of a Judean monarch. Thus, in

334 2 Kgs 23:29. For a discussion of Josiah's death, see Hasegawa 2017. Josiah's rebellion against the Assyrians remains conjectural, and the historical reality of Josiah's religious reforms also remain a subject of debate, on which see, for instance, Fried 2002 and Monroe 2011.

335 2 Kgs: 22–23. Indeed, some scholars have put forward the so-called 'Double Redaction' theory, which holds that an initial edition of the Deuteronomistic History was composed under Josiah: Cross 1973, 274–289; Nelson 1981; Nelson 2005. For a skeptical view, see Davies 2010.

336 Morrow 2019, 157.

337 Morrow 2019, 157. Morrow suggests that the loyalty oath may have been to King Manasseh.

338 As argued in Morrow 2019, 157.

this scenario, these actors could be interpreted as rejecting their status as clients and attempting to redirect the loyalty and vigilance of Judah's subjects. Such a possibility would again be a strong indication of an appreciation of the techniques of Esarhaddon's succession covenant, in particular its demands for vigilance, coupled with a rejection of its fundamental aims – to direct loyalty towards the Assyrian crown and away from local groups. The later scribe(s) writing Deuteronomy 13 would then potentially have repurposed this loyalty oath in a similar way, redirecting loyalty from a person to their god while keeping elements of its contents.

Even though the dynamics of each of these three possibilities are different, therefore, each of them requires one or more Judean scribes to consider the contents of Esarhaddon's succession covenant consequential enough to be worth emulating. So too, these possibilities also contain the possibility that the scribe(s) strongly rejected the covenant's demand that the subjects of Judah be loyal to the Assyrian crown. As such, while Deuteronomy 13 may not be a direct response to Esarhaddon's succession covenant, but rather a response to a response, it is nonetheless worth comparing the nature of vigilance in these two compositions. Even if their relationship is not direct, Deuteronomy 13 provides insights into the elements of Esarhaddon's succession covenant that certain scribes in Judah considered worth keeping, and is likely in itself evidence that they rejected the demand to direct their loyalty towards the Assyrian crown exclusively.

The first and last verses (Dtn. 13:1 and 13:19) of Deuteronomy 13 are considered by several scholars to be a framing device added to the three sections of the core of the chapter (Dtn. 13:2–6; 7–12; 13–18),[339] although this is contested by some.[340] Nonetheless, even if this is correct, the identities of the authors of these different portions of the chapter are unknown, and it is entirely possible that the author(s) of these framing sections may have been aware of Esarhaddon's succession covenant or another covenant based on it.[341]

As in § 10 (scenario no. 3, see Table 1) of Esarhaddon's succession covenant, the first verses of Deuteronomy 13 single out prophets as potentially dangerous and thus as necessary objects of vigilance:

> If a prophet, or one who foretells by dreams, appears among you and announces to you a sign or wonder, and if the sign or wonder spoken of takes place, and the prophet says, 'Let us follow other gods' (gods you have not known) 'and let us worship them,' you must not listen to

339 Levinson 2010, E. Otto 2016.

340 For the division of Deuteronomy into these three sections, see, as well as Morrow 2019, 140–143.

341 For a discussion of the stipulations of the succession covenant, see Chapter 3.

> the words of that prophet or dreamer. The Lord your God is testing you to find out whether you love him with all your heart and with all your soul.[342]

In addition, the description presents the possible statement made by such people as verbatim quotations, similarly to scenarios nos. 12–14 of the succession covenant.[343] In the succession covenant, prophets and diviners are mentioned simply as potential groups that may speak 'any evil, improper, ugly word' concerning Ashurbanipal.[344] In contrast to this, the first stipulation in Deuteronomy 13 is concerned specifically with a situation in which a 'sign or wonder' foretold by the prophet or dream-diviner has taken place. It views such an eventuality as a divine test, something that implies a level of divine scrutiny not postulated in the succession covenant. In these stipulations, the imposer, enforcer, and subject of the stipulations is the god Yhwh. Unlike Ashurbanipal, therefore, Yhwh is portrayed as controlling the situation at all stages of the scenario, even the speech and actions against him.[345] Despite this, the mandated action of the scenario is a response by the covenant's human parties:

> That prophet or dreamer must be put to death for inciting rebellion against the Lord your God, who brought you out of Egypt and redeemed you from the land of slavery. That prophet or dreamer tried to turn you from the way the Lord your God commanded you to follow. You must purge the evil from among you.[346]

Those sworn to the covenant are therefore required to put the 'prophet or dreamer' to death. This bears comparison with Esarhaddon's succession covenant, as such mandated actions are also found there. The stipulation to report to an authority, however, is not present in Deuteronomy 13. Thus, the followers of Yhwh are commanded to act on his behalf, directing their attention towards him and away from those who attempt to draw them away from him. It is implied that they are also to remain alert to the possibility of divine scrutiny in the form of tests sent by Yhwh. The terms in which such a divine test is described parallel the demand in the succession covenant that Assyria's subjects 'shall love Ashurbanipal, the great crown prince designate, son of Esarhaddon, king of Assyria, your

---

342 Dtn. 13:2–4. These translations follow the New International Version (NIV).
343 SAA 2, no. 6, §§ 27, 29 and 31. See also Lauinger 2012.
344 SAA 2, no. 6, § 10: 108 f.
345 Eckart Otto has argued that the undermining of state control in this way was an important conceptual step in the direction of 'human rights' (E. Otto 2002).
346 Dtn. 13:6.

lord, like yourselves.'[347] A similar statement is also found in Deuteronomy 6:4–5, and of course the wording of the succession covenant stipulation is similar to that of Leviticus 19:18: 'Do not seek revenge or bear a grudge against anyone among your people, but love your neighbor as yourself. I am the Lord.' These demands, like that pertaining to Ashurbanipal, mandate a particular attitude.[348]

As in Leviticus 19:18, the stipulations of Deuteronomy 13 focus on behaviour towards the local community. This is highlighted by the command that those bound by the Biblical covenant 'purge the evil from among you'. In a manner similar to the succession covenant, too, the stipulations are imposed upon the group as a whole, not the individual. Thus, Deuteronomy 13 makes the Judeans mutually responsible for one another's religious behaviour, specifically for worshipping Yhwh exclusively. The terms of Deuteronomy 13 seek to ensure the appropriate behaviour of the community by perceiving and punishing dissidents.

Following prophets and dream interpreters, the stipulations move on to the possibility that a member of one's own family may seek to worship other gods:

> If your very own brother, or your son or daughter, or the wife you love, or your closest friend secretly entices you, saying, 'Let us go and worship other gods' (gods that neither you nor your ancestors have known, gods of the peoples around you, whether near or far, from one end of the land to the other), do not yield to them or listen to them. Show them no pity. Do not spare them or shield them. You must certainly put them to death. Your hand must be the first in putting them to death, and then the hands of all the people. Stone them to death, because they tried to turn you away from the Lord your God, who brought you out of Egypt, out of the land of slavery. Then all Israel will hear and be afraid, and no one among you will do such an evil thing again.[349]

Here, the importance of the proper worship of Yhwh among the group is stressed as more important than individual familial relationships. The statement very clearly parallels § 10 of the succession covenant, in which it is also explicitly mandated that one report members of one's own family. Again, in Deuteronomy 13, there is no reporting, only direct action. Those bound to the covenant must disregard the perpetrator and must not protect them. The penalty for the family member is once again death, and the description of the manner in which this individual must be killed is significantly more detailed than parallel passages in Esarhaddon's succession covenant. It is worth noting that chapter 13 is one of the most explicitly

---

347 SAA 2, no. 6: § 24, 266–268; Lauinger 2012, iv 8.

348 As discussed in Chapter 3. See the discussion in Watanabe 2014, 164 and Watanabe 2019, 255 f.

349 Dtn 13:7–12.

violent portions of Deuteronomy, and is certainly far more explicitly violent than the stipulations of Esarhaddon's succession covenant.[350]

The perpetrator is to be stoned to death, a highly visible form of punishment and an opportunity for mutual scrutiny among the group. The purpose of such a measure is clear: it is to serve as a deterrent, and will be something 'all Israel will hear' about. Thus, the threat of such a fate is explicitly intended to foster a fear of speaking against Yhwh, even to one's own family, on the grounds that the family member will communicate this to the wider community, which will result in gruesome execution. Such a process aligns closely with the concept of vigilance.

The final portion of Deuteronomy 13 highlights the towns of Judah as possible sites of failure to properly worship Yhwh. Thus, it turns its attention from potential individual bad actors to groups. This is the focus that least clearly mirrors the third scenario of the succession covenant:[351]

> If you hear it said about one of the towns the Lord your God is giving you to live in that troublemakers have arisen among you and have led the people of their town astray, saying, 'Let us go and worship other gods' (gods you have not known), then you must inquire, probe and investigate it thoroughly. And if it is true and it has been proved that this detestable thing has been done among you, you must certainly put to the sword all who live in that town. You must destroy it completely, both its people and its livestock. You are to gather all the plunder of the town into the middle of the public square and completely burn the town and all its plunder as a whole burnt offering to the Lord your God. That town is to remain a ruin forever, never to be rebuilt, and none of the condemned things are to be found in your hands. Then the Lord will turn from his fierce anger, will show you mercy, and will have compassion on you. He will increase your numbers, as he promised on oath to your ancestors – because you obey the Lord your God by keeping all his commands that I am giving you today and doing what is right in his eyes.[352]

This stipulation begins with the demand that, if those bound by the covenant hear reports that a particular settlement is acting wrongly, they must investigate these claims. Here, rather than demanding that people report to an authority, the chapter requires that they verify rumours that they have heard. Nonetheless, if the statement proves to be true, then the punishment is death for the entire population of the settlement, along with the livestock, and destruction of the buildings. This is far harsher than any of the stipulations found in Esarhaddon's succession covenant. In contrast to the stoning described in the previous scenario, which is a

350 Morrow 2019, 227.
351 For the relationship of this section to Dtn. 2–12, see Morrow 2019, 142f.
352 Dtn. 13:13–19.

warning to the rest of the group, the destruction of the town is therefore a signal to Yhwh that the people are pious.

Despite the uncertainty regarding the context in which scribe(s) who composed Deuteronomy 13 were writing,[353] the terms of this chapter can be viewed as existing in conversation with Judah's history as a client state mandated to swear to covenants. It seems reasonable to assume that the only known covenant composition that was certainly designed for display in religious venues across the provinces and client states of Assyria had an outsized impact on this conversation. Whether the scribe(s) were drawing directly on Esarhaddon's succession covenant or an intermediary composition, the apparent repurposing of these Assyrian stipulations by what was probably the political and religious elite of a client state can be interpreted as evidence that the imposition of Esarhaddon's succession covenant gave rise to consequences that the Assyrian crown had probably not envisioned or intended.

8.4 A literary response: wrongful accusation in the *Story of Ahiqar*

The final source discussed in this chapter is the Aramaic *Story of Ahiqar*, a tale that does not explicitly mention a covenant but appears to reference the atmosphere of mistrust that pervaded Esarhaddon's court in the final years of his reign. The *Story of Ahiqar* is one of two Aramaic literary compositions known from papyri dating to the centuries after the fall of the Assyrian Empire and found in Egypt that tell remarkably accurate tales of events that took place at the courts of the last great kings of Assyria: Sennacherib, Esarhaddon and Ashurbanipal.[354] The other composition is known as the *Tale of the Two Brothers* and was written in Aramaic language and Demotic script on Papyrus Amherst 63.[355] This narrative tells of the war between the royal brothers 'Sarbanabal' (Ashurbanipal) and 'Sarmugi' (Šamaš-šumu-ukin) that did indeed occur in 652–648 BC.[356] The tale focuses in particular on the communications between the two brothers and their deteriorating relationship. Central to this are two other figures, namely the Assyrian

353 On this, see also the arguments of Mark George, who considers that Deuteronomy writes 'Israel' into existence, in large part through the repeated demands to 'monitor' and 'enact' the commands in the book (M. George forthcoming). This theory further highlights the importance of the dynamics of vigilance in the text.
354 Fales 2020, 228.
355 Fales 2020 and Van der Toorn 2018, as well as Dalley 2001.
356 Fales 2020, 230–232; Dalley 2001, 156.

turtānu 'commander-in-chief' and the royal protagonists' sister, Saritra. The princess in question is probably Šeru'a-eṭirat,[357] who is known from contemporary documents and who may well have had an active political role at the court of her father and brother.[358] So too, the term *turtānu* 'commander-in-chief' is accurate to the period in which the narrative is set.[359]

In this way, even though it is certainly fictionalized, the tale reveals a fairly high degree of familiarity with the reality of the Assyrian court and royal family during the time it describes. The *Story of Ahiqar* is similar in this regard: Ahiqar himself is an advisor and 'seal-bearer' to Sennacherib and then to Esarhaddon. His professional designation is not an authentic Neo-Assyrian title, although royal seals were in use among Assyrian officials. The term *rab–unqāti* 'seal-bearer' is, however, attested as a Neo-Babylonian title.[360] Ahiqar himself is not attested at the court of Esarhaddon, and may well not have been a historical figure.[361] Despite this, he has various similarities to known courtiers of Esarhaddon, particularly Adad-šumu-uṣur, who is discussed in Chapter 6. As such, the depiction of Ahiqar appears at the very least to be informed by knowledge of the dynamics at play in Esarhaddon's court.

The narrative begins by introducing Ahiqar, a 'wise and experienced scribe',[362] who had served under Sennacherib and then Esarhaddon.[363] In a manner perhaps similar to *The Underworld Vision of an Assyrian Prince*, the father-son dynamic that is portrayed in the narrative as important within the Assyrian royal family is also

---

357 PNA 3/2, 1264 s.v. Šērū'a-ēṭirat.

358 See the discussion of this figure in Fales 2020, 243 f.

359 Fales 2020, 240 f. See also PNA 4/1, 195–197 s.v. *turtānu.*

360 Radner 2008b, 508; Frame 1991, 55–59.

361 Niehr 2007, 7–10. See also Takayoshi Oshima's discussion of instances of the name Ahiqar in cuneiform sources, which show that one or more individuals with that name were active in Assyria in the seventh century BC: Oshima 2017, 144–146. Note also that a late Uruk cuneiform document (165 BC) mentions Ahiqar with the second name Ṭupšar-Ellil-dari, a name that is attested in the colophon of a seventh-century literary manuscript from the Assyrian royal library at Nineveh: 'Nippur, house of Ṭupšar-Ellil-dari' (Beaulieu 2010, 16, esp. fn. 47).

362 Note that my translations closely follow those of Herbert Niehr's German translation of the Elephantine papyrus (2007, 38–52).

363 Note that Frederick Mario Fales states that Sennacherib is presented as Esarhaddon's successor in the Elephantine version of the *Story of Ahiqar* (Fales 2020, 235, fn. 35). Sennacherib is actually described as the father of Esarhaddon in the Elephantine papyrus (Niehr 2007, 38: i 5, see more recently Moore 2022, 245). Indeed, it is a later Syriac manuscript that makes this error (Moore 2022, 247). Note, however, that the Elephantine papyrus (P 13446) does contain mention of 'Esarharib', a conflation of the two royal names which is also found in a manuscript of the Book of Tobit (Moore 2022, 252).

found between the scribe and his son.[364] As Ahiqar does not have any biological sons, he adopts his nephew, Nadin, securing him a position at court. Both the adoption of a son as an heir and family connections allowing a scholar to enter the monarch's entourage are accurate for this period. Although Adad-šumu-uṣur appears to have had his own sons, it is worth noting that both his sons and his nephews were active at Esarhaddon's court. Despite Ahiqar's kindness to Nadin, the latter betrays Ahiqar: 'my son, who was not my son, was planning wrongdoing'.[365] His treachery entails turning Esarhaddon against his former advisor.[366] This section is preserved only fragmentarily, but it is clear that Nadin does this in part by convincing the monarch that Ahiqar has been involved in a widespread conspiracy. The phrase attributed to Esarhaddon, 'Why should he stir up the country against us?',[367] clearly indicates this involvement, and, as Frederick Mario Fales has already pointed out, it seems likely that the statement alludes to the real historical plot against the monarch that was subdued in 670 BC.[368]

Nadin's scheme is so successful that Esarhaddon condemns Ahiqar to death. An officer named Nabû-šumu-iškun is tasked with the execution, but Ahiqar successfully persuades him against carrying it out. Ahiqar does this by reminding Nabû-šumu-iškun that Sennacherib once ordered that Nabû-šumu-iškun himself be put to death. In contrast to Ahiqar himself, it is perhaps possible to link Nabû-šumu-iškun to a contemporary figure of the same name,[369] strengthening the case that the story contains some historically accurate elements. At the time that Sennacherib made this order, Ahiqar did not follow the monarch's orders, instead hiding the officer in his home and claiming to have killed him, 'until in the [n]ext time and after many days, I brought you before King Sennacherib and removed your offenses in front of him'.[370] As the sage recalls, Sennacherib was approving of the action. Ahiqar claims that Esarhaddon will react similarly now that the roles are reversed: 'Esarhaddon is merciful, as you know. In the future he will remember me and will desire my advice. Th[en] you will take me to him and he will make me live.'[371] The officer agrees to go along with Ahiqar's proposal, killing a eunuch slave of his own instead of the sage. As he explains to his two men, they need a body because Esarhaddon will send people to check that they have

364 Niehr 2007, 38: i 1–2.
365 Niehr 2007, 40: ii 30.
366 On the subject of loyalty and betrayal in the text, see Olyan 2020.
367 Niehr 2007, 40: iii 36.
368 Fales 1994, 48, fn. 56.
369 Niehr 2007, 8, incl. fn. 50; PNA 2/2, 888–890 s.v. Nabû-šumu-iškun.
370 Niehr 2007, 41: iv 49–50.
371 Niehr 2007, 41: iv 53–54.

really carried out their task. Nabû-šumu-iškun then hides Ahiqar in his home, taking care of him, while word of his death spreads throughout the land. The narrative breaks off with Esarhaddon requesting confirmation that Ahiqar has been killed. There then follows a list of wise sayings attributed to Ahiqar.[372]

The papyrus on which this narrative is preserved, Berlin P. 13446, bears additional writing. James D. Moore has recently offered a new reading of one of these 'compositional acts' on the papyrus, which is separate to the narrative described above but still concerns Ahiqar. The sentence reads: 'Saying: 24 years belong(ed) to (the) lord of kings, Senn[acherib] the [k]ing. In Kalhu I advised the kingdom/kingship of Assyria.'[373] As Moore points out, this highly-specific statement – that Ahiqar served Sennacherib from Kalhu – adds yet another parallel to the family of Adad-šumu-uṣur, which was closely associated with that city.[374] The length of Sennacherib's reign given here is also correct.

The papyrus itself is the first known attestation of the tradition of the wise sage Ahiqar. It was found in Elephantine in Egypt and dates to the end of the fifth century BC.[375] While this is the earliest surviving instance of this tale, it is clear that this is one iteration of a long and varied string of narratives concerning this figure.[376] In terms of viewing the *Story of Ahiqar* as evidence of an immediate response to the reality of Esarhaddon's reign, the various points of accuracy concerning the situation of scholars at Assyrian court in general and the profile of Sennacherib and Esarhaddon's advisors in particular suggest that the tale originates close to the Assyrian court. At the very least, the initial promulgators of this tale had a deep knowledge of the context in which they set their narrative.

In addition, it is clear that the *Story of Ahiqar* in the version found in Elephantine has some Babylonian influences, such as the inclusion of a Babylonian professional title. It is perhaps also worth noting that the historical Nabû-šumu-iškun may have been a Babylonian.[377] There is also much later evidence of a cuneiform Ahiqar tradition located in Babylonia: a cuneiform tablet, drawn up in 165 BC in Uruk, lists ancient kings and their sages and includes Esarhaddon and Ṭupšar-Ellil-dari, also known by his Aramaic name, Ahiqar.[378] Even though it is not possi-

372 Niehr 2007, 42–52: vi 79–xiv 222.

373 Moore 2022, 242.

374 James D. Moore focuses in particular on Adad-šumu-uṣur's father, Nabû-zuqup-kenu (2022, 255). See also the discussion of this family in Chapter 1.3.3 and Chapter 6.1.2.

375 Niehr 2007, 1–7.

376 See, for instance, Seth A. Bledsoe's characterization of Ahiqar as an 'international bestseller' (Bledsoe 2021, 55–60).

377 Niehr 2007, 8, incl. fn. 50.

378 Robson 2019, 186 and 202, fn. 177.

ble to know whether or not the Ahiqar tradition had taken hold in Babylonia while that region was still an Assyrian client state, it perhaps implies an interest in this period.

Beyond Babylonia, the find context of Berlin P. 13446 in Elephantine is revealing, as it shows just how far the tradition had spread by the Persian period. The precise find context of the papyri is unfortunately not known, although the existence of many Aramaic documents from Elephantine dating to this period concern a Judean community stationed there among the Achaemenid soldiers.[379] The precise identity of this community is the subject of much discussion,[380] but it is certainly interesting that the character of Ahiqar also appears in the apocryphal Book of Tobit as the nephew of Tobit.[381]

The exact identity of the scribes who originated a version of the *Story of Ahiqar* is unknown, but it is nonetheless still possible to argue that it appears to respond in some ways to the political situation towards the end of Esarhaddon's reign. While the *Story of Ahiqar* does not contain explicit references to covenant, or indeed points of direct intertextuality with the covenant composition, it bears comparison with narratives that do fit that definition, as well as with the covenant composition itself.

As already mentioned, the *Story of Ahiqar* has some similarities to the *Underworld Vision* discussed in Chapter 7.1.4:[382] the tale includes both Sennacherib and his progeny, and mirrors this by involving intergenerational dynamics in a family of scholars. So too, if the linking of that narrative to Sennacherib's murder is correct, they appear to take place in the same period. The fact that the tale compares the actions of Sennacherib to those of Esarhaddon is, of course, also similar to *The Sin of Sargon,* which seeks to determine the correct path for Esarhaddon by examining the mistakes of his father and grandfather.[383] Another clear parallel between all three narratives is the suspicion of scholars and advisors: the deceptive haruspices and scribes of *The Sin of Sargon*, the corrupt scribe in *The Underworld Vision of an Assyrian Prince,* and the wrongfully accused Ahiqar and actually deceptive Nadin in the *Story of Ahiqar.*

---

379 Bledsoe 2021, 22–24.

380 On the information that the literature found at Elephantine provides about the 'historical and cultural context' of the Judean colony, see Kratz 2022.

381 Dimant 2018, 176. Note that the *Story of Ahiqar* also has points of similarity with the Book of Esther and Joseph's story in Genesis (the latter of which also has points of comparison with *Esarhaddon's Apology:* Frahm 2016).

382 See the discussion in Chapter 7.1.4.

383 See the discussion in Chapter 6.2.1.

The interest in the role of scholars in Assyria, and particularly in their ability to deceive the monarch and lead him astray unites these three narratives. Nonetheless, the manner in which the *Story of Ahiqar* deals with the topic differs in various interesting ways from the others. Firstly, there does not seem to be same preoccupation with the will of the gods that is evident in the other two compositions. As far as it is possible to tell from the fragmentary section of the document, although Ahiqar is accused of acting wrongly, there is no indication that he is accused of misleading the king concerning the gods. Instead, he appears to be charged with instigating a plot against the monarch. Beyond this, significantly, the accusation against Ahiqar is false: he has done no such thing. In the other two narratives, meanwhile, the accusations are true. In this regard, the situation in Ahiqar is more complex than those presented in the other two compositions: the scholar under suspicion is not guilty, while the one who is not suspected is himself leading the monarch astray. It is worth noting, however, that the guilty advisor, Nadin, is also not treacherous in the manner of the other two compositions. His actions are concerned only with humans, as his treachery takes the form of a false accusation made against his uncle.

Indeed, the actions of the two scholars here perhaps bear closer comparison with the stipulations of Esarhaddon's succession covenant than do those of the advisors in the other literary compositions. Even though the covenant is not mentioned, the actions of Nadin mirror its stipulations in various ways. Firstly, the manner in which the young advisor acts, from Esarhaddon's point of view, appears to align neatly with the demand that subjects report plots against the crown. The reality of Nadin's actions, however, is more reminiscent of the injunctions against fomenting strife between members of the royal family:

> (As for) the positions which Esarhaddon, king of Assyria, their father, assigned them, you shall not speak in the presence of Ashurbanipal, the great crown prince designate, (trying to make him) remove them from their positions.[384]

Although Ahiqar is not presented as a member of the royal family in the tale, he is portrayed as very powerful and close to the crown. Furthermore, by his own estimation, he is valuable to Esarhaddon and will be needed by him again. By attempting to oust Ahiqar from his position, therefore, Nadin endangers the stability of the crown. These associations between the covenant stipulations and *Story of Ahiqar* are, of course, not sufficiently strong to assert that the latter composition deliberately references the former. Despite this, centuries after the death of Esarhaddon, a tale was still being transmitted in which courtiers reported on each other in the

384 SAA 2, no. 6 § 31: 369–372; Lauinger 2012, v 34–36.

wake of a conspiracy against that very monarch. I consider it reasonable to take this as evidence of an extended afterlife of the discourse that took place in the wake of the conspiracy of 671/670 BC, which was itself heavily shaped by the recent imposition of the succession covenant.

One particularly interesting aspect of the *Story of Ahiqar* is its apparently critical stance towards Esarhaddon's purges. The narrative seems to share the stance of someone like Urdu-Nanaya: it portrays Esarhaddon's vigilance as directed towards the wrong people. In Urdu-Nanaya's letter, he argues that the monarch's distrust is unnecessary, whereas in the *Story of Ahiqar* it ought to be focused on Nadin, rather than the faithful Ahiqar. In an interesting further step, Ahiqar persuades Nabû-šumu-iškun that not executing him will serve to protect Esarhaddon from his own rash behaviour: the king will surely regret his decision and then be glad that the sage is not dead. So too, the problem of wrongfully condemning one's staff is not limited to Esarhaddon in the tale, as Ahiqar reminds Nabû-šumu-iškun that Sennacherib did the same thing to him. Thus Esarhaddon, while misled initially by an advisor, is presented as someone who sometimes needs to be deceived for his own good, like his father before him. Such a message would probably not have met with Esarhaddon's approval, and certainly goes against the stress in the succession covenant on the importance of reporting to the crown. In this way, it is possible to read the *Story of Ahiqar* not merely as a response to the climate of vigilance in the court of Esarhaddon's final years, but as an indictment of it.

8.5 Conclusions

The evidence presented in this chapter can be separated into two clear categories. On the one hand, the royal correspondence sent to Esarhaddon from the client states provides evidence of the manner in which Babylonian and Assyrian subjects of the crown interpreted their duty to report. In Babylonia, some inhabitants of Babylon and Borsippa considered it their duty to report dangerous diviners to Šamaš-šumu-ukin, the crown prince of Babylon. Several letters from Uruk, meanwhile, imply that some inhabitants of that location linked the stipulation to report found in the covenant to another order to report sent by the Assyrian crown. Another letter indicates that a Babylonian linked the covenant with the concept of keeping the *maṣṣartu* 'watch'.

The correspondence of Itti-Šamaš-balaṭu, meanwhile, provides an insight into the preoccupations of an Assyrian who, apparently against his will, was stationed in a client state, and tasked with monitoring the locals. Close reading of Itti-Šamaš-balaṭu's letters indicate that, while he carries out this assignment, his focus – like that of the succession covenant itself – is directed inwards, specifically towards Esarhaddon's own courtiers, whom he accuses of dishonest activities. In this way, the vigilance that Itti-Šamaš-balaṭu exhibits can be argued to go beyond his immediate task of reporting on the client ruler and his people, and to accord more closely with the stipulations of the succession covenant. Itti-Šamaš-balaṭu's ability to quote the composition with utmost accuracy suggests that the local client ruler was taking his own duties to the Assyrian crown seriously, at least in as far as concerned displaying the tablet according to Esarhaddon's demands.

On the other hand, the *Story of Ahiqar* and Deuteronomy 13 may provide insights into some rather more critical views regarding the covenant and its duty of vigilance. Deuteronomy 13 highlights the need for a duty of vigilance, but stresses that this vigilance should be limited to the local community of worshippers of Yhwh. It is unclear whether the scribe who wrote this text made this conceptual leap as a direct or indirect response to Esarhaddon's succession covenant, but it seems likely that the scribes of Judah repurposed the terms of the composition in order to serve their own purposes. Such an act would have altered the purpose of the vigilance of Judah's inhabitants, directing it away from the Assyrian crown and towards local concerns. Such an act was a rejection of the aims of Esarhaddon's succession covenant, even though it also embraced its methods.

The *Story of Ahiqar* shows that a narrative arose that centred on an innocent scholar, falsely accused by a rival and rashly condemned by Esarhaddon. Such a narrative takes up hints found in Esarhaddon's royal correspondence, for instance, that some individuals considered his regime of vigilance excessive and counterproductive. The *Story of Ahiqar* illustrates that knowledge – and possibly disapproval – of the high level of distrust that Esarhaddon had for his own officials spread beyond Esarhaddon's close circle at some point. Both Deuteronomy 13 and the *Story of Ahiqar* can therefore be viewed as evidence of critical responses to the imposition of Esarhaddon's succession covenant. Their specific criticisms are vastly different, however, as the former seems to consider the duty of vigilance to be useful, but misdirected, while the latter seems to suggest that the duty of vigilance was excessive and open to abuse.

Chapter 9: Conclusions. Creating an empire of informers

This study has examined Esarhaddon's succession covenant through the prism of vigilance, employing a model of call and response. The first chapter posited that the text of Esarhaddon's succession covenant laid out an idealized vision of the spatial and social structure of the Assyrian Empire, which those who composed it hoped to make a reality through its enactment.

As discussed in Chapter 3, the subjects of Assyria were required to become alert on behalf of the crown, and to direct their attention towards potential dangers and duplicitous actors, generally high-ranking individuals or groups. Such plotters were expected to take advantage of periods of weakness of the Assyrian crown, in particular the period directly after the death of the sitting monarch. Those bound by the covenant were required to act on the dangers to the crown that they perceived: a detailed list of mandated and forbidden responses to these scenarios serves to elucidate the crown's expectations. In some instances, it was necessary to intervene directly, while in others the subjects were required to report to Ashurbanipal, the crown prince of Assyria. In this way, the projected loyal subject was not dissimilar from the ideal of the 'seeing/saying citizen' that Joshua Reeves traces in the modern US context, although the Assyrian case is more interested in the sense of sound than of sight.[1]

The model Assyrian subject had to do more than merely report, however: in some cases, a subject was required to engage in what approaches vigilantism. This may seem counterintuitive, as vigilantism is defined as 'the practice of ordinary people in a place taking unofficial action to prevent crime or to catch and punish people believed to be criminals'.[2] As the covenant composition mandates intercession on behalf of the crown, such acts are state sanctioned, and can thus perhaps be considered 'official'. Nonetheless, the other aspects of the definition – that ordinary people be the primary agents, that the action prevent crimes or punish people believe to be criminals – accord well with the stipulations of Esarhaddon's succession covenant. Furthermore, while those bound by the covenant are under a general obligation to act in these situations, they are not always necessarily required to seek permission from the crown to act in the moment that these theoretical scenarios become reality. Indeed, part of the intent of the stipulations themselves may have been to allow for immediate action by eliminating

1 Reeves 2017, 3 and passim.

2 According to the online Cambridge Dictionary, s.v. vigilantism.

 https://doi.org/10.1515/9783111323435-012

this potentially dangerously protracted step. As such, the act, while officially sanctioned, can arguably be construed as 'unofficial' in that it was apparently not necessary to notify the crown before taking action. In this way, the mandated intervention of subjects may be characterized as a form of state-sponsored vigilantism.

Beyond acting, a subject was required to adopt particular attitudes: he was expected to obey the demands of the covenant stipulations *ina gammurti libbikunu* 'with your (pl.) whole heart'. In this way, the terms of the succession covenant sought to instill a form of vigilance directed towards the self, as such demands necessitate monitoring one's own thoughts and feelings. This type of demand in the composition culminates in the mandate that subjects of the Assyrian crown 'love Ashurbanipal like your own lives' (Chapter 3.2). Participation in the covenant was supposed to come from within: so too, obeying its stipulations with the goal of protecting the crown prince of Assyria, Ashurbanipal, was supposed to be the result of valuing and preserving him in the same way and at the same level as subjects did their own lives. The sociologist Arlie Hochschild, who has worked extensively on emotions in the contemporary US context, notes when discussing the cultural contingency of human emotion that one judges the appropriateness of one's own emotions based on 'feeling rules' that differ between cultures.[3] Using this concept, the clause can be regarded as taking what was presumably an established Assyrian feeling rule, that one loves oneself and wishes to preserve one's own life, and using it as a reference point to attempt impose a new feeling rule from the top down: one ought to love the crown prince of Assyria in the same way.

This dynamic, attempting to impose from above something that is designed to function either internally or in a bottom-up fashion, is found in all three forms of mandated action in the text of Esarhaddon's succession covenant: reporting, acting, and feeling. As Chapter 4 has argued, Esarhaddon and his advisors also promoted the concept of covenant more broadly, stressing its role in legitimate Assyrian succession in *Esarhaddon's Apology*, and its inescapable nature in *Esarhaddon's Letter to the God Aššur.* The geographical distribution of the manuscripts of these texts highlights Esarhaddon's particular interest in enacting the covenant on inhabitants of the Assyrian provinces, as opposed to the client states, and, within the provincial system, in key cities in the core region, especially Nineveh and Ashur.

Chapter 5, meanwhile, continued to examine the manner in which the imposition of Esarhaddon's succession covenant straddled the line between universal application and pragmatic selection of particular locations and groups as particularly important to the covenant's success. Once again, it seems that the covenant

3 Recently, Hochschild 2013, 6. The concept is introduced in Hochschild 1979.

was enacted more thoroughly on those in provincial capitals, and in particular those located in social proximity to the crown. Nonetheless, it is worth stressing that the scribes who drew up the manuscripts of the succession covenant went to the not inconsiderable effort of writing up versions of the covenant tablet even for fairly politically insignificant client rulers, such as the city-lords of western Iran. This is in itself evidence that the claims that Esarhaddon's succession covenant applied to all subjects of the Assyrian crown were serious. So too, the covenant was designed in part for long-term use and effectiveness, and allowed for further such documents and demands. The manner in which Esarhaddon and his advisors promoted and implemented covenant, therefore, seems to suggest that they acknowledged that the duty of vigilance was more relevant for – and perhaps more feasible to instill in – some groups than others. Despite this, however, it is clear that they simultaneously wished the terms of the covenant to apply to every subject of the crown, without exception, regardless of geographical location, administrative zone or social position.

The second part of this study sought to identify written documents that can be construed as responding to Esarhaddon's succession covenant, in particular its call to vigilance, across different administrative and geographical zones. These texts take many forms: letters to the monarch and the crown princes, literary compositions, a prophecy compilation, archival tablets, legal documents and the stipulations of Deuteronomy 13. The evidence, as the three chapters of Part 2 have illustrated, is distributed unevenly, and is skewed towards the king's court (Chapter 6), the city of Ashur, some locations in the western provinces, particularly Harran and Guzana (Chapter 7), and Babylonia (Chapter 8). The evidence from the small Levantine client states of Arwad and Judah (Chapter 8), however, indicates that elite figures, or those with ties to the Assyrian crown in these locations, also knew of the covenant and in some cases responded to it.

To what extent, then, do these responses provide evidence that Esarhaddon's call to vigilance of 672 BC was successful? The sources discussed in Chapters 6–8 indicate that the implementation of Esarhaddon's succession covenant did provoke responses from a wide variety of places and of a range of types. Nonetheless, this is not the same as establishing that Esarhaddon effectively enacted the duty of vigilance laid out in his succession covenant. When assessing the extent to which he and his advisors achieved their self-imposed agenda, it is necessary to consider two of the core concepts of this study: space and responsibilization. Did Esarhaddon's succession covenant provoke responses in the geographical, administrative and social areas that were its focus? Did the groups in these areas take responsibility for watching out for particular scenarios and responding to them in the way that the crown stipulated? This final chapter of this study will attempt to respond to these questions.

9.1 Space: geographical, administrative, social

The second part of this study examined the source material according to its provenance or, in the case of some letters where this is not certain, the locations to which they pertain. The aim of this was to mirror the spatial dynamics of the empire as they are set out in Esarhaddon's succession covenant itself. As is explored extensively in Chapter 2, the succession covenant was enacted on each individual client state and province. The first legal party mentioned in each covenant tablet was either the ruler of the client state, or the governor of the province, respectively. These first legal parties were followed by a top-down list of members of the client state ruler's family, or members of the provincial administration, ending in both cases with the claim that the covenant applied universally. This 'preamble' appears to be the only portion of the covenant composition for which more than one version was drafted. In this way, the succession covenant seems to view province versus client state as the key difference between the administrative units of the empire.

From a geographical point of view, particular locations appear to be privileged within the covenant composition. In the list of divine witnesses and the adjuration, the gods of particular settlements are mentioned specifically, perhaps implying a greater interest in the covenant's thorough implementation in these locations: the Assyrian core region, northern Babylonia and Harran. The curses support this impression, while adding a focus on the western geographical zone of the empire, encompassing provinces but also client states through references to deities worshipped in that region (Chapter 2.3).

From a social perspective, the stipulations of the covenant appear to focus on groups that are close to the crown, most notably members of Esarhaddon and Ashurbanipal's own family (Chapter 3). Indeed, one of the central objectives of the stipulations seems to be to encourage the subjects of the Assyrian monarch to resist and report high-ranking groups and individuals, eroding loyalty to them and redirecting it towards the crown. The framing of the preamble, meanwhile, highlights high-ranking members of the provincial administration and the ruling families of the client states. While these lists do not correlate perfectly with social proximity to the king, it is reasonable to assume that members of the administrative and non-Assyrian royal elite would have had easier social access to the monarch than lowlier individuals and groups. Nonetheless, both the preamble and the stipulations also emphasize the universality of the bond of the covenant. The preamble explicitly includes all of Esarhaddon's subjects as treaty partners to the succession covenant. The stipulations, meanwhile, do this by using the second person plural to address the collective, while also implying that everyone bound by the covenant is socially close enough to the crown prince to report to him personally.

The complex spatial dynamics of the covenant itself are reflected in the evidence discussed in Part 2 of this study. Responses to the covenant come from both the provinces and the client states, covering both administrative zones (see Figure 6). In terms of geographical distribution, one finds particularly strong evidence from the most significant settlements in the Assyrian heartland, Nineveh, Kalhu and Ashur, as well as Harran, and from Babylonia and the western client states. Responses seem to stem from people at various degrees of social remove from the monarch. Some of Esarhaddon's closest advisors mention the covenant in their letters to the king, and individuals close to the king likely composed literary narratives exploring the role of covenant and the duty of vigilance in Esarhaddon's reign (Chapter 6). In addition to this, however, it appears that the elders of Guzana, and various private legal parties from across the provinces, who were probably not in direct social contact with the monarch, also alluded to the 'covenant of the king' from the reign of Esarhaddon onwards, indicating that the concept permeated groups at a greater social remove from the crown (Chapter 7).

In this way, one can claim that Esarhaddon's succession covenant was successful in reaching populations in all of the administrative zones, geographical regions and social circles that it aimed to influence. Nonetheless, the nature of these responses does not always align with the terms of the covenant composition itself, suggesting that attempts to responsibilize the populations of these spaces were not universally successful.

9.2 Responsibilization: successes

What can be considered successful responsibilization? The stipulation and oath sections of the covenant composition sought to transfer responsibility for the security of the crown's position to the subjects of Assyria. There are definite indications that responsibilization of this kind took place in certain instances.

The easiest successful form of responsibilization to identify is the transfer of seeing/hearing and saying responsibility in the royal correspondence. In the king's entourage, the scholar Adad-šumu-uṣur cites the succession covenant when passing on a report that had been made to him (Chapter 6.1.2). As the details of the report itself are broken, it is unfortunately unclear to what extent it aligns with the scenarios laid out in the covenant. Pertaining to the provinces, the informer Nabû-rehtu-uṣur constitutes a clear example of the successful transfer of seeing/saying responsibility: he reports a scenario with close parallels to more than one of the covenant stipulations (Chapter 7.2.1). He also states in his letters that his duty to report is tied to his status as a *bēl adê ša šarri* 'keeper of the covenant of the king', indicating that he comprehends the connection between the situation described in his letters and his own responsibility to protect the crown.

Beyond the royal correspondence, the literary composition known as *The Underworld Vision of an Assyrian Prince* also explores seeing/saying responsibility through the character of the scribe and his decision in the final lines of the text to report what he had seen (Chapter 7.1.4). This narrative may have been composed at Esarhaddon's court, or perhaps in the city where it was found, Ashur. The story likely concerns not Esarhaddon's succession covenant, but that of Esarhaddon's father, Sennacherib. Nonetheless, as the importance of Sennacherib's succession covenant was stressed under Esarhaddon in anticipation of his imposition of his own covenant, the reference can be interpreted as evidence of the efficacy of royal communication pertaining to *adê* and the duty of vigilance. The scribe in the *Underworld Vision* identifies the actions of the prince, Kummaya, as inappropriate, and reports them to the palace. In this way, the text is a tale of the successful responsibilization of a delinquent character, a scribe who has taken bribes. The tale both explores the responsibility to report and further advertises that responsibility, emphasizing its importance in the context of Esarhaddon's own accession to the throne.

The texts discussed in Part 2 provide little in the way of evidence about subjects of the Assyrian crown attempting to prevent or stop seditious activities by intervening beyond simple reporting. Nabû-ušallim from Ashur is perhaps one such example, if his letter does indeed implicitly reference the succession covenant, which is by no means clear (Chapter 7.1.2). Nabû-ušallim claims that he has not joined those plotting against the king in Ashur, despite their demands that he do so. As their actions accord closely with the scenarios described in the succession covenant, his resistance aligns with the demands of the composition. As the seditious group in Ashur included soldiers, Nabû-ušallim's resistance presumably put him in some danger. If his refusal to comply with the plotters is connected to his obligation to Esarhaddon's succession covenant, then his apparent willingness to put himself in a dangerous situation in order to comply with the terms of the covenant points to an impressive degree of responsibilization. Nonetheless, it is noteworthy that the strongest evidence of direct intervention in a scenario similar to those described in the covenant stipulations is a case of refusal to join, as opposed to the active resistance that is frequently demanded in the mandated actions.

The letters of Itti-Šamaš-balaṭu, a representative of the Assyrian crown at the court of the client ruler of Arwad, may also allude to the dangers of his resistance of the corruption of members of the king's entourage and their merchants, whom he claims are 'systematically scaring me' (Chapter 8.2). This statement is not made in the letter in which Itti-Šamaš-balaṭu quotes the succession covenant, but in another of his missives to Esarhaddon. Nonetheless, it is likely that both letters are

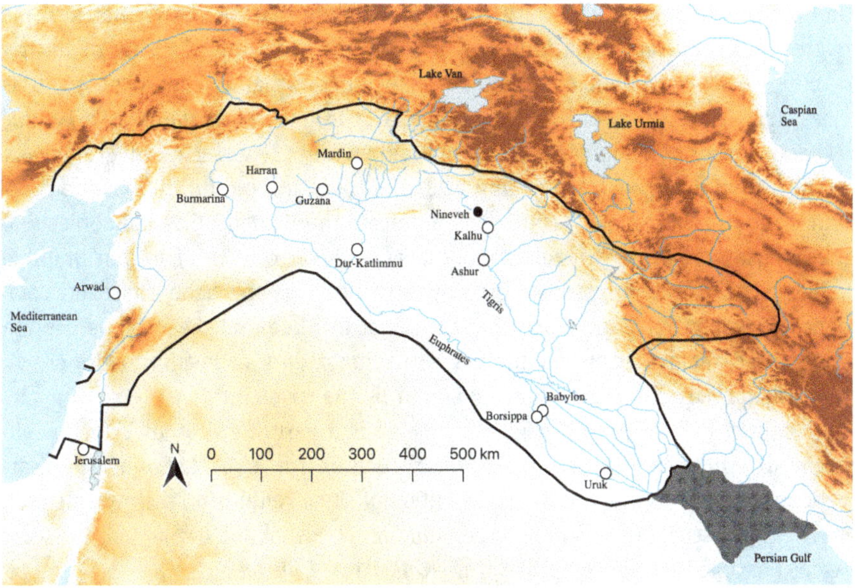

Figure 6: Map of locations of responses to Esarhaddon's succession covenant from the court, provinces and client states. The unknown site of Mallanate is not shown. The capital, Nineveh, is marked in black. The shaded area indicates the ancient coastline.

referencing the same situation. If so, then this is another case of an Assyrian subject, this time located in a client state, refusing to comply with disloyal activity that contravenes the covenant despite it posing a serious risk to him. However, in this instance, the similarity of the actual scenario to those described in the covenant text is less clear, indicating that responsibilization in this case was less precisely targeted. So too, it is probably relevant that neither Itti-Šamaš-balaṭu nor Nabû-ušallim state explicitly that they are seeking to act in accordance with the covenant stipulations by doing what they are doing. This contrasts sharply with the instances of seeing/hearing and saying responsibilization, in which various informers claim that they are reporting in an attempt to uphold their duty to the covenant. While this may in part be due to the nature of the available evidence, it may also indicate that the imposition of the succession covenant was more successful at responsibilising subjects of the Assyrian crown to report than it was at responsibilising them to intervene directly on its behalf. One may conjecture that this is linked to the relative risks of these two forms of action, with the latter likely entailing a higher degree of danger.

As regards the demands that the subjects of the Assyrian crown internalize their duty to the crown, acting on the crown prince's behalf wholeheartedly, and even loving him (Chapter 3.2), there is evidence that the crown was to some degree successful at responsibilising its subjects. While the phrasing of the responses to the covenant does not perfectly mirror that of the covenant, it is clear that reference to *adê* and in particularly to one's own status as a *bēl adê ša šarri* 'keeper of the king's covenant' became a way of expressing general loyalty to the crown (Chapter 7.1.1, 7.2.2 and 7.2.3). This appears to indicate that the covenant's imposition did successfully communicate the demand that the subjects of the Assyrian crown be wholeheartedly loyal to the crown. The fact that multiple subjects of the crown expressed their loyalty in this way probably constitutes evidence of responsibilization, as it indicates that it influenced their self-perception and caused them to view themselves as having a duty of loyalty to the crown that was directly connected to being bound by the covenant.

Some of the responses included in the second part of this study refer to Sennacherib's succession covenant, either as well as or instead of that of his son Esarhaddon (Chapter 7.2.1 and 7.2.2). In the letters of Nabû-rehtu-uṣur, the informer references Sennacherib's covenant in conjunction with that of Esarhaddon. So too, it seems possible that the composer of the *Underworld Vision* narrative was prompted to write a narrative that mentioned Sennacherib's covenant in the wake of the imposition of Esarhaddon's succession covenant (Chapter 7.1.4). *The Sin of Sargon*, while it does not explicitly mention Sennacherib's succession covenant, focuses in part on the circumstances of his death as an opportunity to direct vigilance and suspicion towards the scholarly milieu (Chapter 6.2.1). These references can perhaps be interpreted as evidence of the impact of the general elevation of the importance of covenant that took place under Esarhaddon, by means of the succession covenant itself but also through texts such as *Esarhaddon's Letter to the God Aššur*, and in particular the composition and dissemination of *Esarhaddon's Apology* (Chapter 4). That several of Esarhaddon's subjects reflected on and stressed their duty to Sennacherib's covenant during this time can be considered evidence that Esarhaddon and his advisors successfully responsibilized some of Assyria's subjects using not only his own succession covenant but also by stressing their responsibility to Esarhaddon and to obeying the terms of Sennacherib's succession covenant. As it was clearly part of Esarhaddon's broader strategy to raise the profile of his father's succession covenant in order to support his own position and that of his chosen successor, this can be viewed as a success on his part.

9.3 Responsibilization: limitations and misunderstandings

Despite the evidence of successful responsibilization, there is ample suggestion in the sources that in various cases responsibilization took place only imperfectly.

Indeed, even in instances where subjects appear to have been successfully responsibilized, there is often evidence of a mismatch between the stipulations of Esarhaddon's succession covenant and the manner in which his subjects interpreted their duties. Even the scholar Adad-šumu-uṣur, while he claims that he is following the stipulations of the covenant, merely paraphrases the mandate to report to Ashurbanipal: he does not cite it in full (Chapter 6.1.2). This is typical of references to the covenant in the royal correspondence. With the exception of one letter from Itti-Šamaš-balaṭu, the Assyrian delegate at Arwad, who quotes the curses rather than the stipulations, there are no exact quotations of the covenant in Esarhaddon's correspondence (Chapter 8.2). Despite this, there are several instances of those writing to the monarch claiming that they are quoting directly from the covenant, and instead paraphrasing it. These summaries exhibit the knowledge that all subjects of the monarch are supposed to watch out for threats and report them. It is particularly notable, however, that even Adad-šumu-uṣur, one of Esarhaddon's closest advisors, did not follow the stipulations of the succession covenant with perfect accuracy, making significant deviations from the covenant while claiming to adhere to its terms. The Babylonian correspondents of the king and Šamš-šumu-ukin, the crown prince of Babylon, are similarly vague when it comes to the details of the covenant composition, paraphrasing the demand that subjects report to the crown, and possibly confusing the covenant stipulations with a separate royal order (Chapter 8.1). Beyond this, all but one of the letters analyzed in Chapters 6–8 are addressed to Esarhaddon rather than Ashurbanipal (Chapter 7.2.2). One of the letters, written by Šamaš-šumu-ukin, even shows that some Babylonian subjects of the Assyrian crown chose to report to him, the crown prince of Babylon, rather than to Ashurbanipal, as they should have done according to the covenant (Chapter 8.1.1). This is particularly telling, as the terms of the covenant seek in no small part to protect Ashurbanipal from his brothers (Chapter 2.1 and Chapter 3).

Nonetheless, this need not be taken as evidence that the imposition of Esarhaddon's succession covenant failed to responsibilize at least some of Assyria's subjects. As argued above, the references to the covenant in these letters, accompanied as they often are by information that would have been of interest to the crown, indicate the success of the covenant's implementation. What this does suggest, however, is that the implementation of the covenant succeeded in responsibilising Esarhaddon's subjects to become alert on behalf of the crown generally, rather than to follow the covenant composition's more specific orders to the letter. As discussed above, they do not seem to have been as aware of their duty to intercede on the part of the crown, again indicating that their responsibilization did not accord perfectly with the terms of the succession covenant itself.

In addition to this, it appears that in multiple cases the subjects of Assyria merged the demands of Esarhaddon's succession covenant with those of other royally-commissioned writings, such as previous covenants and royal orders. In one case, the writer also seems to connect it with the king's *maṣṣartu* 'watch' (Chapter 8.1.3). This may well have been encouraged by Esarhaddon and his advisors, and again should not be taken as an indication of failure. It does, however, counteract the covenant composition's own depiction of itself. In the covenant text, it is presented as a legally-binding document that is capable of imposing a duty of vigilance alone, which must be followed perfectly on pain of divine punishment. This differs substantially from the way that it seems to have been interpreted even in the royal correspondence, when talking about it with the king himself, mere years after its imposition.

Beyond the confines of immediate social interaction with the crown, the connection between the covenant stipulations and the responses to the covenant is less evident. The *Underworld Vision* is certainly indicative of responsibilization: it is a narrative in which a scribe takes responsibility for the welfare of the crown by reporting what he has heard, as well as appearing to have the didactic aim of communicating this responsibility to others (Chapter 7.1.4). Nonetheless, the situation in the narrative does not neatly accord with the stipulations of Esarhaddon's succession covenant, or even with what is known of Sennacherib's succession covenant. The scribe, it seems, has already accepted bribes and quite probably helped the seditious prince in his plot against the monarch. According to the covenant, he should have been struck down by divine wrath. This does not happen, however: Nergal, god of the underworld, elects to spare both him and the prince, and the scribe instead changes his ways, going and reporting to the palace. This situation is significantly more complex than the covenant compositions would allow for, and appears to adapt the duty of vigilance to its own narrative purposes. The prophecy compilation discussed in Chapter 6.2.2 also appears to reimagine the role of the gods in a covenant, as compared to the text of Esarhaddon's succession covenant.

Beyond even this, the legal clauses mentioning the *adê ša šarri* 'king's covenant' found in private documents seem to indicate a distance between the understanding of covenant in this context and the specific stipulations Esarhaddon's succession covenant (Chapter 7.3). The legal parties and scribes who used these clauses appear to have viewed a royal covenant as an agent in its own right, able to punish people for misbehaviour beyond the contravention of a covenant's particular stipulations. They seem to have believed that the king's covenant would also act on their behalf, interceding to prevent people from reneging on legal agreements. In this way, these people apparently sought to use the power of the covenant to protect themselves. Whether or not they themselves considered it necessary to follow the written stipulations of the king's covenant is less clear. It can be stated with confidence that these people recognized the concept of covenant as important, and considered the covenant powerful enough to enforce certain behaviour. This implies that they would themselves have sought to obey it. Nevertheless, it is not clear what exactly they would have believed this to entail.

In a similar way, the discovery of an excerpt of a covenant composition in the N4 archive of Ashur indicates that the inhabitants of this private dwelling took this later covenant seriously (Chapter 7.1.3). Despite this, if the excerpt is representative, they seem to have focused more on its curses than its stipulations. As such, it is unclear to what extent they understood the specifics of the covenant's demands, or even considered them important. This may also tie in with the various letters in which it is not entirely clear whether the actions that a particular correspondent describes accord with the scenarios described in the covenant composition: the letters of Itti-Šamaš-balaṭu, the king's representative at Arwad, are but one example of this (Chapter 8.2).

Conversely, there are cases, such as the letter sent to Esarhaddon by the informer Nabû-ušallim, in which the writer does not explicitly cite the succession covenant (Chapter 7.1.2). It is relevant that, of the hundreds of letters sent to Esarhaddon, only fifteen explicitly respond to the covenant. The letter of Nabû-ušallim, and many others, could be taken as following the order to report to the crown in the scenarios laid out in the succession covenant. While I have discussed Nabû-ušallim's letter on the grounds that it contains clear links to the covenant composition in various places, it is difficult to assess whether the writers of these letters perceived themselves as responding to Esarhaddon's succession covenant. In short, it is unclear whether or not they had been responsibilized by its imposition. More broadly, the relative lack of explicit responses to the Esarhaddon's succession covenant can perhaps be viewed as evidence of the limited success of the covenant's implementation.

9.4 Responsibilization: failures

Several of the sources that can, in my view, be taken as 'responses' to the imposition of Esarhaddon's succession covenant provide evidence that it failed to responsibilize various individuals and groups. In the king's own entourage, it appears that even the chief *asû*-healer, Urdu-Nanaya, was skeptical of the duty to report to the crown (Chapter 6.1.1). While he argues that the covenant has been successful, he appears to caution against the culture of vigilance that it promotes. Similarly, an early version of the *Story of Ahiqar*, found at Elephantine, may also refer to the perceived excesses of the monitoring of one another that took place at Esarhaddon's court (Chapter 8.4).

Beyond this implicit criticism, it appears that some outright rejected the demands of the succession covenant. In the client kingdom of Judah, it seems that portions of the succession covenant were repurposed, possibly directly for a covenant with the local god, Yhwh, or perhaps for a monarch of Judah (Chapter 8.3). This reuse of the covenant stipulations can certainly be seen as embracing the policy of attempting to harness the attention of one's subjects in one's own interest. However, it appears to reject the demand that the people of Judah do so for the good of the Assyrian monarch, their imperial overlord, rather attempting to redirect their attention and loyalty towards either the local ruler or the local god. In this way, the implementation of Esarhaddon's succession covenant seems to have prompted resistance to the Assyrian crown – the very opposite of what it aimed to achieve.

It is worth stressing that King Esarhaddon himself may at times have perceived the implementation of his succession covenant as having failed, or at least have temporarily worried that it had done so. The evidence discussed in Chapter 6 in particular suggests that, within Esarhaddon's inner circle, steps were taken to reassure the monarch that he enjoyed divine support and that the imposition of his covenant had indeed been successful. That such measures were necessary at all suggests that Esarhaddon himself had doubts on this subject. The cause of his doubt seems to have been the occurrence shortly after the covenant's imposition of widespread resistance to his succession arrangements. While various people reported this to him, indicating that they had been successfully responsibilized, the plotters appear to have engaged in various activities that are strictly prohibited in the covenant composition. Nabû-ušallim's letter from Ashur, for instance, states that an apparently large group had sworn allegiance to the conspirators by means of a covenant (Chapter 7.1.2). This act is forbidden in the covenant composition, and yet only a year or so later, it appears that some people were willing to do it anyway. As the plotters sought to seize the throne, it is possible that – far from protecting the designated successor, Ashurbanipal – the wide publicization of Esarhaddon's choice of crown prince sparked resistance. As such, the fact that several people reported on the seditious activities that they perceived reinforces the idea that Esarhaddon's succession covenant was successful; and yet, if the succession covenant had been more successful, they would have had less to report to the crown in the first place.

9.5 Spatial dynamics of responsibilization

Although the evidence presented in Part 2 of this study was arranged along spatial lines, it is not clear that the extent of the Esarhaddon's succession covenant's responsibilization of Assyria's subjects can be neatly divided according to space. In broad strokes, the conclusions of each chapter of Part 2 can be characterized as follows. In Chapter 6, the members of the king's entourage were at the greatest geographical and social proximity to the monarch. In administrative terms, they were located in the provinces, but their status as people who were likely privy to, and perhaps involved in, the creation and dissemination of the covenant composition warrants their discussion in a separate category. The evidence from this group, I argue, in large part suggests that the monarch and his close advisors were preoccupied in the aftermath of the 671/670 BC conspiracy with the question of whether or not the imposition of the covenant had failed. The focus of this discussion was not logistical, but rather hinged on the question of the extent to which the king himself enjoyed divine support. These documents indicate that the succession covenant informed the manner in which events in the latter part of Esarhaddon's reign were viewed, implying that Esarhaddon's advisors and the king himself had to some degree internalized its message. The scholar Adad-šumu-uṣur appears to have considered it his duty to pass on information to the monarch, a duty decreed by the covenant, pointing to his successful – if imperfect – responsibilization (Chapter 6.1.2). His close colleague, Urdu-Nanaya, however, despite attributing the successful prevention of a plot against the king to the covenant, appears rather critical of the culture of vigilance that it demands (Chapter 6.1.1).

The evidence from the provinces, discussed in Chapter 7, largely falls into the category of partially successful responsibilization. The letters of Nabû-rehtu-uṣur from Harran could perhaps be taken as seeking to assure the monarch that his covenant is effective and that it will continue to be so, in a manner that accords with the discussion analyzed in Chapter 6 (Chapter 7.2.1). The source material often eludes analysis of the precise extent to which the people responding to the covenant understood the details of its stipulations, and thus the extent to which they can be said to have been successfully responsibilized by it. So too, the plotters of 671/670 BC were clearly active in the provinces, meaning that it was arguably the zone in which the covenant's imposition provoked the strongest backlash. The relationship between the gods and the covenant, as well as the status of the covenant as a god-like being in its own right, feature prominently in much of the source material, such as the legal documents and archival material (Chapter 7.1.3, 7.1.4 and 7.3). As in Chapter 6, the manner in which these things were depicted sometimes implies that the subjects in question are considering the relevance of the covenant to them, suggesting a degree of monitoring directed towards the self.

Chapter 8 indicates that certain subjects in the client states of Arwad and Babylonia were successfully responsibilized to a similar degree to the court scholar Adad-šumu-uṣur, one of the closest people to Esarhaddon both physically and socially (Chapter 6.1.2 and Chapter 8.1 and 8.2). Indeed, Itti-Šamaš-balaṭu is the correspondent of Esarhaddon discussed in Part 2 of this study who was probably located furthest from the monarch geographically, given that he is posted at the court of the client king of Arwad on an island off the modern Syrian coast, and yet his letter is the only surviving one that quotes the covenant accurately (Chapter 8.2). The Biblical Book of Deuteronomy similarly reveals a deep knowledge of the covenant composition in the client states, in the case of Judah (Chapter 8.4), while the *Story of Ahiqar* hints at intimate familiarity with the climate at court during the final years of Esarhaddon's reign (Chapter 8.4). These sources indicate that the covenant and its attempts to responsibilize Assyria's subjects also provoked criticism in the client states. Nevertheless, Deuteronomy 13, while hostile to the aims of the covenant, clearly embraces its methods. Meanwhile, the *Story of Ahiqar*, while it appears disapproving, does portray Esarhaddon's court as a place gripped by a culture of suspicion, in which subjects reported on their peers to the monarch.

In this way, all of the three zones discussed in the second part of this study appear to have provoked a wide variety of responses, pointing to successful, limited and failed responsibilization in each category. What, if anything, then, does the spatial distribution of the evidence tell us about responsibilization across the Assyrian Empire?

It is clear that social proximity to the monarch was of considerable importance in ensuring responsibilization. The most direct responses to the demands of the covenant, and thus the clearest indications of successful responsibilization, are found in Esarhaddon's correspondence. These, by definition, all come from individuals who were in direct contact with the crown. As discussed in Chapter 7, it seems probable that even the anonymous denunciations sent to Esarhaddon came from people who were known to him. So too, Itti-Šamaš-balaṭu, located in far-away Arwad though he was, was socially close to the monarch: he appears to be aware of some details of court affairs and was actively campaigning to return to the heartland (Chapter 8.2). Similarly, the literary texts that respond in some way to the implementation of the succession covenant were all composed by people who were, at the very least, familiar with the dynamics of the Assyrian royal family and the court. Even if they were not themselves in the king's social circle, therefore, they are unlikely to have been particularly far removed from it. While these literary texts do not always unequivocally suggest successful responsibilization, they do imply an understanding of the covenant's aims. Deuteronomy 13, the strongest evidence for failed responsibilization, also comes from a milieu that would have been close to the king of Judah, who would himself have been the first named treaty partner in Judah's manuscript of the covenant composition (Chapter 8.3). He would presumably have been in social contact with Esarhaddon, and certainly could have contacted the Assyrian crown if he wished to do so.

It ought to be taken into account, however, that people at close social proximity to the Assyrian monarch are better attested in the existing source material than those living in the Assyrian Empire at a greater social remove from the monarch. Nevertheless, it would be consistent with the covenant composition's own framing if those who were socially connected to the monarch, regardless of their geographical distance from or proximity to him, were most responsibilized by the covenant's implementation. Although the covenant professes to apply universally, the hierarchical list of treaty partners with which both versions begin belies this. In addition to this, the stipulation to report to the crown would have been far more easily achievable for those in contact with the monarch, or with the social status to write to him. While the right of direct appeal to the king may have existed in this period,[4] the available sources do not provide evidence that subjects of Assyria actually made use of this in order to follow the covenant's stipulations by reporting to the crown. Whether this is due to the patchy nature of the surviving sources is difficult to decide.

Considering this distribution of the available source material, it is remarkable that evidence of knowledge of, and responses to, Esarhaddon's succession covenant was not limited only to those who had direct social connections with the monarch. So too, those who were in direct contact with him appear to have used references to covenant and considered the ramifications of covenant beyond their interactions with the crown. While it is perhaps less clear how they interpreted it, other people seem to have engaged with the covenant, and this is evidence of the crown's success in implementing it. Nonetheless, it is frequently impossible to ascertain the degree to which these references that take place at a greater remove from the king are evidence of successful responsibilization by the covenant. When the elders of Guzana claim, when speaking to their local governor, to be loyal to Esarhaddon because they are keepers of the king's covenant (Chapter 7.2.3), what precisely do they consider this to entail? What do they believe their duties to the monarch and his covenant to be? Similarly, if parties in a legal transaction (Chapter 7.3), or their scribes, include a penalty clause mentioning the king's covenant, does this show that they are familiar with the stipulations of that covenant? What is abundantly clear from the materials discussed in this study is that Esarhaddon's at-

4 See Postgate 1974a, Postgate 1980, Garelli 1989 and Radner 2003a, 887.

tempt in 672 BC to instill a culture of vigilance among his subjects had long-lasting repercussions that went far beyond those directly involved with the Assyrian crown at the time.

Abbreviations

AHw	Von Soden, Wolfram (1965–1981) *Akkadisches Handwörterbuch unter Benutzung des lexikalischen Nachlasses von Bruno Meissner (1868–1947). Bände 1–3.* Wiesbaden: Harrassowitz.
CAD	(1956–2010) *The Assyrian Dictionary of the Oriental Institute of the University of Chicago.* Chicago/Glückstadt: The Oriental Institute of the University of Chicago/J. J. Augustin Verlagsbuchhandlung.
CDA	Black, Jeremy, George, Andrew and Postgate, J. Nicholas (2000) *A Concise Dictionary of Akkadian.* Wiesbaden: Harrassowitz.
CTN 2	Postgate, J. Nicholas (1973) *The Governor's Palace Archive.* London: British School of Archaeology in Iraq. (Cuneiform Texts from Nimrud 2).
CTN 6	Herbordt, Suzanne, et al. (2019) *Documents from the Nabu Temple and from Private Houses on the Citadel.* London: British Institute for the Study of Iraq (Cuneiform Texts from Nimrud, 6).
PNA 1/1	Radner, Karen (ed.) (1998) *The Prosopography of the Neo-Assyrian Empire. Volume 1, Part I: A.* Helsinki: The Neo-Assyrian Text Corpus Project.
PNA 1/2	Radner, Karen (ed.) (1999) *The Prosopography of the Neo-Assyrian Empire. Volume 1, Part II: B-G*, Helsinki: The Neo-Assyrian Text Corpus Project.
PNA 2/1	Baker, Heather D. (ed.) (2000) *The Prosopography of the Neo-Assyrian Empire. Volume 2, Part I: Ḫ-K.* Helsinki: The Neo-Assyrian Text Corpus Project.
PNA 2/2	Baker, Heather D. (ed.) (2001) *The Prosopography of the Neo-Assyrian Empire. Volume 2, Part II: L-N.* Helsinki: The Neo-Assyrian Text Corpus Project.
PNA 3/1	Baker, Heather D. (ed.) (2002) *The Prosopography of the Neo-Assyrian Empire. Volume 3, Part I: P-Ṣ.* Helsinki: The Neo-Assyrian Text Corpus Project.
PNA 3/2	Baker, Heather D. (ed.) (2011) *The Prosopography of the Neo-Assyrian Empire. Volume 3, Part II: Š-Z.* Helsinki: The Neo-Assyrian Text Corpus Project.
PNA 4/1	Baker, Heather D. (ed.) (2017) *Neo-Assyrian Specialists: Crafts, Offices, and other Professional Designations. The Prosopography of the Neo-Assyrian Empire. Volume 4, Part I: Index of Professions.* Helsinki: The Neo-Assyrian Text Corpus Project.
RIMA 3	Grayson, A. Kirk (1996) *Assyrian Rulers of the Early First Millennium BC II: 858–745 BC.* Toronto: University of Toronto Press (The Royal Inscriptions of Mesopotamia. Assyrian Periods. Volume 3).
RINAP 1	Tadmor, Hayim and Yamada, Shigeo (2011) *The Royal Inscriptions of Tiglath-pileser III (744–727 BC) and Shalmanese V (726–722 BC), Kings of Assyria.* Winona Lake, IN: Eisenbrauns (The Royal Inscriptions of the Neo-Assyrian Period, 1).
RINAP 2	Frame, Grant (2020) *The Royal Inscriptions of Sargon II, King of Assyria (721–705 BC).* University Park, PA: Eisenbrauns (The Royal Inscriptions of the Neo-Assyrian Period, 2).
RINAP 3/1	Grayson, A. Kirk and Novotny, Jamie (2012) *The Royal Inscriptions of Sennacherib, King of Assyria (704–681 BC), Part 1.* Winona Lake, IN: Eisenbrauns (The Royal Inscriptions of the Neo-Assyrian Period, 3/1).
RINAP 3/2	Grayson, A. Kirk and Novotny, Jamie (2014) *The Royal Inscriptions of Sennacherib, King of Assyria (704–681 BC), Part 2.* Winona Lake, IN: Eisenbrauns (The Royal Inscriptions of the Neo-Assyrian Period, 3/2).
RINAP 4	Leichty, Erle (2011) *The Royal Inscriptions of Esarhaddon, King of Assyria (680–669 BC).* Winona Lake, IN: Eisenbrauns (The Royal Inscriptions of the Neo-Assyrian Period, 4).

 https://doi.org/10.1515/9783111323435-013

RINAP 5/1 Novotny, Jamie and Jeffers, Joshua (2018) *The Royal Inscriptions of Ashurbanipal (668–631 BC), Aššur-etel-ilāni (630–627 BC), and Sîn-šarra-iškun (626–612 BC), Kings of Assyria, Part 1.* Winona Lake, IN: Eisenbrauns (The Royal Inscriptions of the Neo-Assyrian Period, 5/1).

RlA (1928–) *Reallexikon der Assyriologie und Vorderasiatischen Archäologie.* Leipzig/Berlin/New York: Walter de Gruyter.

SAA 1 Parpola, Simo (1987) *The Correspondence of Sargon II, Part I: Letters from Assyria and the West.* Helsinki: Helsinki University Press (State Archives of Assyria, 1).

SAA 2 Parpola, Simo and Watanabe, Kazuko (1988) *Neo-Assyrian Treaties and Loyalty Oaths.* Helsinki: Helsinki University Press (State Archives of Assyria, 2).

SAA 3 Livingstone, Alasdair (1989) *Court Poetry and Literary Miscellanea.* Helsinki: Helsinki University Press (State Archives of Assyria, 3).

SAA 4 Starr, Ivan (1990) *Queries to the Sungod: Divination and Politics in Sargonid Assyria.* Helsinki: Helsinki University Press (State Archives of Assyria, 4).

SAA 5 Lanfranchi, Giovanni Battista and Parpola, Simo (1990) *The Correspondence of Sargon II, Part II: Letters from the Northern and Northeastern Provinces.* Helsinki: Helsinki University Press (State Archives of Assyria, 5).

SAA 6 Kwasman, Theodore and Parpola, Simo (1991) *Legal Transactions of the Royal Court of Nineveh, Part I. Tiglath-Pileser III through Esarhaddon.* Helsinki: Helsinki University Press (State Archives of Assyria, 6).

SAA 7 Fales, F. Mario and Postgate, J. Nicholas (1992) *Imperial Administrative Records, Part I: Palace and Temple Administration.* Helsinki: Helsinki University Press (State Archives of Assyria, 7).

SAA 8 Hunger, Hermann (1992) *Astrological Reports to Assyrian Kings.* Helsinki: Helsinki University Press (State Archives of Assyria, 8).

SAA 9 Parpola, Simo (1997) *Assyrian Prophecies.* Helsinki: Helsinki University Press (State Archives of Assyria, 9).

SAA 10 Parpola, Simo (1993) *Letters from Assyrian and Babylonian Scholars.* Helsinki: Helsinki University Press (State Archives of Assyria, 10).

SAA 13 Cole, Steven W. and Machinist, Peter (1998) *Letters from Assyrian and Babylonian Priests to Kings Esarhaddon and Assurbanipal.* Helsinki: Helsinki University Press (State Archives of Assyria, 13).

SAA 14 Mattila, Raija (2002) *Legal Transactions of the Royal Court of Nineveh, Part II: Assurbanipal through Sin-šarru-iškun.* Helsinki: Helsinki University Press (State Archives of Assyria, 14).

SAA 15 Fuchs, Andreas and Parpola, Simo (2001) *The Correspondence of Sargon II, Part III: Letters from Babylonia and the Eastern Provinces.* Helsinki: Helsinki University Press (State Archives of Assyria, 14).

SAA 16 Luukko, Mikko and Van Buylaere, Greta (2002) *The Political Correspondence of Esarhaddon.* Helsinki: Helsinki University Press (State Archives of Assyria, 16).

SAA 18 Reynolds, Frances (2003) *The Babylonian Correspondence of Esarhaddon and Letters to Assurbanipal and Sin-šarru-iškun from Northern and Central Babylonia.* Helsinki: Helsinki University Press (State Archives of Assyria, 18).

SAA 19 Luukko, Mikko (2012) *The Correspondence of Tiglath-Pileser III and Sargon II from Calah/Nimrud.* Helsinki: The Neo-Assyrian Text Corpus Project (State Archives of Assyria, 19).

SAA 20 Parpola, Simo (2017) *Assyrian Royal Rituals and Cultic Texts.* Helsinki: The Neo-Assyrian Text Corpus Project (State Archives of Assyria, 20).

SAA 21 Parpola, Simo (2018) *The Correspondence of Assurbanipal, Part I: Letters from Assyria, Babylonia, and Vassal States.* Winona Lake, IN: The Neo-Assyrian Text Corpus Project (State Archives of Assyria, 21).

StAT 2 Donbaz, Veysel and Parpola, Simo (2001) *Neo-Assyrian Legal Texts in Istanbul.* Saarbrücken: SDV Saarbrücker Druckerei und Verlag (Studien zu den Assur-Texten, 2).

Bibliography

Adalı, Selim F. (2017) 'The Anatolian and Iranian Frontiers: Analyzing the Foreign Policy of the Assyrian Empire under Esarhaddon', in Olga Drewnowska and Małgorzata Sandowicz (eds.) *Fortune and Misfortune in the Ancient Near East: Proceedings of the 60th Rencontre Assyriologique Internationale at Warsaw 21–25 July 2014.* Winona Lake, IN: Eisenbrauns, pp. 307–324.

Allen, Spencer L. (2013) 'Rearranging the Gods in Esarhaddon's Succession Treaty (SAA 2 6:414–465)', *Die Welt des Orients*, 43(1), pp. 1–24.

Arnold, Bill T. and Shockey, Brian T. (2022) 'Deuteronomy 13 and the Succession Treaty of Esarhaddon: A Fresh Investigation', in James K. Hoffmeier et al. (eds.) *'Now These Records are Ancient': Studies in Ancient Near Eastern and Biblical History, Language and Culture in Honor of K. Lawson Younger, Jr.* Münster: Zaphon (Ägypten und Altes Testament, 114), pp. 1–14.

Ataç, Mehmet-Ali (2004) 'The "Underworld Vision" of the Ninevite Intellectual Milieu', *Iraq*, 66, pp. 67–76.

Bach, Johannes (2018) 'A Transtextual View on the "Underworld Vision of an Assyrian Prince"', in Strahil V. Panayotov and Luděk Vacín (eds.) *Mesopotamian Medicine and Magic: Studies in Honor of Markham J. Geller.* Leiden/Boston: Brill (Ancient Magic and Divination, 14), pp. 69–92.

Bach, Johannes and Fink, Sebastian (2022) *The King as a Nodal Point of Neo-Assyrian Identity. Kasion 8.* Münster: Zaphon.

Bagg, Ariel M. (2007) *Répertoire Géographique das Textes Cunéiformes VII/1. Die Orts- und Gewässernamen der neuassyrischen Zeit. Teil 1: Die Levante.* Wiesbaden: Dr. Ludwig Reichert Verlag.

Bagg, Ariel M. (2011) *Die Assyrer und das Westland: Studien zur historischen Geographie und Herrschaftspraxis in der Levante im 1. Jt. v. u. Z.* Leuven/Paris/Walpole, MA: Uigeverij Peeters & Departement Oosterse Studies (Orientalia Lovaniensia Analecta, 216).

Bagg, Ariel M. (2017) *Répertoire Géographique des Textes Cunéiformes VII/2. Die Orṭs- und Gewässernamen der neuassyrischen Zeit Teil 2: Zentralassyrien und benachbarte Gebiete, Ägypten und die arabische Halbinsel.* Wiesbaden: Dr. Ludwig Reichert Verlag.

Bagg, Ariel M. (2020) *Répertoire Géographique des Textes Cunéiformes VII/3. Die Orts- und Gewässernamen der neuassyrischen Zeit Teil 3: Babylonien, Urarṭu und die östlichen Gebiete.* Wiesbaden: Dr. Ludwig Reichert Verlag.

Balatti, Silvia (2017) *The Mountain Peoples in the Ancient Near East. The Case of the Zagros in the First Millennium BCE.* Wiesbaden: Harrassowitz (Classica et Orientalia, 18).

Baker, Heather D. and Gross, Melanie M. (2015) 'Doing the King's Work: Perceptions of Service in the Assyrian Royal Correspondence', in Stephan Procházka, Lucian Reinfandt, and Sven Tost (eds.) *Official Epistolography and the Language(s) of Power. Proceedings of the First International Conference of the Research Network Imperium & Officium.* Wien: Österreichische Akademie der Wissenschaften (Comparative Studies in Ancient Bureaucracy and Officialdom, 8), pp. 73–90.

Barmash, Pamela (2020) *The Laws of Hammurabi: At the Confluence of Royal and Scribal Traditions.* Oxford: Oxford University Press.

Barcina, Cristina (2016) 'The Display of Esarhaddon's Succession Treaty at Kalḫu as a Means of Internal Political Control', *Antiguo Oriente*, 14, pp. 11–52.

Barcina, Cristina (2017) 'The Conceptualization of the *Akitu* under the Sargonids: Some Reflections', *State Archives of Assyria Bulletin*, 23, pp. 91–129.

 https://doi.org/10.1515/9783111323435-014

Barjamovic, Gojko (2011) 'Pride, Pomp and Circumstance: Palace, Court and Household In Assyria 879 – 612 BCE', in Jeroen Duindam, Tülay Artan, and Metin Kunt (eds.) *Royal Courts in Dynastic States and Empires.* (Rulers & Elites, 1), pp. 25 – 61.

Bartelmus, Alexa (2007) '*Talīmu* and the Relationship between Assurbanipal and Šamaš-šumu-ukīn', *State Archives of Assyria Bulletin*, 16, pp. 287 – 302.

Beaulieu, Paul-Alain (2010) 'The Afterlife of Assyrian Scholarship in Hellenistic Babylonia', in Jeffrey Stackert, Barbara N. Porter, and David P. Wright (eds.) *Gazing on the Deep: Ancient Near Eastern and Other Studies in Honor of Tzvi Abusch.* Bethesda, MD: Eisenbrauns, pp. 1 – 18.

Bendix, Reinhard (1960) *Max Weber: An Intellectual Portrait.* Garden City, NY: Doubleday.

Bennett, Ellie (2019) '"I am a Man": Masculinities in the Titulary of the Neo-Assyrian Kings in the Royal Inscriptions', *KASKAL*, 16, pp. 373 – 392.

Black, Jeremy and Green, Anthony (2004) *Gods, Demons and Symbols of Ancient Mesopotamia: An Illustrated Dictionary.* London: The British Museum Press.

Bledsoe, Seth A. (2021) *The Wisdom of the Aramaic Book of Ahiqar: Unravelling a Discourse of Uncertainty and Distress.* Leiden/Boston: Brill (Supplements to the Journal for the Study of Judaism, 199).

Böschen, Stefan (2018) 'Reflexive Responsibilisierung – feldtheoretisch ausgeleuchtet', in Anna Henkel et al. (eds.) *Reflexive Responsibilisierung: Verantwortung für nachhaltige Entwicklung.* Bielefeld: transcript (Sozialtheorie), pp. 247 – 266.

Botta, Paul É. (1849) *Monument de Ninive.* Paris: Imprimerie Nationale.

Brendecke, Arndt (2009) *Imperium und Empirie. Funktionen des Wissens in der spanischen Kolonialherrschaft.* Köln: Böhlau Verlag.

Brendecke, Arndt (2016) *The Empirical Empire: Spanish Colonial Rule and the Politics of Knowledge.* Berlin/Boston: De Gruyter.

Breuer, Stefan (2019) 'The Relevance of Weber's Conception and Typology of Herrschaft', in Edith Hanke, Lawrence Scaff, and Sam Whimster (eds.) *The Oxford Handbook of Max Weber.* Oxford: Oxford University Press, pp. 237 – 257.

Brosius, Maria (2021) *A History of Ancient Persia: The Achaemenid Empire.* Hoboken, NJ: Wiley Blackwell (Blackwell Companions to the Ancient World).

Cambridge Dictionary, https://dictionary.cambridge.org/dictionary/english [last access: 20. 12. 2022].

Cancik-Kirschbaum, Eva (2015) *Die Assyrer. Geschichte, Gesellschaft, Kultur.* Munich: C. H. Beck.

Cifola, Barbara (1995) *Analysis of Variants in the Assyrian Royal Titulary from the Origins to Tiglath-Pileser III.* Naples: Instituto Universitario Orientale.

Cogan, Mordechai (1977) 'Ashurbanipal Prism F: Notes on Scribal Techniques and Editorial Procedures', *Journal of Cuneiform Studies*, 29(2), pp. 97 – 107.

Cogan, Mordechai (2005) 'Some Text Critical Issues in Hebrew Bible from an Assyriological Perspective', *Textus*, 22, pp. 1 – 20.

Cross, Frank M. (1973) *Canaanite Myth and Hebrew Epic: Essays in the History of the Religion of Israel.* Cambridge, MA: Harvard University Press.

Crouch, Carly L. (2014) *Israel and the Assyrians. Deuteronomy, the Succession Treaty of Esarhaddon, and the Nature of Subversion.* Atlanta, GA: SBL Press (Ancient Near East Monographs, 8).

Dalley, Stephanie M. (2001) 'Assyrian Court Narratives in Aramaic and Egyptian: Historical Fiction', in T. Abusch et al. (eds.) *Historiography of the Cuneiform World.* Bethesda, MD: Eisenbrauns (CRRAI, 45), pp. 149 – 161.

Dalley, Stephanie M. and Siddall, Luis R. (2021) 'A Conspiracy to Murder Sennacherib? A Revision of SAA 18 100 in the Light of a Recent Join.', *Iraq*, 83, pp. 45 – 56.

Davies, Philip R. (2010) 'The Deuteronomistic History and "Double Redaction"', in Kurt L. Noll and Brooks Schramm (eds.) *Raising Up a Faithful Exegete: Essays in Honor of Richard D. Nelson.* Winona Lake, IN: Eisenbrauns, pp. 51–59.

Dercksen, Jan G. (2004) *Old Assyrian Institutions.* Leiden: Nederlands Instituut vor het Nabije Oosten (MOS Studies, 4).

Dezső, Tamás (2006) 'Šubria and the Assyrian Empire', *Acta Antiqua Academiae Scientiarum Hungaricae*, 46, pp. 33–38.

Dezső, Tamás (2016) *The Assyrian Army II. Recruitment and Logistics.* Budapest: Eötvös University Press (Antiqua et Orientalia, 6).

Dimant, Devorah (2018) 'Tobit and Ahiqar', in J. Harold Ellens et al. (eds.) *Wisdom Poured Out Like Water: Studies on Jewish and Christian Antiquity in Honor of Gabriele Boccaccini.* Berlin/Boston: De Gruyter, pp. 276–291.

Dubovský, Peter (2004) 'Neo-Assyrian Warfare: Logistics and Weaponry during the Campaigns of Tiglath-pileser III', *Anodos: Studies of the Ancient World,* 4–5, pp. 61–67.

Dubovský, Peter (2012) 'King's Direct Control: Neo-Assyrian *Qēpu* Officials', in G. Wilhelm (ed.) *Representation, and Symbols of Power in the Ancient Near East.* Winona Lake, IN: Eisenbrauns, pp. 449–460.

Düring, Bleda S. (2021) 'Book Review of The Neo-Assyrian Empire in the Southwest: Imperial Domination and Its Consequences By Avraham Faust. Oxford: Oxford University Press 2021. Pp. 400. $115. ISBN 9780198841630 (cloth).', *American Journal of Archaeology,* 125(4), https://doi.org/doi:10.3764/ajaonline1254.During [last access: 11.01.2023].

Edenburg, Cynthia and Müller, Reinhard (2019) 'Literary Connections and Social Contexts: Approaches to Deuteronomy in Light of the Assyrian *adê*-Tradition', *Hebrew Bible and Ancient Israel*, 8(2), pp. 73–86.

Ephʿal, Israel and Tadmor, Hayim (2006) 'Observations on two inscriptions of Esarhaddon: Prism Nineveh A and the letter to the god', in Yairah Amit et al. (eds.) *Essays on ancient Israel in its Near Eastern context: A Tribute to Nadav Na'aman.* Winona Lake, IN: Eisenbrauns, pp. 155–170.

Ermidoro, Stefania (2017) 'Ruling over Time. The Calendar in the Neo-Assyrian Royal Propaganda', *State Archives of Assyria Bulletin*, 23, pp. 131–156.

Fadhil, Anmar A. (2012) *Eine kleine Tontafelbibliothek aus Assur (Ass. 15426).* Ruprecht-Karls-Universität Heidelberg: PhD dissertation.

Faist, Betina (2012) 'Der neuassyrische Kaufvertrag im Spannungsfeld zwischen Formular und konkretem Fall', *Zeitschrift für altorientalische und biblische Rechtsgeschichte*, 18, pp. 209–219.

Faist, Betina (2015) 'Der Eid im neuassyrischen Gerichtsverfahren', in Heinz Barta, Martin Lang, and Robert Rollinger (eds.) *Prozessrecht und Eid. Recht und Rechtsfindung in antiken Kulturen. Teil 1.* Wiesbaden: Harrassowitz (Philippikka, 86), pp. 63–78.

Faist, Betina (2020) *Assyrische Rechtsprechung im 1. Jahrtausend v. Chr. Münster: Zaphon* (dubsar, 15).

Faist, Betina and Klengel-Brandt, Evelyn (2010) 'Die Siegel der Stadtvorsteher von Assur', in Ş. Dönmez (ed.) *Veysel Donbaz'a sunulan yazılar. DUB.SAR É.DUB.BA.A. Studies presented in honour of Veysel Donbaz.* Istanbul: Ege Publications, pp. 115–134.

Fales, F. Mario (1988) 'Prosopography of the Neo-Assyrian Empire, 2: The Many Faces of Nabû-šarru-uṣur', *State Archives of Assyria Bulletin,* 2(2), pp. 105–124.

Fales, F. Mario (1991) 'Narrative and Ideological Variations in the Account of Sargon's Eighth Campaign', in Mordechai Cogan and Israel Eph'al (eds.) *Ah, Assyria...: Studies in Assyrian History and Ancient Near Eastern Historiography Presented to Hayim Tadmor.* Jerusalem: Magnes Press (ScrHier, 33), pp. 129–147.

Fales, F. Mario (1994) 'Riflessioni sull' Ahiqar di Elefantina', in *Orientis Antiqui Miscellanea*, 1. Roma: IPOCAN – Istituto per l'Oriente Carlo Alfonso Nallino, pp. 39–60.
Fales, F. Mario (1999) 'Assyrian Royal Inscriptions: Newer Horizons', *State Archives of Assyria Bulletin*, 13, pp. 115–144.
Fales, F. Mario (2011) '*Maṣṣartu*: The Observation of Astronomical Phenomena in Assyria', in Enrico M. Corsini (ed.) *The Inspiration of Astronomical Phenomena*, VI. San Francisco: Astronomical Society of the Pacific, pp. 361–370.
Fales, F. Mario (2012) 'After Ta'yinat. The New Status of Esarhaddon's *adê* for Assyrian Political History', *Revue d'Assyriologie*, 106, pp. 133–158.
Fales, F. Mario (2017a) 'Assyrian Legal Traditions', in Eckart Frahm (ed.) *A Companion to Assyria.* Malden, MA: Wiley, pp. 398–422.
Fales, F. Mario (2017b) 'Phoenicia in the Neo-Assyrian Period: An Updated Overview', *State Archives of Assyria Bulletin*, 23, pp. 181–295.
Fales, F. Mario (2020) 'Saritra and the others: A Neo-Assyrian view of Papyrus Amherst 63', in Maria E. Balza et al. (eds.) *Città e parole argilla e pietra: studi offerti a Clelia Mora da allievi, colleghi e amici.* Bari-S. Spirito: Edipuglia (Biblioteca di Athenaeum, 65), pp. 225–251.
Fales, F. Mario (2023a) 'Love and Kindness in the Assyrian State', in Karen Sonik and Ulrike Steinert (eds.) *The Routledge Handbook of Emotions in the Ancient Near East.* London/New York: Routledge, pp. 695–724.
Fales, F. Mario (2023b) 'The Assyrian Empire: Perspectives on Culture and Society', in Karen Radner, Nadine Moeller, and Daniel T. Potts (eds.) *The Oxford History of the Ancient Near East. Volume IV: The Age of Assyria.* Oxford: Oxford University Press, pp. 425–519.
Fales, F. Mario, Bachelot, Luc and Attardo, Ezio (1996) 'An Aramaic Tablet from Tell Shioukh Fawqani, Syria', *Semitica*, 46, pp. 81–121.
Fincke, Jeanette C. (2004) 'The British Museum's Ashurbanipal Library Project', *Iraq*, 66, pp. 55–60.
Finkel, Irving (2021) *The First Ghosts: Most Ancient of Legacies.* London: Hodder and Stoughton.
Finn, Jennifer (2017) *Much Ado about Marduk: Questioning Discourses of Royalty in First Millennium Mesopotamian Literature.* Berlin/Boston: De Gruyter (Studies in Ancient Near Eastern Records, 16).
Foster, Benjamin (2005) *Before the Muses: An Anthology of Akkadian Literature.* 3rd edn. Bethesda, MD: CDL Press.
Foster, Benjamin (2007) *Akkadian Literature of the Late Period.* Münster: Ugarit-Verlag (Guides to the Mespotamian Textual Record, 2).
Frahm, Eckart (1997) *Einleitung in die Sanherib-Inschriften.* Wien: Institut für Orientalik der Üniversität (Archiv für Orientforschung Beiheft, 26).
Frahm, Eckart (1999) 'Nabû-zuqup-kenu, das Gilgamesch-Epos und der Tod Sargons II.', *Journal of Cuneiform Studies*, 51, pp. 73–90.
Frahm, Eckart (2009a) *Historische und historisch-literarische Texte. Keilschrifttexte aus Assur literarischen Inhalts 3.* Wiesbaden: Harrassowitz (WVDOG, 121).
Frahm, Eckart (2009b) 'Warum die Brüder Böses planten: Überlegungen zu einer alten Crux Asarhaddons "Ninive-A" Inschrift', in Werner Arnold et al. (eds.) *Philologisches und Historisches zwischen Anatolien und Sokotra: Analecta Semitica in Memoriam Alexander Sima.* Wiesbaden: Harrassowitz, pp. 27–49.
Frahm, Eckart (2010) 'Hochverrat in Assur', in Stefan M. Maul and Nils P. Heeßel (eds.) *Assur-Forschungen: Arbeiten aus der Forschungsstelle "Edition literarischer Keilschrifttexte aus Assur" der Heidelberger Akademie der Wissenschaften.* Wiesbaden: Harrassowitz, pp. 89–139.

Frahm, Eckart (2014) 'Family Matters: Psychohistorical Reflections on Sennacherib and His Times', in Seth Richardson and Isaac Kalimi (eds.) *Sennacherib at the Gates of Jerusalem. Story, History and Historiography.* Leiden: Brill (Culture and History of the Ancient Near East, 71), pp. 161–222.

Frahm, Eckart (2016) '"And His Brothers Were Jealous of Him": Surprising Parallels between Joseph and King Esarhaddon', *Biblical Archaeology Review*, 42(3), pp. 43–64.

Frahm, Eckart (2017a) 'Assyria and the South: Babylonia', in Eckart Frahm (ed.) *A Companion to Assyria.* Malden, MA: Wiley, pp. 286–298.

Frahm, Eckart (2017b) 'The Neo-Assyrian Period (ca. 1000–609 BCE)', in Eckart Frahm (ed.) *A Companion to Assyria.* Malden, MA: Wiley, pp. 161–208.

Frahm, Eckart (2019) 'The Neo-Assyrian Royal Inscriptions as Text: History, Ideology, and Intertextuality', in Giovanni Battista Lanfranchi, Raija Mattila, and Robert Rollinger (eds.) *Writing Neo-Assyrian History: Sources, Problems, and Approaches.* Helsinki: The Neo-Assyrian Text Corpus Project (State Archives of Assyria Studies, 29), pp. 139–159.

Frame, Grant (1991) 'Nabonidus, Nabû-šarra-uṣur, and the Eanna temple', *Zeitschrift für Assyriologie und vorderasiatische Archäologie*, 81, pp. 37–86.

Frame, Grant (1992) *Babylonia 689–627 B.C. A Political History.* Leiden: Nederlands Instituut vor het Nabije Oosten (PIHANS, LXIX).

Frame, Grant (2008) 'Babylon: Assyria's Problem and Assyria's Prize', *Journal of the Canadian Society for Mesopotamian Studies*, 3, pp. 21–31.

Frankena, Rintje (1965) 'The Vassal-Treaties of Esarhaddon and the Dating of Deuteronomy', *Oudtestamentische studiën (Old Testament Studies)*, 14, pp. 122–154.

Fried, Lisbeth S. (2002) 'The High Places (*Bāmôt*) and the Reforms of Hezekiah and Josiah: An Archaeological Investigation', *Journal of the American Oriental Society*, 122(3), pp. 437–465.

Fuchs, Andreas (1994) *Die Inschriften Sargons II. aus Khorsabad.* Göttingen: Cuvillier Verlag.

Fuchs, Andreas (2008) 'Der Turtan Šamši-ilu und die große Zeit der assyrischen Großen (830–746)', *Die Welt des Orients*, 38, pp. 61–145.

Fuchs, Andreas (2012) 'Urarṭu in der Zeit', in S. Kroll et al. (eds.) *Biainili-Urartu, 12.–14. Oktober 2007, München.* Leuven: Peeters (Acta Iranica, 51), pp. 135–161.

Fuchs, Andreas (2023) 'The Medes and the Kingdom of Mannea', in Karen Radner, Nadine Moeller, and Daniel T. Potts (eds.) *The Oxford History of the Ancient Near East. Volume IV: The Age of Assyria.* Oxford: Oxford University Press, pp. 674–768.

Gadebusch Bondio, Mariacarla et al. (2023) *Techniken der Responsibilisierung. Historische und gegenwartsbezogene Studien.* Hannover: Wehrhahn.

Garelli, Paul (1989) 'L'appel au roi sous l'empire assyrien', in Marc Lebeau and Philippe Talon (eds.) *Reflets des Deux Fleuves. Volume de melanges offerts à André Finet.* Leuven: Peeters (Akkadica Supplementum, 6), pp. 45–46.

Garelli, Paul (1991) 'The Achievement of Tiglath-pileser III: Novelty or Continuity?', in Mordechai Cogan and Israel Eph'al (eds.) *Ah, Assyria...: Studies in Assyrian History and Ancient Near Eastern Historiography Presented to Hayim Tadmor.* Jerusalem: Magnes Press (Scripta Hierosolymitana, 33), pp. 46–51.

George, Andrew (1986) 'Sennacherib and the Tablet of Destinies', *Iraq*, 48, pp. 133–146.

George, Andrew (1993) *House Most High: The Temples of Ancient Mesopotamia.* Winona Lake, IN: Eisenbrauns (Mesopotamian Civilizations, 5).

George, Mark (forthcoming) *How Deuteronomy Created Israel: Technologies of the Self, Government, and Writing.* London/New York: Routledge.

Grayson, A. Kirk (1975) *Assyrian and Babylonian Chronicles.* New York: Locust Valley (Texts from Cuneiform Sources, 5).

Grayson, A. Kirk (1980) 'Assyria and Babylonia', *Orientalia, NOVA SERIES*, 49(2), pp. 140 – 194.

Grayson, A. Kirk (1987) 'Akkadian Treaties of the Seventh Century BC', *Journal of Cuneiform Studies*, 39, pp. 127 – 160.

Gross, Melanie M. (2014) 'Ḫarrān als kulturelles Zentrum in der altorientalischen Geschichte und sein Weiterleben', in Lea Müller-Funk et al. (eds.) *Kulturelle Schnittstelle: Mesopotamien, Anatolien, Kurdistan: Geschichte. Sprachen. Gegenwart.* Vienna: Selbstverlag des Instituts für Orientalistik der Universität Wien, pp. 139 – 154.

Gross, Melanie M. (2020) *At the Heart of an Empire: The Royal Household in the Neo-Assyrian Period.* Leuven: Peeters (Orientalia Lovaniensia Analecta, 292).

Gross, Melanie M. and Pirngruber, Reinhard (2014) 'On Courtiers in the Neo-Assyrian Empire: *ša-rēši* and *mazzāz pāni*', *Altorientalische Forschungen*, 41(2), pp. 161 – 175.

Günbati, Cahit et al. (2020) *Kahramanmaraş'ta Bulunmuş Yeni Asurca Tabletler.* Ankara: Türk Tarih Kurumu.

Harrison, Timothy P. (2014) 'Articulating Neo-Assyrian Imperialism at Tell Tayinat', in Matthew T. Rutz and Morag M. Kersel (eds.) *Archaeologies of Text. Archaeology, Technology, and Ethics.* Oxford/ Philadelphia: Oxbow Books.

Harrison, Timothy P. and Osborne, James F. (2012) 'Building XVI and the Neo-Assyrian Sacred Precinct at Tell Tayinat', *Journal of Cuneiform Studies*, 64, pp. 125 – 143.

Hasegawa, Shuichi (2017) 'Josiah's Death: Its Reception History as Reflected in the Books of Kings and Chronicles', *Zeitschrift für die alttestamentliche Wissenschaft*, 129(4), pp. 522 – 535.

Hätinen, Aino (2021) *The Moon God Sîn in Neo-Assyrian and Neo-Babylonian Times.* Münster: Zaphon (dubsar, 20).

Hauser, Stefan R. (2012) *Status, Tod und Ritual: Stadt- und Sozialstruktur Assurs in neuassyrischer Zeit.* Wiesbaden: Harrassowitz (Abhandlungen der Deutschen Orient-Gesellschaft, 26).

Hinds, Lyn and Grabosky, Peter (2010) 'Responsibilisation Revisited: From Concept to Attribution in Crime Control', *Security Journal*, 23, pp. 95 – 113.

Hipp, Krzysztof (2015) 'Fugitives in the State Archives of Assyria', *State Archives of Assyria Bulletin*, 21, pp. 47 – 77.

Hochschild, Arlie (1979) 'Emotion Work, Feeling Rules, and Social Structure', *American Journal of Sociology*, 85(3), pp. 551 – 575.

Hochschild, Arlie (2013) *So How's the Family?: And Other Essays.* Berkeley and Los Angeles, CA: University of California Press.

Huehnergard, John (2011) *A Grammar of Akkadian.* 3rd edn. Winona Lake, IN: Eisenbrauns (Harvard Semitic Studies, 45).

Hussein, Auday (2020) 'Crown Prince or Prince? The Translation of *mār šarri* and Its Impact on the Succession in the Neo-Assyrian Period', *State Archives of Assyria Bulletin*, 26, pp. 58 – 88.

Jeffers, Joshua (2018) 'Neo-Assyrian Scribal Practices: The Editorial History of Ashurbanipal's Prism F Inscription', *Zeitschrift für Assyriologie und vorderasiatische Archäologie*, 108(2), pp. 209 – 225.

Jursa, Michael and Radner, Karen (1996) 'Keilschrifttexte aus Jerusalem', *Archiv für Orientforschung*, 42(3), pp. 89 – 108.

Kantorowicz, Ernst H. (1985) *Kaiser Friedrich der Zweite.* 6d edn. Stuttgart: Klett-Cotta.

Karlsson, Mattias (2020) 'From Sumer to Assyria: The Term "Black-headed People" in Assyrian Texts', *Akkadica*, 141(2), pp. 127 – 139.

Kertai, David (2015) *The Architecture of Late Assyrian Royal Palaces.* Oxford: Oxford University Press.

Kitz, Anne M. (2007) 'Curses and Cursing in the Ancient Near East', *Religion Compass*, 1(6), pp. 615–627.

Knapp, Andrew (2015) *Royal Apologetic in the Ancient Near East.* Atlanta, GA: SBL Press (Writings from the Ancient World Supplement Series, 4).

Knapp, Andrew (2016) 'The Sitz im Leben of Esarhaddon's Apology', *Journal of Cuneiform Studies*, 68, pp. 181–195.

Knapp, Andrew (2020) 'The Murderer of Sennacherib, yet Again: The Case against Esarhaddon.', *Journal of the American Oriental Society*, 140(1), pp. 165–182.

Knoppers, Gary N. (2010) 'Theories of the Redaction(s) of Kings', in Baruch Halpern and André Lemaire (eds.) *The Book of Kings: Sources, Composition, Historiography and Reception.* Leiden/Boston: Brill (Supplements to Vetus Testamentum, 129), pp. 69–88.

Koch, Christoph (2008) *Vertrag, Treueid und Bund.* Berlin/New York: Walter de Gruyter (Beihefte zur Zeitschrift für die alttestamentliche Wissenschaft, 383).

Kölbel, Ralf et al. (2021) 'Responsibilisierung', *Working Paper des SFB 1369 'Vigilanzkulturen'*, 2, pp. 4–19.

Kratz, Reinhard G. (2022) 'Aḥiqar and Bisitun: The Literature of the Judeans at Elephantine', in Reinhard G. Kratz and Bernd U. Schipper (eds.) *Elephantine in Context: Studies on the History, Religion and Literature of the Judeans in Persian Period Egypt.* Tübingen: Mohr Siebeck (Forschungen zum Alten Testament, 155), pp. 301–322.

Kvanvig, Helge S. (1981) 'An Akkadian Vision as Background for Daniel 7', *Studia Theologica*, 35, pp. 85–89.

Kvanvig, Helge S. (1988) *Roots of Apocalyptic: The Mesopotamian Background of the Enoch Figure and of the Son of Man.* Neukirchen-Vluyn: Neukirchener Verlag.

Lambert, Wilfred G. (1983) 'The God Aššur', *Iraq*, 45(1), pp. 82–86.

Lanfranchi, Giovanni Battista (2003) 'The Assyrian Expansion in the Zagros and the Local Ruling elites', in Giovanni Battista Lanfranchi, Michael Roaf, and Robert Rollinger (eds.) *Continuity of Empire (?): Assyria, Media, Persia.* Padova: S.A.R.G.O.N. Editrice e Libreria, pp. 79–118.

Latour, Bruno (1987) *Science in Action: How to Follow Scientists and Engineers through Society.* Cambridge, MA: Harvard University Press.

Lauinger, Jacob (2011) 'Some Preliminary Thoughts on the Tablet Collection in Building XVI', *The Canadian Society for Meopotamian Studies Journal*, 6(Fall), pp. 5–14.

Lauinger, Jacob (2012) 'Esarhaddon's Succession Treaty at Tell Tayinat: Text and Commentary', *Journal of Cuneiform Studies*, 64, pp. 87–123.

Lauinger, Jacob (2013) 'The Neo-Assyrian *adê:* Treaty, Oath, or Something Else?', *Zeitschrift für altorientalische und biblische Rechtsgeschichte*, 19, pp. 99–115.

Lauinger, Jacob (2015) 'Neo-Assyrian Scribes, "Esarhaddon's Succession Treaty", and the Dynamics of Textual Mass Production', in Paul Delnero and Jacob Lauinger (eds.) *Texts and Contexts. The Circulation and Transmission of Cuneiform Texts in Social Space.* Berlin/Boston: De Gruyter (Studies in Ancient Near Eastern Records, 9), pp. 337–347.

Lauinger, Jacob (2019) 'Literary Connections and Social Contexts: Approaches to Deuteronomy in Light of the Assyrian *adê*-Tradition', *Hebrew Bible and Ancient Israel*, 8(2), pp. 87–100.

Lauinger, Jacob (2021) 'Observing Neo-Assyrian Scribes at Work', in Rodney Ast et al. (eds.) *Observing the Scribe at Work: Scribal Practice in the Ancient World.* Leuven/Paris/Bristol, CT: Peeters (Orientalia Lovaniensia analecta, 301), pp. 177–185.

Leichty, Erle (1991) 'Esarhaddon's "Letter to the Gods"', in Mordechai Cogan and Israel Eph'al (eds.) *Ah, Assyria... Studies in Assyrian History and Ancient Near Eastern Historiography presented to Hayim Tadmor.* Jerusalem: The Magnes Press, pp. 52–57.

Leichty, Erle (2007) 'Esarhaddon's Exile: Some Speculative History', in Martha T. Roth et al. (eds.) *From the Workshop of the Chicago Assyrian Dictionary Volume 2: Studies Presented to Robert D. Biggs.* Chicago, IL: The Oriental Institute of the University of Chicago (Assyriological Studies, 27), pp. 189–191.

Lenzi, Alan (2008) *Secrecy and the Gods: Secret Knowledge in Ancient Mesopotamia and Biblical Israel.* Helsinki: The Neo-Assyrian Text Corpus Project (State Archives of Assyria Studies, 19).

Levinson, Bernard M. (2010) 'Esarhaddon's Succession Treaty as the Source for the Canon Formula in Deuteronomy 13:1', *Journal of the American Oriental Society*, 130, pp. 337–347.

Liverani, Mario (1995) 'The Medes at Esarhaddon's Court', *Journal of Cuneiform Studies*, 47, pp. 57–62.

Liverani, Mario (2014a) *The Ancient Near East: History, Society and Economy.* Translated by Soraia Tabatabai. London/New York: Routledge.

Liverani, Mario (2014b) 'The King and His Audience', in Salvatore Gaspa et al. (eds.) *From Source to History: Studies on Ancient Near Eastern Worlds and Beyond. Dedicated to Giovanni Battista Lanfranchi on the Occasion of His 65th Birthday on June 23, 2014.* Münster: Ugarit-Verlag (Alter Orient und Altes Testament, 412), pp. 373–385.

Liverani, Mario (2017) *Assyria: The Imperial Mission.* Winona Lake, IN: Eisenbrauns (Mesopotamian Civilisations, 21).

Livingstone, Alasdair (2017) 'Babylonian Hemerologies and Menologies', in Donald Harper and Marc Kalinowski (eds.) *Books of Fate and Popular Culture in Early China. The Daybook Manuscripts of the Warring States, Qin, and Han.* Leiden/Boston: Brill (Handbook of Oriental Studies. Section 4 China, 33), pp. 408–436.

Livingstone, David N. (2003) *Putting Science in its Place: Geographies of Scientific Knowledge.* Chicago, IL: University of Chicago Press.

Loktionov, Alexandre A. (2016) 'An "Egyptianising" Underworld Judging an Assyrian Prince? New Perspectives on VAT 10057', *Journal of Ancient Near Eastern History*, 3(1), pp. 39–55.

Lundström, Steven (2009) *Die Königsgrüfte im Alten Palast von Assur.* Wiesbaden: Harrassowitz (WVDOG, 123).

Luukko, Mikko (2007) 'The Administrative Roles of the "Chief Scribe" and the "Palace Scribe" in the Neo-Assyrian Period', *State Archives of Assyria Bulletin*, 16, pp. 227–256.

Luukko, Mikko (2018) 'Anonymous Neo-Assyrian Denunciations in a Wider Context', in Shigeo Yamada (ed.) *Neo-Assyrian Sources in Context: Thematic Studies of Texts, History and Culture.* Helsinki: The Neo-Assyrian Text Corpus Project (State Archives of Assyria Studies, 28), pp. 163–184.

Machinist, Peter (1984) 'The Assyrians and Their Babylonian Problem: Some Reflections', in *Jahrbuch des Wissenschaftskollegs zu Berlin*, pp. 353–364.

Mattila, Raija (2000) *The King's Magnates: A Study of the Highest Officials of the Neo-Assyrian Empire.* Helsinki: The Neo-Assyrian Text Corpus Project (State Archives of Assyria Studies, 11).

Maul, Stefan M. (2010) 'Die Tontafelbibliothek aus dem sogenannten "Haus des Beschwörungspriesters"', in Stefan M. Maul and Nils P. Heeßel (eds.) *Assur-Forschungen: Arbeiten aus der Forschungsstelle 'Edition Literarische Keilschrifttexte aus Assur' der Heidelberger Akademie der Wissenschaften.* Wiesbaden: Harrassowitz, pp. 189–228.

Maul, Stefan M. and Manasterska, Sara (2023) *Schreiberübungen aus neuassyrischer Zeit: Mit Beiträgen von Anmar A. Fadhil und einer internationalen Forschergruppe. Keilschrifttexte aus Assur literarischen Inhalts 15.* Wiesbaden: Harrassowitz (WVDOG, 162).

Maul, Stefan M. and Miglus, Peter A. (2020) 'Erforschung des *ekal māšarti* auf Tell Nebi Yunus in Ninive 2018 – 2019', *Zeitschrift für Orient-Archäologie*, 13, pp. 128 – 213.

May, Natalie N. (2018) 'The Scholar and Politics: Nabû-zuqup-kēnu, his Colophons and the Ideology of Sargon II', in *Proceedings of the International Conference Dedicated to the Centenary of Igor Mikhailovich Diakonoff (1915 – 1999).* St. Petersburg: The State Hermitage Publishers (Transactions of the State Hermitage Museum, 95), pp. 110 – 164.

Mayer, Walter (1998) 'Der Weg auf den Thon Assurs: Sukzession und Usurpation im assyrischen Königshaus', in Manfried Dietrich et al. (eds.) *'Und Mose schrieb dieses Lied auf': Studien zum Alten Testament und zum Alten Orient. Festschrift für Osward Loretz zur Vollendung seines 70. Lebensjahres mit Beiträgen von Freunden, Schülern und Kollegen.* Münster: Ugarit-Verlag (Alter Orient und Altes Testament, 250), pp. 533 – 555.

Melville, Sarah C. (1999) *The Role of Naqia/Zakutu in Sargonid Politics.* Helsinki: The Neo-Assyrian Text Corpus Project (State Archives of Assyria Studies, 9).

Mofidi-Nasrabadi, Behzad (1999) *Untersuchungen zu den Bestattungssitten in Mesopotamien in der ersten Hälfte des ersten Jahrtausends v. Chr.* Mainz: von Zabern (Baghdader Forschungen, 23).

Monroe, Lauren A. S. (2011) *Josiah's Reform and the Dynamics of Defilement: Israelite Rites of Violence and the Making of a Biblical Text.* Oxford: Oxford University Press.

Moore, James D. (2022) '"Ahikariana" New Readings of Berlin P. 13446 and Developments in Ahiqar Research', in Reinhard G. Kratz and Bernd U. Schipper (eds.) *Elephantine in Context: Studies on the History, Religion and Literature of the Judeans in Persian Period Egypt.* Tübingen: Mohr Siebeck (Forschungen zum Alten Testament, 155), pp. 237 – 263.

Morrow, William (2009) 'The Paradox of Deuteronomy 13. A Post-Colonial Reading', in Reinhard Achenbach and Martin Arneth (eds.) *'Gerechtigkeit und Recht zu üben' (Gen 18,19). Studien zuraltorientalischen und biblischen Rechtsgeschichte, zur Religionsgeschichte Israels und zur Religionssoziologie. FS E. Otto.* Wiesbaden: Harrassowitz (BZAR, 13), pp. 227 – 239.

Morrow, William (2019) 'Have Attempts to Establish the Dependency of Deuteronomy on the Esarhaddon Succession Treaty (EST) Failed?', *Hebrew Bible and Ancient Israel*, 8(2), pp. 133 – 158.

Na'aman, Nadav (1974) 'Sennacherib's 'Letter to God' on His Campaign to Judah', *Bulletin of the American Schools of Oriental Research*, 214, pp. 25 – 39.

Na'aman, Nadav (2006) 'Sennacherib's Sons' Flight to Urartu', *N.A.B.U.*, 2006(1), pp. 4 – 5.

Nelson, Richard D. (1981) *The Double Redaction of the Deuteronomistic History.* Sheffield: JSOT Press (Journal for the Study of the Old Testament Supplement Series, 18).

Nelson, Richard D. (2005) 'The Double Redaction of the Deuteronomistic History: The Case Is Still Compelling', *Journal for the Study of the Old Testament*, 29, pp. 319 – 337.

Niehr, Herbert (2007) *Weisheitliche, magische und legendarische Erzählungen: Aramäischer Aḥiqar.* Gütersloh: Gütersloher Verlagshaus (Jüdische Schriften aus hellenistisch-römischer Zeit, 2).

Nissinen, Martti (1998) *References to Prophecy in Neo-Assyrian Sources.* Helsinki: The Neo-Assyrian Text Corpus Project (State Archives of Assyria Studies, 7).

Nissinen, Martti (2003) *Prophets and Prophecy in the Ancient Near East.* Atlanta, GA: Society of Biblical Literature (Writings from the Ancient World, 12).

Nissinen, Martti (2019) 'Religious Texts as Historical Sources: Assyrian Prophecies as Sources of Esarhaddon's Nineveh A Inscription', in Giovanni Battista Lanfranchi, Raija Mattila, and Robert Rollinger (eds.) *Writing Neo-Assyrian History: Sources, Problems and Approaches. Proceedings of an*

International Conference Held at the University of Helsink on September 22–25, 2014. Helsinki: The Neo-Assyrian Text Corpus Project (State Archives of Assyria Studies, 29), pp. 183–193.

Noth, Martin (1957) *Überlieferungsgeschichtliche Studien: die sammelnden und bearbeitenden Geschichtswerke im Alten Testament.* 2nd edn. Tübingen: M. Niemeyer.

Noth, Martin (1991) *The Deuteronomistic History.* 2nd edn. Sheffield: JSOT Press (Journal for the Study of the Old Testament Supplement Series, 15).

Novotny, Jamie (2014) '"I Did Not Alter the Site Where The Temple Stood". Thoughts on Esarhaddon's Rebuilding of the Aššur Temple', *Journal of Cuneiform Studies*, 66, pp. 91–112.

Novotny, Jamie (2018) 'A Previously Unrecognized Version of Esarhaddon's "Annals"', *Zeitschrift für Assyriologie und vorderasiatische Archäologie*, 108(2), pp. 203–208.

Novotny, Jamie (2020) 'Royal Assyrian Building Activities in the Northwestern Provincial Centre of Ḫarrān', in Shuichi Hasegawa and Karen Radner (eds.) *The Reach of the Assyrian and Babylonian Empires: Case studies in Eastern and Western Peripheries.* Wiesbaden: Harrassowitz (Studia Chaburensia, 8), pp. 73–94.

Novotny, Jamie (2023) 'The Assyrian Empire in Contact with the World', in Karen Radner, Nadine Moeller, and Daniel T. Potts (eds.) *The Oxford History of the Ancient Near East. Volume IV: The Age of Assyria.* Oxford: Oxford University Press, pp. 352–424.

Novotny, Jamie and Singletary, Jennifer (2009) 'Family Ties. Assurbanipal's Family Revisited', in Mikko Luukko, Raija Mattila, and Saana Svärd (eds.) *Of God(s), Trees, Kings, and Scholars: Neo-Assyrian and Related Studies in Honour of Simo Parpola.* Helsinki: Finnish Oriental Society (Studia Orientalia, 106), pp. 167–177.

Oates, Joan and Oates, David (2001) *Nimrud: An Assyrian Imperial City Revealed.* London: British Institute for the Study of Iraq.

Oded, Bustenay (1979) *Mass Deportations and Deportees in the Neo-Assyrian Empire.* Wiesbaden: Ludwig Reichert Verlag.

Olyan, Saul M. (2020) 'The Literary Dynamic of Loyalty and Betrayal in the Aramaic Ahiqar Narrative', *Journal of Near Eastern Studies*, 79(2), pp. 261–269.

Oppenheim, A. Leo (1960) 'The City of Assur in 714 B.C.', *Journal of Near Eastern Studies*, 19(2), pp. 133–147.

Oppenheim, A. Leo (1968) 'The Eyes of the Lord', *Journal of the American Oriental Society*, 88/1, pp. 173–180.

Oppenheim, A. Leo (1979) 'Neo-Assyrian and Neo-Babylonian Empires', in Harold D. Lasswell, Daniel Lerner, and Hans Speier (eds.) *Propaganda and Communication in World History. Volume I: The Symbolic Instrument in Early Times.* Honolulu: University of Hawai'i Press, pp. 111–144.

'ORACC: Ancient Mesopotamian Gods and Goddesses', http://oracc.museum.upenn.edu/amgg/index.html [last access: 10.01.2023].

'ORACC: The Royal Inscriptions of Assyria online'. http://oracc.museum.upenn.edu/riao/index.html [last access: 10.01.2023].

Oshima, Takayoshi (2017) 'How Mesopotamian was Ahiqar the Wise? A Search for Ahiqar in Cuneiform Texts', in Angelika Berlejung, Aren M. Maeir, and Andreas Schüle (eds.) *Wandering Arameans: Arameans Outside Syria. Textual and Archaeological Perspectives.* Wiesbaden: Harrassowitz (Leipziger Altorienalische Studien, 5), pp. 141–167.

Otto, Adelheid (2015) 'Neo-Assyrian capital cities: from imperial hedquarters to cosmopolitan cities', in Norman Yoffee (ed.) *The Cambridge World History: Volume III Early Cities in Comparative Perpective. 4000 BCE–1200 CE.* Cambridge: Cambridge University Press, pp. 469–490.

Otto, Eckart (1998) 'Die Ursprünge der Bundestheologie im Alten Testament und im Alten Orient', *Zeitschrift für Altorientalische und Biblische Rechtsgeschichte / Journal for Ancient Near Eastern and Biblical Law*, 4, pp. 1 – 84.

Otto, Eckart (1999) *Das Deuteronomium: Politische Theologie und Rechtsreform in Juda und Assyrien.* Berlin: Walter de Gruyter.

Otto, Eckart (2002) *Gottes Recht als Menschenrecht. Rechts- und literaturhistorische Studien zum Deuteronomium.* Wiesbaden: Harrassowitz (BZAR, 2).

Otto, Eckart (2009) 'Die Geburt des Mose. Die Mose-Figur also Gegenentwurf zur neuassyrischen Königsideologie im 7. Jh. v. Chr.', in Eckart Otto (ed.) *Die Tora. Studien zum Pentateuch. Gesammelte Aufsätze.* Wiesbaden: Harrassowitz (BZAR, 9), pp. 9 – 45.

Otto, Eckart (2016) *Deuteronomium 12 – 34: Erster Teilband 12,1 – 23,15.* Freiburg: Verlag Herder (Herders Theologischer Kommentar zum Alten Testament).

Otto, Eckart (2017) *Deuteronomium 12 – 34: Zweiter Teilband 23,16 – 34,12.* Freiburg: Verlag Herder (Herders Theologischer Kommentar zum Alten Testament).

Pakkala, Juha (2009) 'The Date of the Oldest Edition of Deuteronomy', *Zeitschrift für die alttestamentliche Wissenschaft*, 121, pp. 388 – 401.

Pakkala, Juha (2019) 'The Influence of Treaties on Deuteronomy, Exclusive Monolatry, and Covenant Theology', *Hebrew Bible and Ancient Israel*, 8(2), pp. 159 – 183.

Parker, Bradley J. (2001) *The Mechanics of Empire: The Northern Frontier of Assyria as a Case Study in Imperial Dynamics.* Helsinki: The Neo-Assyrian Text Corpus Project.

Parpola, Simo (1972) 'A Letter from Šamaš-šumu-ukīn to Esarhaddon', *Iraq*, 34(1), pp. 21 – 34.

Parpola, Simo (1980) 'The Murderer of Sennacherib', in Bendt Alster (ed.) *Death in Mesopotamia: Papers Read at the XXVIe Rencontre Assyriologique Internationale.* Copenhagen: Akademisk, pp. 171 – 82.

Parpola, Simo (1983) *Letters from Assyrian Scholars to the Kings Esarhaddon and Assurbanipal. Part II: Commentary and Appendices.* Neukirchen-Vluyn: Neukirchener Verlag (Alter Orient und Altes Testament: Veröffentlichungen zur Kultur und Geschichte des Alten Orients und des Alten Testaments, 5/2).

Pedersén, Olof (1986) *Archives and Libraries in the City of Assur: A Survey of the Material from the German Excavations. Part II.* Uppsala: Uppsala University Press (Studia Semitica Upsaliensia, 8).

Pedersén, Olof (1998) *Archives and Libraries in the Ancient Near East 1500 – 300 B.C.* Bethesda, MD: CDL Press.

Piccin, Michela and Worthington, Martin (2015) 'Schizophrenia and the Problem of Suffering in the Ludlul Hymn to Marduk', *Revue d'assyriologie et d'archéologie orientale*, 109, pp. 113 – 124.

Piccin, Michela (2020) 'Assyrian Treaties: "Patchwork" Texts', *Journal of Ancient Civilizations*, 35(1), pp. 33 – 70.

Ponchia, Simonetta (2014) 'The Neo-Assyrian *Adê* Protocol and the Administration of the Empire', in Salvatore Gaspa et al. (eds.) *From Source to History: Studies on Ancient Near Eastern Worlds and Beyond. Dedicated to Giovanni Battista Lanfranchi on the Occasion of His 65th Birthday on June 23, 2014.* Münster: Ugarit-Verlag (Alter Orient und Altes Testament, 412), pp. 501 – 525.

Pongratz-Leisten, Beate (1999) *Herrschaftswissen in Mesopotamien. Formen der Kommunikation zwischen Gott und König im 2. und 1. Jahrtausend v. Chr.* Helsinki: The Neo-Assyrian Text Corpus Project (State Archives of Assyria Studies, 10).

Pongratz-Leisten, Beate (2013) 'All the King's Men: Authority, Kingship, and the Rise of the Elites in Assyria', in Jane A. Hill, Philip Jones, and Antonio J. Morales (eds.) *Experiencing Power,*

Generating Authority: Cosmos, Politics, and the Ideology of Kingship in Ancient Egypt and Mesopotamia. Philadephia: University of Pennsylvania Press, pp. 285–309.

Pongratz-Leisten, Beate (2015) *Religion and Ideology in Assyria.* Berlin/Boston: De Gruyter (Studies in Ancient Near Eastern Records, 6).

Pongratz-Leisten, Beate (n.d.) '"The Writing of the God" and the Textualization of Neo-Assyrian Prophecy', http://nyu.academia.edu/BeatePongratzLeisten/Papers/338208/_The_Writing_of_the_God_and_the_Textualization_of_Neo_Assyrian_Prophecy, pp. 1–29 [last access: 11.01.2023].

Porter, Barbara N. (1993) *Images, Power, and Politics: Figurative Aspects of Esarhaddon's Babylonian Policy.* Philadephia: American Philosophical Society (Memoirs of the American Philosophical Society, 208).

Porter, Barbara N. (2004) 'Ishtar of Nineveh and Her Collaborator, Ishtar of Arbela, in the Reign of Assurbanipal', *Iraq*, 66, pp. 41–44.

Postgate, J. Nicholas (1974a) 'Royal Exercise of Justice under the Assyrian Empire', in Paul Garelli (ed.) *Le palais et la royauté.* Paris: Geuthner (CRRAI, 19), pp. 417–426.

Postgate, J. Nicholas (1974b) *Taxation and Conscription in the Assyrian Empire.* Rome: Biblical Institute Press (Studia Pohl: Series Maior, 3).

Postgate, J. Nicholas (1979) 'The Economic Structure of the Assyrian Empire', in Mogens T. Larsen (ed.) *Power and Propaganda; A Symposium on Ancient Empires.* Copenhagen: Akademisk (Mesopotamia Copenhagen Studies in Assyriology, 7), pp. 193–221.

Postgate, J. Nicholas (1980) '"Princepts Iudex" in Assyria', *Revue d'Assyriologie et d'archéologie orientale*, 74(2), pp. 180–182.

Postgate, J. Nicholas (1989) 'Ownership and Exploitation of Land in Assyria in the 1st Millennium B.C.', in Marc Lebeau and Philippe Talon (eds.) *Reflets des deux fleuves: Volume de melanges offerts a Andre Finet.* Leuven, pp. 141–152.

Postgate, J. Nicholas (1992) 'The Land of Assur and the Yoke of Assur', *World Archaeology*, 23(3), pp. 247–263.

Postgate, J. Nicholas (1995) 'Assyria: The Home Provinces', in Mario Liverani (ed.) *Neo-Assyrian Geography.* Rome: University of Rome, 'La Sapienza' (Quaderni di Geografia Storica, 5), pp. 1–17.

Postgate, J. Nicholas (2007) 'The Invisible Hierarchy: Assyrian Military and Civilian Administration in the 8th and 7th Centuries BC', in J. Nicholas Postgate (ed.) *The Land of Assur & the Yoke of Assur: Studies on Assyria, 1971–2005.* Oxford: Oxbow Press, pp. 331–360.

Potts, Daniel T. (2014) *Nomadism in Iran: from Antiquity to the Modern Era.* Oxford: Oxford University Press.

Radner, Karen (1997) *Die neuassyrischen Privatrechtsurkunden als Quelle für Mensch und Umwelt.* Helsinki: The Neo-Assyrian Text Corpus Project (State Archives of Assyria Studies, 6).

Radner, Karen (2002) *Die neuassyrischen Texte aus Tall Šēḫ Ḥamad. Mit Beiträgen von Wolfgang Röllig zu den aramäischen Beischriften.* Berlin: Dietrich Reimer Verlag (Berichte der Ausgrabung Tell Schech Hamad, 6).

Radner, Karen (2003a) 'Neo-Assyrian Period', in Raymond Westbrook (ed.) *A History of Ancient Near Eastern Law.* Leiden: Brill (Handbook of Oriental Studies. Section 1 The Near and Middle East), pp. 883–910.

Radner, Karen (2003b) 'Salmanassar V. in den Nimrud Letters', *Archiv für Orientforschung*, 50, pp. 95–104.

Radner, Karen (2003c) 'The Trials of Esarhaddon. The Conspiracy of 670 BC.', *Isimu: Revista sobre Oriente Proximo y Egipto en la antiguedad*, 6, pp. 165–184.

Radner, Karen (2005) *Die Macht des Namens. Altorientalische Strategien zur Selbsterhaltung.* Wiesbaden: Harrassowitz (SANTAG Arbeiten und Untersuchungen zur Keilschriftkunde, 8).
Radner, Karen (2006) 'Assyrische *ṭuppi adê* als Vorbild für Deuteronomium 28, 22 – 44?', in Jan C. Gertz et al. (eds.) *Die deuteronomistischen Geschichtswerke.* Berlin/New York: Walter de Gruyter (Beihefte zur Zeitschrift für die alttestamentliche Wissenschaft, 365), pp. 351 – 378.
Radner, Karen (2007) 'Abgaben an den König von Assyrien aus dem In- und Ausland', in Hilmar Klinkott, Sabina Kubish and Renate Müller-Wollermann (eds.) *Geschenke und Steuern, Zölle und Tribute: Antike Abgabenformen in Anspruch und Wirklichkeit.* Leiden: Brill (Culture and History of the Ancient Near East, 29), pp. 213 – 230.
Radner, Karen (2008a) 'Esarhaddon's Expedition from Palestine to Egypt in 671 BCE. A Trek through Negev and Sinai', in Dominik Bonatz, Rainer M. Czichon, and F. Janoscha Kreppner (eds.) *Fundstellen: Gesammelte Schriften zur Archäologie und Geschichte Altvorderasiens ad honorem Hartmut Kühne.* Wiesbaden: Harrassowitz, pp. 305 – 314.
Radner, Karen (2008b) 'The Delegation of Power. Neo-Assyrian Bureau Seals', in Pierre Briant, Wouter F. M. Henkelman, and Matthew W. Stolper (eds.) *L'archive des Fortifications de Persépolis. État des questions et perspectives de recherches. Actes du colloque organisé au Collège de France par la 'Chaire d'Histoire et Civilisation du Monde Achéménide et de l'Empire d'Alexandre' et le 'Réseau International d'Ètudes et de Recherches Achéménides', 3. – 4. November 2006.* Paris: Éditions de Boccard (Persika, 12), pp. 481 – 515.
Radner, Karen (2009) 'The Assyrian King and his Scholars. The Syro-Anatolian and the Egyptian Schools', *Studia Orientalia*, 106, pp. 221 – 238.
Radner, Karen (2010a) 'Assyrian and Non-Assyrian Kingship in the First Millennium BC', in Giovanni Battista Lanfranchi and Robert Rollinger (eds.) *Concepts of Kingship in Antiquity.* Padova: S.A.R.G.O.N. Editrice e Libreria (History of the Ancient Near East / Monographs, 11), pp. 15 – 24.
Radner, Karen (2010b) 'Gatekeepers and Lock Masters. The Control of Access in the Neo-Assyrian Palaces', in Heather D. Baker, Eleanor Robson, and Gabor Zólyomi (eds.) *Your Praise is Sweet: A Memorial Volume for Jeremy Black from Students, Colleagues and Friends.* London: British Institute for the Study of Iraq, pp. 269 – 280.
Radner, Karen (2011a) 'Royal Decision-Making. Kings, Magnates, and Scholars', in Karen Radner and Eleanor Robson (eds.) *The Oxford Handbook of Cuneiform Culture.* Oxford: Oxford University Press (Oxford Handbooks in Classics and Ancient History), pp. 358 – 379.
Radner, Karen (2011b) 'Schreiberkonventionen im assyrischen Reich. Sprachen und Schriftsysteme', in Johannes Renger (ed.) *Assur – Gott, Stadt und Land: 5. Internationales Colloquium der Deutschen Orient-Gesellschaft 18. – 21. Februar 2004 in Berlin.* Wiesbaden: Harrassowitz, pp. 385 – 403.
Radner, Karen (2011c) 'The Assur-Nineveh-Arbela Triangle. Central Assyria in the Neo-Assyrian Period', in Peter A. Miglus and Simone Mühl (eds.) *Between the Cultures: The Central Tigris Region from the 3rd to the 1st Millennium BC, 22. – 24. Januar 2009, Heidelberg.* Heidelberg: Heidelberger Orientverlag, pp. 321 – 329.
Radner, Karen (2012) 'Between a Rock and a Hard Place. Musasir, Kumme, Ukku and Subria – the Buffer States between Assyria and Urartu', in Stephan Kroll et al. (eds.) *Biainili-Urartu, 12. – 14. Oktober 2007, München.* Leuven: Peeters (Acta Iranica, 51), pp. 243 – 264.
Radner, Karen (2014a) 'An Imperial Communication Network: The State Correspondence of the Neo-Assyrian Empire', in Karen Radner (ed.) *State Correspondence in the Ancient World: From New Kingdom Egypt to the Roman Empire.* New York: Oxford University Press (Oxford Studies in Early Empires), pp. 64 – 93.

Radner, Karen (2014b) 'The Neo-Assyrian Empire', in Robert Rollinger and Michael Gehler, (eds.)) *Imperien und Reiche in der Weltgeschichte: Epochenübergreifende und globalhistorische Vergleiche.* Wiesbaden: Harrassowitz, pp. 101–119.

Radner, Karen (2015a) *Ancient Assyria: A Very Short Introduction.* Oxford: Oxford University Press.

Radner, Karen (2015b) 'Royal Pen Pals: The Kings of Assyria in Correspondence with Officials, Clients and Total Strangers (8th and 7th centuries BC)', in Stephan Procházka, Lucian Reinfandt, and Sven Tost (eds.) *Official Epistolography and the Language(s) of Power.* Vienna: Verlag der Österreichischen Akademie der Wissenschaften (Papyrologica Vindobonensia, 8), pp. 127–143.

Radner, Karen (2016) 'Revolts in the Assyrian Empire: Succession Wars, Rebellions against a False King and Independence Movements.', in John J. Collins and Joseph G. Manning (eds.) *Revolt and Resistance in the Ancient Classical World and the Near East: In the Crucible of Empire.* Leiden: Brill (Culture and History of the Ancient Near East, 85), pp. 41–54.

Radner, Karen (2017a) 'Assur's "Second Temple Period": The Restoration of the Cult of Assur, c. 538 BC', in Christoph Levin and Reinhard Müller (eds.) *Herrschaftslegitimation in vorderorientalischen Reichen der Eisenzeit.* Tübingen: Mohr Siebeck (Orientalische Religionen in der Antike, 21), pp. 77–96.

Radner, Karen (2017b) 'Economy, Society, and Daily Life in the Neo-Assyrian Period', in Eckart Frahm (ed.) *A Companion to Assyria*. Malden, MA: Wiley, pp. 209–228.

Radner, Karen (2017c) *Mesopotamien. Die frühen Hochkulturen an Euphrat und Tigris.* Munich: C. H. Beck.

Radner, Karen (2018a) 'Last Emperor or Crown Prince Forever? Aššur-uballiṭ II of Assyria according to Archival Sources', in Shigeo Yamada (ed.) *Neo-Assyrian Sources in Context: Thematic Studies of Texts, History and Culture.* Winona Lake, IN: Eisenbrauns (State Archives of Assyria Studies, 28), pp. 135–142.

Radner, Karen (2018b) 'The "Lost Tribes of Israel" in the Context of the Resettlement Programme of the Assyrian Empire', in Karen Radner, Christoph Levin, and Shuichi Hasegawa (eds.) *The Last Days of the Kingdom of Israel.* Berlin: De Gruyter (Beihefte zur Zeitschrift für die alttestamentliche Wissenschaft, 511), pp. 101–123.

Radner, Karen (2019) 'Neo-Assyrian Treaties as a Source for the Historian: Bonds of Friendship, the Vigilant Subject and the Vengeful King's Treaty', in Giovanni Battista Lanfranchi, Raija Mattila, and Robert Rollinger (eds.) *Writing Neo-Assyrian History: Sources, Problems, and Approaches.* Helsinki: The Neo-Assyrian Text Corpus Project (State Archives of Assyria Studies, 29), pp. 309–328.

Radner, Karen (2021) 'Diglossia and the Neo-Assyrian Empire's Akkadian and Aramaic Text Production', in Louis C. Jonker, Angelika Berlejung, and Izak Cornelius (eds.) *Multilingualism in Ancient Contexts: Perspectives from Ancient Near Eastern and Early Christian Contexts.* Stellenbosch: African Sun Media, pp. 146–181.

Reade, Julian E. (1986) 'Archaeology and the Kuyunjik archives', in Klaas R. Veenhof (ed.) *Cuneiform Archives and Libraries. Papers Read at the 30e Rencontre Assyriologique Internationale, Leiden 4–8 July 1983.* Leiden/Istanbul: Nederlands Instituut vor het Nabije Oosten (Uitgaven van het Nederlands Historisch-Archaeologisch Instituut te Istanbul, 57), pp. 213–222.

Reculeau, Hervé (2022) 'Assyria in the Late Bronze Age', in Karen Radner, Nadine Moeller and Daniel T. Potts (eds.) *The Oxford History of The Ancient Near East. From the Hyksos to the Late Second Millennium BC.* Oxford: Oxford University Press, pp. 707–800.

Reeves, Joshua (2017) *Citizen Spies: The Long Rise of America's Surveillance Society.* New York: New York University Press.

Roaf, Michael (2021) 'Cyaxares in Assyria', *N.A.B.U.*, 2021(4), pp. 277–279.

Robson, Eleanor (2008) 'Mesopotamian Medicine and Religion: Current Debates, New Perspectives', *Religion Compass*, 2(4), pp. 455–483.

Robson, Eleanor (2011) 'Empirical Scholarship in the Neo-Assyrian Court', in Gebhard J. Selz and Klaus Wagensommer (ed.) *The Empirical Dimension of Ancient Near Eastern Studies / Die empirische Dimension altorientalischer Forschungen.* Berlin/Vienna: Lit Verlag (Wiener Offene Orientalistik 6), pp. 603–629.

Robson, Eleanor (2013) 'Reading the Libraries of Assyria and Babylonia', in Jason König, Katerina Oikonomopoulou, and Greg Woolf (eds.) *Ancient Libraries.* Cambridge: Cambridge University Press, pp. 38–56.

Robson, Eleanor (2019) *Ancient Knowledge Networks: A Social Geography of Cuneiform Scholarship in First-Millennium Assyria and Babylonia.* London: UCL Press.

Röllig, Wolfgang (2014) *Die aramäischen Texte aus Tall Šēḫ Ḥamad / Dūr-Katlimmu / Magdalu.* Wiesbaden: Harrassowitz (Berichte der Ausgrabung Tell Schech Hamad, 17).

Rubin, Zachary M. (2021) *The Scribal God Nabû in Ancient Assyrian Religion and Ideology*, Dissertation. Brown University.

Rüterswörden, Udo (2006) *Das Buch Deuteronomium.* Stuttgart: Katholisches Bibelwerk (Neuer Stuttgarter Kommentar Altes Testament, 4).

Sanders, Seth L. (2009) 'The First Tour of Hell: From Neo-Assyrian Propaganda to Early Jewish Revelation', *Journal of Ancient Near Eastern Religions*, 9(2), pp. 151–169.

Sano, Katsuji (2020) *Die Deportationspraxis in neuassyrischer Zeit.* Münster: Ugarit-Verlag (Alter Orient und Altes Testament, 466).

Šašková, Kateřina (2010a) 'Adad-šumu-uṣur and his Family in the Service of Assyrian Kings', in Petr Charvát and Petra M. Vičková (eds.) *Who Was King? Who Was Not King? The Rulers and the Ruled in the Ancient Near East.* Prague: Institute of Archaeology of the Academy of Sciences of the Czech Republic, pp. 113–132.

Šašková, Kateřina (2010b) 'Esarhaddon's Accession to the Assyrian Throne', in Kateřina Šašková, Lukáš Pecha, and Petr Charvát (eds.) *Shepherds of the Black-headed People: The Royal Office vis-à-vis Godhead in Ancient Mesopotamia.* Plzeň: Západočeská univerzita, pp. 147–177.

Schwemer, Daniel (2001) *Die Wettergottgestalten Mesopotamiens und Nordsyriens im Zeitalter der Keilschriftkulturen. Materialien und Studien nach den schriftlichen Quellen.* Wiesbaden: Harrassowitz.

Schwemer, Daniel (2011) 'Magic Rituals: Conceptualization and Performance', in Karen Radner and Eleanor Robson (eds.) *The Oxford Handbook of Cuneiform Culture.* Oxford: Oxford University Press, pp. 418–442.

Schwemer, Daniel (2015) 'The Ancient Near East', in David J. Collins (ed.) *The Cambridge History of Magic and Witchcraft in the West: From Antiquity to the Present.* Cambridge University Press, pp. 17–51.

Scurlock, JoAnn (1999) 'Physician, Exorcist, Conjurer, Magician: A Tale of Two Healing Professionals', in Tzvi Abusch and Karel van der Toorn (eds.) *Mesopotamian Magic: Textual, Historical, and Interpretative Perspectives.* Groningen: Styx Publications (Ancient Magic and Divination, 1), pp. 69–79.

Scurlock, JoAnn (2012) 'Getting Smashed at the Victory Celebration or What Happened to Esarhaddon's so-called Vassal Treaties and Why', in Natalie N. May (ed.) *Iconoclasm and Text Destruction in the Ancient Near East and Beyond.* Chicago, IL: The Oriental Institute of the University of Chicago, pp. 175–186.

'SFB 1369 "Vigilanzkulturen"', https://www.en.sfb1369.uni-muenchen.de/the-crc/index.html [last access: 11.01.2023].

Steymans, Hans Ulrich (1995) *Deuteronomium 28 und die adê zur Thronfolgeregelung Asarhaddons: Segen und Fluch im Alten Orient und in Israel.* Göttingen: Vandenhoeck Ruprecht (Biblicus et Orientalis, 145).

Steymans, Hans Ulrich (2003) 'Die neuassyrische Vertragsrhetorik der "Vassal Treaties of Esarhaddon" und das Deuteronomium', in Georg Braulik (ed.) *Das Deuteronomium.* Frankfurt a. M.: Peter Lang, pp. 89–152.

Steymans, Hans Ulrich (2004) 'Asarhaddon und die Fürsten im Osten: Der gesellschaftspolitische Hintergrund seiner Thronfolgeregelung', in Friedrich Schipper (ed.) *Zwischen Euphrat und Tigris: Österreichische Forschungen zum Alten Orient.* Vienna: LIT, pp. 61–85.

Steymans, Hans Ulrich (2006) 'Die literarische und historische Bedeutung der Thronfolgevereidigungen Asarhaddons', in Jan C. Gertz et al. (eds.) *Die deuteronomistischen Geschichtswerke.* Berlin/New York: Walter de Gruyter (Beihefte zur Zeitschrift für die alttestamentliche Wissenschaft, 365), pp. 331–349.

Steymans, Hans Ulrich (2013) 'Deuteronomy 28 and Tell Tayinat', *Verbum et ecclesia*, 34(2), pp. 1–13.

Steymans, Hans Ulrich (2019) 'Deuteronomy 13 in Comparison with Hittite, Aramaic and Assyrian Treaties', *Hebrew Bible and Ancient Israel*, 8(2), pp. 101–132.

Stökl, Jonathan (2015) 'Prophecy and the Royal Court in the Ancient Near East', *Religion Compass*, 9(3), pp. 55–65.

Tadmor, Hayim (1983) 'Autobiographical Apology in the Royal Assyrian Literature', in Hayim Tadmor and Moshe Weinfeld (eds.) *History, Historiography, and Interpretation: Studies in Biblical and Cuneiform Literatures.* Jerusalem: Magnes Press, pp. 36–57.

Tadmor, Hayim (1997) 'Propaganda, Literature, Historiography: Cracking the Code of the Assyrian Royal Inscriptions', in Simo Parpola and Robert Whiting (eds.) *Assyria 1995.* Helsinki: The Neo-Assyrian Text Corpus Project, pp. 325–338.

Tadmor, Hayim (2004) 'An Assyrian Victory Chant and Related Matters', in Grant Frame (ed.) *From the Upper Sea to the Lower Sea: studies on the history of Assyria and Babylonia in honour of A. K. Grayson.* Istanbul/Leiden: Nederlands Inst. voor het Nabije Oosten (PIHANS, 101), pp. 269–276.

Tadmor, Hayim, Landsberger, Benno and Parpola, Simo (1989) 'The Sin of Sargon and Sennacherib's Last Will', *State Archives of Assyria Bulletin*, 3, pp. 3–51.

Thomas, Fredy (1993) 'Sargon II., der Sohn Tiglat-pilesers III.', in Manfried Dietrich and Oswald Loretz (eds.) *Mesopotamica – Ugaritica – Biblica. Festschrift für Kurt Bergerhof zur Vollendung seines 70. Lebensjahres am 7. Mai 1992.* Neukirchen-Vluyn: Neukirchener Verlag (Alter Orient und Altes Testament, 232), pp. 465–470.

Toptaş, Koray and Akyüz, Faruk (2021) 'A Neo-Assyrian Sale Contract from the Province of the Chief Cupbearer (*rab-šaqê*) kept at the Hasankeyf Museum (Batman)', *Zeitschrift für Assyriologie*, 111(1), pp. 77–87.

Tsetskhladze, Gocha R. (ed.) (2021) *Archaeology and History of Urartu (Biainili).* Leuven/Paris/Bristol, CT: Peeters.

Tushingham, Poppy (2019) 'Uniformity versus Regional Variation in the Legal and Scribal Practices of the Neo-Assyrian Empire', *Zeitschrift für altorientalische und biblische Rechtsgeschichte*, 25, pp. 29–53.

Tushingham, Poppy (2023) 'The *adê* covenants of the Assyrian Empire: Imposed by Humans, Enforced by the Gods', in Christian A. Eberhart and Wolfgang Kraus (eds.) *Covenant – Concepts of Berit, Diatheke, and Testamentum.* Tübingen: Mohr-Siebeck, pp. 43–66.

Ur, Jason (2017) 'Physical and Cultural Landscapes of Assyria', in Eckart Frahm (ed.) *A Companion to Assyria.* Malden, MA: Wiley, pp. 13–35.

Valk, Jonathan (2020) 'Crime and Punishment: Deportation in the Levant in the Age of Assyrian Hegemony', *Bulletin of the American Schools of Oriental Research*, 384, pp. 77–103.

Van der Toorn, Karel (2018) *Papyrus Amherst 63.* Münster: Ugarit-Verlag (Alter Orient und Altes Testament, 448).

Veenhof, Klaas R. (2017) 'The Old Assyrian Period (20th–18th Century BCE)', in Eckart Frahm (ed.) *A Companion to Assyria.* Malden, MA: Wiley (Blackwell Companions to the Ancient World), pp. 57–79.

Vera Chamaza, Galo W. (1992) 'Sargon II's Ascent to the Throne: the Political Situation', *State Archives of Assyria Bulletin*, 6, pp. 21–33.

Vera Chamaza, Galo W. (2002) *Die Omnipotenz Aššurs: Entwicklungen in der Aššur-Theologie unter den Sargoniden Sargon II., Sanherib und Asarhaddon.* Münster: Ugarit-Verlag (Alter Orient und Altes Testament, 295).

Von Soden, Wolfram (1936) 'Die Unterweltsvision eines assyrischen Kronprinzen', *Zeitschrift für Assyriologie und vorderasiatische Archä*ologie, 43, pp. 1–32.

Von Soden, Wolfram (1995) *Grundriss der akkadischen Grammatik.* 3rd edn. Roma: Editrice Pontificio Istituto Biblico (Analecta Orientalia, 33).

Wallenfels, Ronald (2022) 'K. 2673 and Esarhaddon's Succession Treaty Seal C: A Disambiguation', *N.A.B.U.*, 2022(3), pp. 254–258.

Watanabe, Kazuko (1985) 'Die Siegelung der 'Vassallenverträge Asarhaddons' durch den Gott Aššur,' *Baghdader Mitteilungen*, 16, pp, 377–392, Pl. 33.

Watanabe, Kazuko (1987) *Die adê-Vereidigung anläßlich der Thronfolgeregelung Asarhaddons.* Berlin: Gebr. Mann (Baghdader Mitteilungen, Beiheft 3).

Watanabe, Kazuko (2014) 'Esarhaddon's Succession Oath Documents Reconsidered in Light of the Tayinat Version', *Orient*, 49, pp. 145–170.

Watanabe, Kazuko (2015) 'Innovations in Esarhaddon's Succession Oath Documents Considered from the Viewpoint of the Documents' Structure', *State Archives of Assyria Bulletin*, 21, pp. 173–215.

Watanabe, Kazuko (2017) 'A Study of Assyrian Cultural Policy as Expressed in Esarhaddon's Succession Oath Documents', in Amitai Barucki-Unna et al. (eds.) *'Now It Happened in Those Days': Studies in Biblical, Assyrian, and Other Ancient Near Eastern Historiography Presented to Mordechai Cogan on His 75th Birthday.* Winona Lake, IN: Eisenbrauns, pp. 473–492.

Watanabe, Kazuko (2019) 'Aššurbanipal and His Brothers Considered from the References in Esarhaddon's Succession Oath Documents', in Ichiro Nakata et al. (eds.) *Prince of the Orient: Ancient Near Eastern Studies in Memory of H.I.H. Prince Takahito Mikasa.* Tokyo: The Society for Near Eastern Studies in Japan (Orient: Supplement, 1), pp. 237–257.

Watanabe, Kazuko (2020) 'Adoration of Oath Documents in Assyrian Religion and its Development', *Orient*, 55, pp. 71–86.

Watanabe, Kazuko (2021) 'Diversity and Tradition in Esarhaddon's Reform', in Ada Taggar-Cohen (ed.) *International Cultural Diversity in the Ancient Near East: Archaeological and Textual Approaches – Proceedings of International Conference on the Ancient Near Eastern World held at Doshisha University April 13–14, 2019.* Kyoto: Doshisha University, Center for interdisciplinary study of the monotheistic religions, pp. 23–52.

Waters, Matthew W. (1999a) 'ABL 268 and Tammaritu', *Archív orientální*, 67, pp. 72–74.

Waters, Matthew W. (1999b) 'Te'umman in the Neo-Assyrian Correspondence', *Journal of the American Oriental Society*, 119, pp. 473–477.

Waters, Matthew W. (2022) 'Peace in Pieces. Making Peace in Elam', in Giovanni Battista Lanfranchi, Simonetta Ponchia, and Robert Rollinger (eds.) *Making Peace in the Ancient World. Proceedings of the 7th Melammu Workshop, Padova, 5 – 7 November 2018.* Münster: Zaphon (Melammu Workshops and Monographs, 5), pp. 249 – 263.

Weaver, Ann M. (2004) 'The "Sin of Sargon" and Esarhaddon's Reconception of Sennacherib: A Study in Divine Will, Human Politics and Royal Ideology', *Iraq*, 66, pp. 61 – 66.

Weber, Max (1923) *Gesammelte Aufsätze zur Religionssoziologie. Band 2: Hinduismus und Buddhismus.* 2nd ed. Tübingen: Mohr Siebeck.

Weber, Max (1980) *Wirtschaft und Gesellschaft. Grundriß der verstehenden Soziologie.* Johannes Winckelmann (ed.), 5th ed. Tübingen: Mohr Siebeck.

Weidner, Ernst F. (1939) 'Assurbânipal in Assur', *Archiv für Orientforschung*, 13, pp. 204 – 218, plates XI – XVI.

Westbrook, Raymond (2005) 'Patronage in the Ancient Near East', *Journal of the Economic and Social History of the Orient*, 48(2), pp. 210 – 233.

Westenholz, Joan G. (1997) *Legends of the Kings of Akkade.* Winona Lake, IN: Eisenbrauns (Mesopotamian Civilizations, 7).

Wiseman, Donald J. (1958) 'The Vassal-Treaties of Esarhaddon', *Iraq*, 20, pp. 1 – 99, plates 1 – 53.

Zawadzki, Stefan (1994) 'The Revolt of 746 BC and the Coming of Tiglath-pileser III to the Throne', *State Archives of Assyria Bulletin*, 8, pp. 53 – 54.

Index of Personal Names

[...]-šarru-uṣur 5
Abdâ 172
Adad-šumu-uṣur 24, 24 fn. 95, 28, 144, 148–152, 149 fn. 36, 149 fn. 38, 164, 173, 179, 208, 210, 235–236, 236 fn. 374, 245, 249, 253–254
Ahiqar 27, 179, 204, 234–239, 234 fn. 361
Akkullanu 170 fn. 89
Aplaya 206, 208–209, 214–215
Ashurbanipal 2–3, 3 fn. 8, 6–7, 15, 17 fn. 69, 24, 24 fn. 95, 33, 36–38, 36 fn. 12, 37 fn. 16, 38 fn. 18, 42–43, 55, 57–94, 65 fn. 97, 80 fn. 155, 86 fn. 171, 89 fn. 181, 91 fn. 186, 98–99, 109, 120–122, 124–128, 129 fn. 320, 130, 133, 135 fn. 347, 137, 147 fn. 27, 151, 160 fn. 69, 161, 161 fn. 71, 163, 170, 179, 179 fn. 142, 180–181, 183, 185–186, 189–190, 194 fn. 203, 196, 200–201, 205, 205 fn. 232, 207, 209, 218, 221–222, 224, 227, 230–231, 233, 238, 241–242, 244, 249, 252
Ashurnasirpal II 23, 183
Aššur-da''in-aplu 99–100
Aššur-etel-ilani-mukin-apli 4 fn. 12. *See also* Esarhaddon
Aššur-nadin-šumi 206
Aššur-nerari V 49, 52
Aššur-uballiṭ II 199
Aššur-zeru-ibni 185, 193
Auwa 196, 196 fn. 213

Ba'al (king of Tyre) 34 fn. 2, 35 fn. 11, 39, 222–223, 222 fn. 309
Bel-eṭir 206, 209, 209 fn. 247
Bel-ibni 1 fn. 3
Bel-iddina 206
Bel-ušallim 207, 216
Binunî 170–171

Darius I 16 fn. 60

Esarhaddon 2–8, 4 fn. 12, 5 fns. 14–16, 5 fn. 18, 6 fns. 18–19, 15, 20–25, 22 fn. 85, 33–42, 35 fn. 8, 35 fn. 11, 38 fn. 18, 45, 45 fn. 49, 47, 47–48 fn. 60, 57, 59–65, 64 fn. 95, 68–74, 76–77, 81–84, 86–88, 90–91, 93, 95–117, 105 fn. 229, 106 fn. 232, 107 fn. 235, 110 fn. 151, 120–121, 123, 126, 128, 134 fn. 341, 141–143, 141 fn. 2, 145–149, 151–156, 155 fn. 54, 159–166, 171–172, 180, 183, 183 fn. 160, 185, 185 fn. 163, 187–188, 190, 192–193, 196, 202–203, 205, 208–209, 211, 213–215, 217–224, 233–240, 234 fn. 363, 242–243, 246, 248, 250–252, 254–255
Ešarra-ḫammat 6

Gabbu-ilani-ereš 23, 24 fn. 93, 149

Humbareš 39–40
Hu-Teššub 112 fn. 256

Ik-Teššub 110–112, 112 fn. 256
Ikkaru 143
Ikkilû (Yakin-Lû) 217–220, 222–223
Ina-teši-eṭir 212
Issar-šumu-ereš 23–25, 24 fn. 95, 25 fn. 97, 123, 123 fn. 284, 124–126, 129, 141, 148, 148 fn. 30, 152
Itti-Marduk-balaṭu 206, 209, 210–214, 210 fn. 256
Itti-Šamaš-balaṭu 204, 209, 216–223, 217 fn. 289, 239–240, 246–247, 249, 251, 254

Joseph 237 fn. 381
Josiah 225, 227–228, 228 fns. 334–335

Kummaya 177–180, 179 fn. 142, 246

Manasseh 17 fn. 69, 224, 228 fn. 337
Marduk-aplu-iddina 156
Mar-Issar 192 fn. 192
Mati'-ilu 49, 52
Moses 226, 226 fn. 328
Mullissu-hammat 196

https://doi.org/10.1515/9783111323435-015

Nabopolassar 210
Nabû-ahhe-ereš 205
Nabû-kina-[...] 211 fn. 261
Nabû-naṣir 210
Nabû-rehtu-uṣur 184–189, 191, 193, 200, 208, 208 fn. 243, 213, 248, 253
Nabû-šarru-uṣur 218, 218 fn. 293
Nabû-šumu-iddina 129, 135 fn. 347
Nabû-šumu-iškun 235–236, 239
Nabû-ušabši 209, 210 fn. 256
Nabû-ušallim 168–169, 171–173, 208, 246, 251–252
Nabû-zer-kitti-lišir 107, 112, 112 fn. 259
Nabû-zeru-lešir 24–25
Nabû-zuqup-kenu 24, 24 fn. 95, 148, 236 fn. 374
Nadin 235, 237–239
Nadinu 214–215
Naqi'a 7, 74 fn. 132, 145, 214
Nergal-belu-uṣur 170

Pabba'u 196

Rusâ I 108 fn. 240, 109

Šamaš-metu-uballiṭ 179
Šamaš-šumu-ukin 6, 8, 38, 38 fn. 18, 53, 57, 65, 74, 83–84, 98, 190 fn. 185, 196, 201, 203, 205–210, 207 fn. 241, 212, 233, 239, 249
Šamaš-zeru-iqiša 206
Šamši-Adad V 99–101
Sarmugi 233. *See also* Šamaš-šumu-ukin
Sarbanabal 233. *See also* Ashurbanipal
Sargon II 4, 4 fn. 10, 24, 26, 29, 47, 47 fn. 60, 51, 71, 90 fn. 182, 96, 101–104, 101 fn. 215, 102, 102 fn. 219, 108–112, 108 fns. 240–242, 109, 109 fns. 244–245, 112 fn. 256, 114, 128, 128 fn. 310, 148, 153–157, 154 fn. 47, 186 fn. 168, 189
Šaridu 205, 206 fn. 234

Saritra 234. *See* Šeru'a-eṭirat.
Sasî 7, 172, 186–187, 186 fn. 168, 190, 220
Šeru'a-eṭirat 234. *See also* Saritra.
Še'-'ušnî 199
Sennacherib 1, 3 fn. 8, 4, 4 fns. 11–12, 5, 5 fn. 14–15, 22, 24, 35, 35 fn. 11, 37, 39, 41, 41 fn. 31, 44–45, 44 fn. 40, 45 fn. 49, 48 fn. 60, 55 fn. 80, 69, 70 fn. 118, 71, 73 fn. 130, 74 fn. 130, 74 fn. 132, 76 fn. 141, 81, 90 fn. 182, 95–96, 99–106, 102 fn. 219, 108, 108 fn. 241, 110, 113, 114, 126, 126 fn. 301, 126 fn. 306, 148, 153–156, 153 fn. 44, 158, 160, 162, 176 fn. 127, 178–182, 181 fn. 152, 186–187, 186 fn. 168, 189, 191, 198, 200–201, 213, 233–237, 234 fn. 363, 239, 246, 248, 250
Shalmaneser III 26, 99–100
Shalmaneser V 101, 102 fn. 219, 103
Sîn-ahu-uṣur 109, 109 fn. 245
Sîn-nadin-apli 6 fn. 18
Sîn-per'u-ukin 150
Sîn-šarru-iškun 101 fn. 214, 169, 174, 175 fn. 117, 182, 199, 201,
Šulmu-šarri 194 fn. 203
Šumaya 24 fn. 95

Tapnahti 196
Tiglath-pileser III 4, 4 fn. 10, 41, 51, 101, 101 fn. 215, 102 fn. 219, 166, 166 fn. 81
Tobit 237
Tukulti-Ninurta II 23
Ṭupšar-Ellil-dari 234 fn. 361, 236

Ululayu 102 fn. 219. *See* Shalmaneser V
Urdu-Gula 24 fn. 95, 179, 179 fn. 144
Urdu-Mullissu 5, 180–181
Urdu-Nabû 206, 209
Urdu-Nanaya 28, 143–152, 145 fn. 18, 164, 239, 251, 253

Zera-ukin 212

Index of Divine Names

Adad 47, 51, 80, 104, 155
Anath-Bethel 51
Anu 50, 216
Aramiš (Aramis) 51
Aššur 22, 22 fn. 85, 34, 43–47, 44 fns. 41–42, 45 fn. 45, 80, 82, 101, 103–104, 109–110, 113, 126, 130, 144, 146–147, 154–157, 161–162, 168–171, 173–174, 178, 189, 193, 197, 214, 218

Ba'al (god) 51
Bel 103, 126, 137, 146–147, 197 fn. 218, 218. *See also* Marduk
Bethel 51

Ea 120, 178, 216
Ea-šarru 168, 171
Ellil 216. *See also* Enlil
Enlil 48, 108 fn. 240, 156. *See also* Ellil
Ereškigal 178

Gula 120, 122, 175 fn. 120

Hadad 51
Haldi 110

Ištar 47, 47 fn. 57, 103, 119, 162–163, 187 fn. 170, 189, 196, 209–210, 218
Išum 178

Jupiter (planet) 46, 206

Karhuha 51
Kubaba 51

Marduk 45, 48, 76 fn. 141, 104, 108 fn. 240, 125–126, 126 fn. 301, 128, 154–157. *See also* Bel
Mars (planet) 46
Mercury (planet) 46
Mullissu 154, 187 fn. 170, 188–189

Nabû 17, 47–48, 103–104, 116, 125–129, 126 fn. 301, 126 fns. 305–306, 134–138, 134 fn. 341, 135 fn. 347, 146–147, 149, 157, 197 fn. 218, 206, 209, 218
Nergal 174–176, 175 fns. 119–120, 178–179, 182, 200, 218, 250
Nikkal 184, 185 fn. 163, 186–189, 187 fn. 170
Nusku 185 fn. 163, 186

Šala 51
Šamaš 6 fn. 18, 50, 80, 103–104, 122, 146–147, 154–155, 179, 197 fn. 218, 218
Šarrat-Ekron 51
Saturn (planet) 46
Sîn 47, 48 fn. 60, 49–50, 50 fn. 70, 103–104, 154, 185 fn. 163, 186, 197 fn. 218
Sirius (star) 46, 206

Tašmetu 125
Teššub 51

Venus (planet) 46

Yhwh 224, 227–228, 230–233, 240, 252

https://doi.org/10.1515/9783111323435-016

Index of Place Names

Adia 99
Akkad 1 fn. 3, 35, 35 fn. 8, 46–47, 74
Amedu 100
Ammon 128
Anatolia 143 fn. 14
Arbela 47, 47 fn. 57, 100, 103, 123–124, 161–162, 196, 218
Armenia 51
Arpad 49, 52
Arrapha 100
Arwad 29, 203, 205, 216–217, 219, 222–223, 243, 246, 249, 251, 254
Ashur (Qal'at Sherqat) 4, 17–18, 23, 25, 29, 44 fn. 42, 45, 47, 95, 97, 100–101, 112–115, 113 fn. 261, 123–124, 126, 126 fn. 306, 135, 155, 161, 166–169, 171–172, 174, 176–177, 176 fn. 126, 181–182, 181 fn. 152, 182 fn. 152, 195–196, 195 fn. 206, 198–202, 208, 214, 242–243, 245–246, 251–252
Aza'i 51

Babylon 1 fn. 3, 35, 38, 38 fn. 18, 45, 47–48, 53, 57, 74, 84, 98, 155, 190 fn. 185, 203, 205–206, 207, 209, 239, 249
Babylonia 1 fn. 3, 3 fn. 8, 6–8, 16 fn. 61, 17 fn. 63, 25, 29, 35, 35 fn. 8, 38–39, 45, 48–51, 48 fn. 62, 53, 101, 103, 154–155, 183, 191, 201, 203–204, 206–208, 210, 214, 216, 225, 236–237, 239, 243–245, 254
Bisitun 16 fn. 60
Bit-Šaširia 99
Bit-Zamani 199
Borsippa 47–48, 205–206, 214 fn. 276, 239
Burmarina 167, 195 fn. 206, 198

Carchemish 51

Dariga 100
Dur-balaṭi 100
Dur-Katlimmu (Tell Sheikh Hamad) 167, 194 fn. 203, 195 fn. 206, 198–199, 201
Dur-Šarruken 108, 129

Eanna 209–211
Egypt 3, 6–7, 16 fn. 60, 128, 141 fn. 2, 144 fn. 14, 180, 196, 204, 218, 230–231, 233, 236
Elam 7, 16 fn. 60, 206, 209, 211–212
Elephantine 16 fn. 60, 27, 204, 234 fns. 362–363, 236–237, 237 fn. 380, 251
Euphrates 183
Ezida 17

Gaza 128
Gimir 183–184, 191
Girnavaz 167, 195 fn. 206, 198. *See also* Mardin
Guzana, (Tell Halaf) 29, 167, 183, 185, 192–193, 195 fn. 206, 198, 200–201, 243, 245, 255

Hanigalbat 106–107, 106 fn. 234
Harmašu 211
Harran (Carrhae, near Şanlıurfa) 5 fn. 14, 7, 29, 47, 47 fn. 60, 48 fn. 60, 49, 50 fn. 70, 107 fn. 235, 155, 167, 183–185, 185 fn. 163, 187, 190, 192, 199–202, 213, 243–245, 253
Hindanu 100
Huzirina 100

Ilhina 195 fn. 206
Imgur-Enlil 99
Iššabri 99

Jerusalem 132, 205
Judah 17 fn. 69, 18, 29, 108, 116, 128, 132, 204, 223–225, 228–229, 232–233, 240, 243, 252, 254–255

Kahat 100
Kalhu (Nimrud) 17–18, 17 fn. 64, 19 fn. 76, 20, 24, 34 fn. 3, 36 fn. 12, 38 fn. 20, 39–40, 42, 47, 51, 116, 118, 122–131, 128 fn. 313, 134, 137–138, 141, 141 fn. 2, 148–149, 166–167, 195 fn. 206, 198, 201–203, 206, 209, 236, 245

https://doi.org/10.1515/9783111323435-017

Karduniaš 74. *See also* Babylonia
Khabur Triangle 183
Kidmuri 218
Kilizu 47, 123–124
Kipšuna 100
Kullania (Kunalia, Kunulua, Tell Tayinat) 17–18, 34 fn. 3, 34 fn. 5, 35 fn. 11, 36 fn. 12, 38 fn. 20, 39, 41, 42 fn. 34, 81 fn. 157, 88 fn. 177, 117 fn. 262, 122, 127, 129–130, 134–135. *See also* Tell Tayinat.
Kurbail (Kurba'il) 51, 100

Levant 2, 51
Lubdu 100

Mallanate 167, 195 fn. 206, 198, 247
Mardin (Girnavaz) 167, 195 fn. 206. *See also* Girnavaz
Mediterranean Sea 121
Memphis 3, 7
Moab 128
Muṣaṣir 109–110

Nabulu 100
Nahšimarti 39
Nebi Yunus 97
Nineveh 1, 3 fn. 8, 18, 21, 21 fn. 80, 24–25, 37, 47, 95, 97, 97 fn. 195, 97 fn. 199, 99, 103, 106–108, 112 fn. 259, 114–115, 123–126, 126 fn. 305, 128, 128 fn. 313, 128 fn. 316, 129, 137, 141, 141 fn. 2, 153, 161, 166–167, 169, 172, 177, 181, 186, 195, 198, 205, 209, 218–219, 234 fn. 361, 242, 245, 247
Nippur 47–48, 234 fn. 361

Orontes Valley 41

Persian Gulf 121
Phoenicia 203, 216–217, 217 fn. 292, 222–223

Qarne 51 fn. 72
Qarnina 51, 51 fn. 72
Que 128, 183–184

Raqmat 100

Sealand 107
Šibaniba 99
Šibhiniš, 99
Ṣimirra 219
Šimu 99
Spain 9, 166
Šuanna 1 fn. 3
Šubria 23, 96, 110–112, 110 fn. 251, 112 fn. 256, 128 fn. 310, 157, 166, 166 fn. 82
Sumer 1 fn. 3, 35 fn. 8, 46–47, 74
Susa 97

Tamnuna 99–100
Tell Tayinat 17–18, 19 fn. 76, 41, 117. *See also* Kullania
Tidu 100
Tigris 23, 110
Til Barsip (Tell Aḥmar) 7 fn. 29, 199
Til-abni 100
Tyre 34 fn. 2, 35 fn. 11, 39, 222, 222 fn. 309

Urakka 100
Urartu 5 fn. 17, 106 fn. 232, 108 fn. 240, 109–111, 109 fn. 247, 110 fn. 251, 156–157
Uruk 203, 205–207, 209–212, 210 fn. 256, 214, 234 fn. 361, 236, 239
Van Van 109
Zaban 100
Zagros 2, 131–132, 132 fn. 330, 138

Index of Ancient Texts

2 Kgs. 19 5 fn. 15, 5 fn. 17
2 Kgs. 22 228 fn. 335
2 Kgs. 23 228 fns. 334–335

ABL 268 209 fn. 253
ACP, no. 28 195 fn. 206
Ashur Archive N4 169, 173–174, 173 fn. 110, 176–177, 182, 200–201, 251
Ashur Archive N6 169, 177, 182, 200
Ashur Archive N31 195–196
Ashurbanipal's Library 3 fn. 8
Ashur Ostracon 26 fn. 98
Aššur Charter. *See* RINAP 2, no. 89

Berlin P. 13446 236–237
Book of Deuteronomy. *See* Dtn.
Book of Esther 237 fn. 381
Book of Genesis 237 fn. 381
Book of Isaiah. *See* Isa.
Book of Leviticus. *See* Leviticus
Book of Kings. *See* 2 Kgs.
Book of Tobit 234 fn. 363, 237

Codex Hammurabi 76 fn. 140
CTN 2, no. 221 195 fn. 206
CTN 6, no. 7 194 fn. 206
Cuthean Legend of Naram-Sîn 154 fn. 48

Dtn. 2–12 232 fn. 351
Dtn. 4–30 226
Dtn. 6 231
Dtn. 13 27, 204, 223–233, 230 fn. 342, 231 fn. 349, 240, 243, 254
Dtn. 28 204, 223–227

Esarhaddon's Apology 21–22, 21 fn. 80, 76 fn. 141, 80 fn. 153, 95–107, 99 fn. 208, 109, 112–115, 128 fn. 313, 138, 149, 157, 160 fn. 69, 161–162, 180–182, 187, 200, 237 fn. 381, 242, 248. *See also* RINAP 4, no. 1
Esarhaddon's Letter to the God Aššur 21–23, 95–96, 98 fn. 201, 107–115, 138, 149, 156–157, 168, 242, 248. *See also* RINAP 4, no. 33
Exorcist's Manual 174 fn. 112

Faist 2020, no. 12 (VAT 20691) 195 fn. 206, 196 fn. 215
Frahm 2009a, no. 66 174 fn. 113
Frahm 2009a, no. 67 4 fn. 13, 41 fn. 31, 55 fn. 80, 74 fn. 130, 74 fn. 132, 174 fn. 113
Frahm 2009a, no. 68 4 fn. 13, 41 fn. 31, 55 fn. 80, 174 fn. 113
Frahm 2009a, no. 69 4 fn. 13, 41 fn. 31, 55 fn. 80, 74 fn. 130, 126 fn. 301, 174 fn. 113, 176 fn. 127
Frahm 2009a, no. 70 17 fn. 65
Frahm 2009a, no. 71 17 fn. 65, 174 fn. 113
Frahm 2010. *See* YBC 11382

George 1993, no. 1239 17 fn. 63
Grayson 1975, no. 1 5 fn. 16, 7 fn. 26, 141 fn. 2, 142 fn. 7
Grayson 1975, no. 14 7 fn. 26, 141 fn. 2, 142 fn. 7
Günbati et al. 2020, no. 11 135 fn. 346
Günbati et al. 2020, no. 12 135 fn. 346
Günbati et al. 2020, no. 13 135 fn. 346
Günbati et al. 2020, no. 26 135 fn. 346
Günbati et al. 2020, no. 33 135 fn. 346

Isa. 37 5 fn. 15, 5 fn. 17

Lauinger 2012 17 fn. 67, 17 fn. 70, 34 fn. 34, 38 fn. 18, 41 fn. 30, 44 fn. 39, 46 fn. 52, 47 fn. 54, 49 fn. 65, 52 fn. 78, 63 fn. 91, 72 fns. 125–126, 76 fn. 141, 78 fn. 146, 78 fn. 148, 80 fn. 156, 81 fns. 157–159, 82 fns. 160–161, 83 fn. 163, 87 fn. 173, 88 fns. 177–178, 89 fn. 179, 90 fn. 183, 91 fn. 185, 91 fn. 187, 92 fn. 188, 117 fn. 263, 118 fn. 267, 118 fn. 269, 119 fn. 270, 125 fn. 298, 130 fn. 322, 133 fn. 333, 133 fn. 335, 133 fn. 338, 173 fn. 107, 174 fn. 116, 188 fns. 172–173, 207 fn. 240, 221 fn. 307,

 https://doi.org/10.1515/9783111323435-018

222 fn. 308, 230 fn. 343, 231 fn. 347, 238 fn. 384
Leviticus 19, 231

ND 4327 20, 34 fn. 3, 36 fn. 12. *See also* SAA 2, no. 6
ND 4336 64. *See also* SAA 2, no. 6
ND 4336C 122 fn. 278. *See also* SAA 2, no. 6
ND 4349U 174 fn. 116. *See also* SAA 2, no. 6
ND 4354D 122 fn. 278. *See also* SAA 2, no. 6
ND 4354F 122 fn. 278. *See also* SAA 2, no. 6
ND 4356 71 fn. 123. *See also* SAA 2, no. 6
Niehr 2007 234 fn. 363, 235 fns. 264–265, 235 fn. 267, 235 fns. 269–271, 236 fn. 372. *See also* Story of Ahiqar
Nissinen 2003, no. 86 161 fn. 73. *See also* SAA 9, no. 3
Nissinen 2003, no. 87 162 fn. 77, 163 fn. 78 *See also* SAA 9, no. 3

Papyrus Amherst 63, 233
Parpola 1983, no. 133 143 fn. 11, 148 fn. 29, 151 fn. 41. *See also* SAA 10, no. 199
Parpola 1983, no. 238 143 fn. 11
Parpola 1983, no. 246 147 fns. 26–27. *See also* SAA 10, no. 315
Parpola 1983, no. 247 143 fns. 11–12, 146 fn. 24. *See also* SAA 10, no. 316
Parpola 1983, nos. 265(+)266(+)267) 145 fn. 18. *See also* SAA 10, no. 327

Radner 2002, no. 128 199 fn. 228
Radner 2002, no. 199 199 fn. 228
RIMA 3, A.0.103.1 99 fn. 209, 100 fn. 210
RINAP 2, no. 1 90 fn. 182
RINAP 2, no. 65 (Sargon II's Letter to the God Aššur) 108 fn. 240, 110 fn. 248, 156 fn. 59, 156 fn. 61.
RINAP 2, no. 89 (Aššur Charter) 101, 101 fn. 217, 104, 104 fn. 226
RINAP 2, no. 2002 109 fn. 245
RINAP 3/1, no. 1 1 fn. 3
RINAP 3/1, no. 2 1 fn. 3
RINAP 3/1, no. 3 1 fn. 3
RINAP 3/1, no. 16 90 fn. 182
RINAP 3/2, no. 212 44 fn. 40
RINAP 3/2, no. 213 1 fn. 3
RINAP 3/2, no. 1015 108 fn. 241
RINAP 4, no. 1 (Nineveh A) 5 fn. 14, 5 fns. 16–17, 21 fns. 79–80, 76 fn. 141, 80 fn. 153, 90 fn. 182, 96 fn. 192, 97 fn. 195, 97 fn. 198, 104 fns. 223–225, 105 fns. 228–230, 106 fn. 231, 106 fn. 233, 107 fn. 237, 112 fns. 258–259. *See also* Esarhaddon's Apology
RINAP 4, no. 2 128 fn. 315
RINAP 4, no. 5 (Nineveh F/S) 21 fn. 80, 97 fn. 195
RINAP 4, no. 6 (Nineveh D/S) 21 fn. 80, 97 fn. 195
RINAP 4, no. 13 4 fn. 12
RINAP 4, no. 33 21 fn. 79, 23 fn. 87, 95 fn. 190, 96 fn. 193, 110 fn. 252, 111 fns. 253–255, 157 fn. 63. *See also* Esarhaddon's Letter to the God Aššur
RINAP 4, no. 48 64 fn. 95
RINAP 4, no. 57 135 fn. 349
RINAP 4, no. 77 122 fn. 277
RINAP 4, no. 93 122 fn. 277
RINAP 4, no. 97 7 fn. 29
RINAP 4, no. 98 7 fn. 29
RINAP 5/1, no. 9 121 fn. 273
RINAP 5/1, no. 11 37 fn. 16

SAA 1, nos. 29–40, 102 fn. 219
SAA 1, no. 110, 128 fn. 312
SAA 2, no. 2, 49 fn. 63, 52 fn. 77
SAA 2, no. 3, 4 fn. 13, 55 fn. 80, 176 fn. 127, 187
SAA 2, no. 4, 175 fn. 120, 187
SAA 2, no. 5 34 fn. 2, 35 fn. 11, 39 fns. 24–25, 175 fn. 120, 222 fns. 309–310, 223 fn. 312
SAA 2, no. 6 17 fn. 64, 34 fn. 4, 35 fn. 9, 36 fn. 13, 37 fn. 14, 38 fn. 18, 39 fn. 23, 44 fn. 39, 46 fn. 52, 47 fn. 54, 49 fn. 65, 52 fn. 78, 57 fn. 89, 63 fn. 89, 64 fn. 92, 65 fns. 99–100, 66 fns. 102–103, 67 fn. 106, 67 fn. 108, 68 fn. 110, 68 fn. 113, 69 fn. 117, 70 fn. 121, 71 fn. 122, 71 fn. 124, 72 fns. 125–126, 73 fns. 128–129, 74 fn. 131, 74 fn. 133, 75 fns. 134–136, 75 fns. 138–139, 76 fn. 141–143, 77 fns. 144–145, 78 fn. 146, 78 fn. 148, 79 fns. 151–152, 80 fn. 154, 80 fn. 156, 81 fns. 157–159, 82

fns. 160–161, 83 fn. 163, 84 fns. 164–166, 85 fn. 167, 85 fn. 169, 86 fn. 170, 86 fn. 172, 87 fn. 171, 87 fns. 173–175, 88 fns. 176–178, 89 fn. 179, 90 fn. 183, 91 fn. 185, 91 fn. 187, 92 fn. 188, 109 fn. 246, 117 fn. 263, 118 fn. 267, 118 fn. 269, 119 fn. 270, 125 fn. 298, 130 fn. 322, 133 fn. 333, 133 fn. 335, 133 fn. 338, 173 fn. 107, 174 fn. 116, 175 fn. 120, 184 fn. 160, 186 fn. 166, 188 fns. 172–174, 207 fn. 240–241, 208 fn. 246, 221 fn. 307, 222 fn. 308, 230 fns. 343–344, 231 fn. 347, 238 fn. 384
SAA 2, no. 7 7 fn. 30
SAA 2, no. 8 8 fn. 31
SAA 2, no. 9 89 fn. 181
SAA 2, no. 11 175 fn. 117
SAA 2, no. 12 174 fns. 114–115
SAA 2, no. 14 7 fn. 30
SAA 3, no. 32 26 fn. 103, 177 fns. 130–132, 178 fns. 133–134, 178 fn. 136, 178 fns. 138–140. *See also* Underworld Vision of an Assyrian Prince
SAA 3, no. 33 26 fn. 103, 153 fn. 42, 153 fn. 45, 154 fns. 48–49, 155 fn. 52. *See also* Sin of Sargon
SAA 4, no. 89 218 fn. 294
SAA 4, nos. 139–148 156 fn. 56
SAA 4, nos. 149–173 156 fn. 56
SAA 4, nos. 183–199 147 fn. 27
SAA 5, nos. 44–45 112 fn. 256
SAA 5, no. 52 128 fn. 310
SAA 5, no. 281 102 fn. 219
SAA 7, no. 9 186 fn. 165
SAA 8, no. 160 150 fn. 38
SAA 8, no. 161 150 fn. 38
SAA 8, no. 162 150 fn. 38
SAA 8, no. 163 50 fn. 38
SAA 8, no. 536 207, 214–215, 216 fn. 287
SAA 9, nos. 1–4 160 fn. 69
SAA 9, no. 3 119 fn. 271, 161 fns. 72–73, 162 fns. 74–75, 162 fn. 77, 163 fn. 78
SAA 9, nos. 5–11 161 fn. 71
SAA 10, no. 1 149 fn. 38
SAA 10, no. 3 149 fn. 38
SAA 10, no. 5 123 fn. 284, 124 fn. 288
SAA 10, no. 6 123 fn. 284, 123 fn. 286, 124 fns. 289–290, 124 fn. 293
SAA 10, no. 7 123 fns. 284–285, 123 fn. 287, 124 fn. 291
SAA 10, no. 24 149 fn. 38
SAA 10, nos. 185–232 149 fn. 38
SAA 10, no. 199 144, 148 fn. 29, 150 fn. 40, 208 fn. 242. *See also* Parpola 1983, no. 133
SAA 10, no. 224 24 fn. 95
SAA 10, no. 256 149 fn. 38
SAA 10, no. 259 149 fn. 38
SAA 10, no. 281 149 fn. 38
SAA 10, no. 294 24 fn. 95
SAA 10, no. 315 147 fn. 26. *See also* Parpola 1983, no. 246
SAA 10, no. 316 146 fns. 22–23, 146 fn. 25. *See also* Parpola 1983, no. 247
SAA 10, no. 327 145 fn. 18. *See also* Parpola 1983, nos. 265(+)266(+)267)
SAA 13, no. 45 168 fn. 84, 169, 170 fns. 91–92, 170 fn. 94, 171 fns. 97–98, 193, 214–215
SAA 13, no. 56 126 fn. 300
SAA 13, no. 70 126 fn. 300
SAA 13, no. 78 126 fn. 300, 135 fn. 347
SAA 13, nos. 81–123 129 fn. 320
SAA 14, no. 155 197 fn. 218
SAA 15, no. 60 128 fn. 310
SAA 16, no. 21 205–206, 205 fn. 231, 206 fn. 234, 206 fn. 238, 207 fn. 239
SAA 16, no. 22 205 fn. 231
SAA 16, no. 23 205 fn. 231
SAA 16, no. 24 205 fn. 231
SAA 16, no. 34 24 fn. 95
SAA 16, no. 35 24 fn. 95
SAA 16, no. 59 161 fn. 71, 184, 185 fn. 162, 186, 186 fn. 168, 187 fn. 169, 188, 208 fn. 243
SAA 16, no. 60 149 fn. 36, 161 fn. 71, 184, 185 fn. 162, 188, 188 fn. 176, 189 fn. 179, 193, 208 fn. 243, 215
SAA 16, no. 61 161 fn. 71, 184, 185 fn. 162, 188–189, 188 fn. 175, 193, 208 fn. 243, 215
SAA 16, no. 63 185, 192, 193 fns. 196–197, 213 fn. 272, 215
SAA 16, no. 69 190
SAA 16, no. 70 190
SAA 16, no. 71 184, 190, 190 fn. 182, 191 fns. 186–187, 192–193, 213, 213 fn. 272

SAA 16, no. 126 217, 217 fn. 289, 218 fn. 297, 219, 221 fn. 305
SAA 16, no. 127 217 fn. 289, 219, 219 fns. 299–301, 220
SAA 16, no. 128 217 fn. 289, 219 fn. 298, 220, 220 fn. 303
SAA 16, no. 129 217 fn. 289
SAA 16, no. 167 149 fn. 38
SAA 18, no. 80 206, 209, 209 fn. 248, 209 fn. 252, 210, 210 fn. 254, 210 fn. 256, 210 fns. 259–260, 211–212, 212 fns. 265–266, 213–214
SAA 18, no. 81 206, 209 fn. 248, 211–215, 211 fns. 261–263
SAA 18, no. 83 207, 209 fn. 248, 212–213, 212 fn. 269, 213 fn. 271, 213 fns. 273–274, 216
SAA 18, no. 100 70 fn. 118
SAA 18, no. 102 207, 214, 214 fns. 276–278, 215 fn. 279
SAA 19, no. 8 102 fn. 219
SAA 19, no. 9 102 fn. 219
SAA 19, no. 10 102 fn. 219
SAA 19, no. 11 102 fn. 219
SAA 19, no. 157 102 fn. 219
SAA 19, no. 158 102 fn. 219
SAA 19, no. 229 186 fn. 167
SAA 20, no. 49 126 fn. 306
SAA 21, no. 28 130 fn. 326
SAA 21, no. 75 130 fn. 326
SAA 21, no. 79 129 fn. 320
Sargon II's Letter to the God Aššur. *See* RINAP 2, no. 65
Sin of Sargon 26, 28, 142, 152–165, 177, 180–183, 189 fn. 180, 192 fn. 194, 208, 237, 248. *See also* SAA 3, no. 33
StAT 2, no. 33 195 fn. 206
StAT 2, no. 145 195 fn. 206
StAT 2, no. 146 195 fn. 206
StAT 2, no. 164 180 fn. 148, 195–197, 195 fn. 206, 196 fn. 212, 197 fn. 220
StAT 2, no. 169 195 fn. 206
StAT 2, no. 184 195 fn. 209
StAT 2, no. 242 195 fn. 206
StAT 2, no. 266 195 fn. 206
StAT 2, no. 272 195 fn. 206
Story of Ahiqar 27, 29, 179, 204, 233–240, 236 fn. 276, 251, 254. *See also* Niehr 2007

T 1801, 34 fn. 3. *See also* Lauinger 2012.
Tale of the Two Brothers. *See* Papyrus Amherst 63
Toptaş and Akyüz 2021 195 fn. 206, 198 fn. 223

Underworld Vision of an Assyrian Prince 26, 167, 169, 177–183, 197, 200, 208, 234, 237, 246, 248, 250. *See also* SAA 3, no. 32

YBC 11382 (Frahm 2010) 168 fn. 85, 169, 171, 172 fn. 101–102, 172 fns. 104–105, 173 fn. 106, 208 fn. 245

www.ingramcontent.com/pod-product-compliance
Lightning Source LLC
Chambersburg PA
CBHW051536230426
43669CB00015B/2623